12/25
STRAND PRICE
$5.00

Springer Earth System Sciences

Series editors

Philippe Blondel, Bath, UK
Eric Guilyardi, Paris, France
Jorge Rabassa, Ushuaia, Argentina
Clive Horwood, Chichester, UK

More information about this series at http://www.springer.com/series/10178

Pablo Bouza · Andrés Bilmes
Editors

Late Cenozoic of Península Valdés, Patagonia, Argentina

An Interdisciplinary Approach

Editors
Pablo Bouza
Instituto Patagónico para el Estudio de los
  Ecosistemas Continentales (CCT
  CONICET-CENPAT)
Puerto Madryn
Argentina

Andrés Bilmes
Instituto Patagónico de Geología y
  Paleontología (CCT CONICET-
  CENPAT)
Puerto Madryn
Argentina

ISSN 2197-9596   ISSN 2197-960X   (electronic)
Springer Earth System Sciences
ISBN 978-3-319-48507-2   ISBN 978-3-319-48508-9   (eBook)
DOI 10.1007/978-3-319-48508-9

Library of Congress Control Number: 2016957320

© Springer International Publishing AG 2017
This work is subject to copyright. All rights are reserved by the Publisher, whether the whole or part of the material is concerned, specifically the rights of translation, reprinting, reuse of illustrations, recitation, broadcasting, reproduction on microfilms or in any other physical way, and transmission or information storage and retrieval, electronic adaptation, computer software, or by similar or dissimilar methodology now known or hereafter developed.
The use of general descriptive names, registered names, trademarks, service marks, etc. in this publication does not imply, even in the absence of a specific statement, that such names are exempt from the relevant protective laws and regulations and therefore free for general use.
The publisher, the authors and the editors are safe to assume that the advice and information in this book are believed to be true and accurate at the date of publication. Neither the publisher nor the authors or the editors give a warranty, express or implied, with respect to the material contained herein or for any errors or omissions that may have been made.

Printed on acid-free paper

This Springer imprint is published by Springer Nature
The registered company is Springer International Publishing AG
The registered company address is: Gewerbestrasse 11, 6330 Cham, Switzerland

Prologue

"The Late Cenozoic of Península Valdés, Patagonia, Argentina: an interdisciplinary approach" by Pablo Bouza and Andrés Bilmes

Jorge Rabassa

Península Valdés is one of the most fascinating and emblematic places of the Argentine Patagonian coast.

This is the place where its faunal resources have given this region an international reputation.

This is the place where every winter, hundreds of whales, particularly southern right whales, sail north from the Antarctic Peninsula, crossing the endless blue of the South Atlantic Ocean, to give birth to their offspring, feed them, train and prepare them to return to the open seas, and then find mates for the unequalled task of building up the new generation.

This is the place where sea elephants enjoy sunshine, sleeping on the beach, where sea lions rest on the rocks or play in the surf, where orcas patiently patrol the intertidal plains looking for dinner, where ñandús (South American ostriches) indefatigably explore their pathway across the immense Patagonian plains, where guanacos (South American camels) wander amidst the dunes and where an unthinkable variety of marine birds fill with life every remote corner of the coast.

This is the place where Magellanic penguins unexpectedly nest in caves in the ground and hide behind the spiny shrubs, surprising tourists who may assume that all penguins live only in contact with ice and snow. Penguins have been here since the early Neogene period, being their adaptation to ice and polar environments a much modern event, and not the other way around.

This is the place where the desert is boundless, the climate is cruel and demanding, the wind is overwhelming and suffocating, sand and dunes pinpoint the landscape, deflation hollows are countless, and ventifacts and yardangs are as common as they are in the Sahara or the Arabian Desert.

This is the place where endless coastal cliffs recall a medieval fortress, where zillion Miocene fossil oysters evoke discussions about the non-biological origin of sea shells in the mountains in the times of Nicholas Steno, and where flat-lying sedimentary beds persist away "to infinity and beyond", Buzz Lightyear *dixit*.

This is the place where Native Americans prospered during the last 13,000 years in spite of the harsh environment, totally adapted to the painful tablelands. This is the place of the arrival point of their unending migration from Siberia, the habitat where modern Tehuelches, then pedestrian hunter-gatherers, adapted to the European horse as recent as the beginnings of the 17th century.

This is the place where the dream of an independent Wales homeland was nurtured.

This is the place where the Welsh colonizers disembarked in 1865. The Welsh Colony. Y Wladychfa Gymreig.

A long way from Cardiff.

The establishment of a new Welsh speaking colony in the province of Chubut lead to the founding of several towns that still today preserve the Welsh flavour in their names: Rawson, Trelew, Madryn, Gaiman, Dolavon. No previous European settlements were long lasted in this region. Language, cultural roots and traditions stand up. Welsh is still taught in some private schools. Names in street signs show remembrance of the Welsh epic settlement. Eisteddfod brings into life, yearly, these legends to the approximately 73,000 Argentine-Welsh living in this territory. The flag of Puerto Madryn, merging the Argentine blue and white flag and the Welsh red dragon, bring homesickness of the far away motherland.

The idea of a Welsh colony in South America started around 1840, and Patagonia was the chosen destiny, supported by lands gifted by the Argentine Government, together with the promise that the Welsh Colony would become an Argentine province when the population exceeded 20,000. In 1862, a Welsh delegation visited this region for the first time. They explored the coast and in the middle of a fierce storm, they landed in a small cove within the Golfo Nuevo (New Gulf) which they named as "Port Madryn", honouring the birthplace of Captain Love Jones Parry in North Wales.

The definitive settlement of the Welsh colony took place on 28th July, 1865, when 153 Welsh men and women arrived on board of a sailboat. One hundred and fifty years later, memories of such heroic period abound, honouring the courage of the pioneers.

This fascinating world of awesome biological, paleontological, anthropological and historical heritage is the scenario where Pablo Bouza and Andrés Bilmes have undertaken the difficult task of exposing a complete and ample knowledge of the natural and cultural history of this extreme region.

This book shows the climatology, tectonics, stratigraphy, geomorphology, paleontology, archaeology and recent history of a fascinating environment. The editors have gathered a long list of distinguished scholars, most of them of continental prestige. The authors have contributed a valuable set of convincing and educating chapters, which provide an updated overview of our present knowledge in these fields.

The labor of the editors has been paramount, gathering the expertise of many relevant scientists and they have compiled a volume full of information, knowledge, references, diagnostic and proposals, which I believe will be quite appreciated by the readers coming from diverse disciplines.

Congratulations then to authors and editors! Their work will undoubtedly be enjoyed, used and remembered by many colleagues, dedicated lecturers and interested readers during forthcoming decades.

Ushuaia, July 2016   Jorge Rabassa

Contents

Climatic, Tectonic, Eustatic, and Volcanic Controls
on the Stratigraphic Record of Península Valdés 1
Andrés Bilmes, Leandro D'Elia, José Cuitiño, Juan Franzese
and Daniel Ariztegui

Geology of Península Valdés 23
Miguel J. Haller

Miocene Marine Transgressions: Paleoenvironments
and Paleobiodiversity 47
José I. Cuitiño, María T. Dozo, Claudia J. del Río, Mónica R. Buono,
Luis Palazzesi, Sabrina Fuentes and Roberto A. Scasso

The Climate of Península Valdés Within a Regional Frame 85
Fernando Coronato, Natalia Pessacg and María del Pilar Alvarez

Late Cenozoic Landforms and Landscape Evolution
of Península Valdés 105
Pablo Bouza, Andrés Bilmes, Héctor del Valle and César Mario Rostagno

Vegetation of Península Valdés: Priority Sites for Conservation 131
Mónica B. Bertiller, Ana M. Beeskow, Paula D. Blanco,
Yanina L. Idaszkin, Gustavo E. Pazos and Leonardo Hardtke

Soil–Geomorphology Relationships and Pedogenic Processes
in Península Valdés 161
Pablo Bouza, Ileana Ríos, César Mario Rostagno and Claudia Saín

Soil Degradation in Peninsula Valdes: Causes, Factors, Processes,
and Assessment Methods 191
Paula D. Blanco, Leonardo A. Hardtke, Cesar M. Rostagno,
Hector F. del Valle and Gabriela I. Metternicht

Groundwater Resources of Península Valdés 215
María del Pilar Alvarez and Mario Alberto Hernández

Archaeology of the Península Valdés: Spatial and Temporal Variability in the Human Use of the Landscape and Geological Resources........ 233
Julieta Gómez Otero, Verónica Schuster and Anahí Banegas

Animal Diversity, Distribution and Conservation................ 263
Ricardo Baldi, Germán Cheli, Daniel E. Udrizar Sauthier, Alejandro Gatto, Gustavo E. Pazos and Luciano Javier Avila

Index 305

Introduction

This book presents extensive and new information on the Late Cenozoic record of one of the most important world heritage sites of southern South America: The Península Valdés. From a multidisciplinary approach, that includes geological, biological, paleontological and archeological perspectives, the book describes the main environmental factors and the land and biological processes of this region from the late Cenozoic to the present. The volume brings together an update of the geology, climate, geomorphology, soils, biodiversity, paleontology, archeology of Península Valdés as well as the human impacts on their ecosystems. It includes many published and unpublished information developed mostly by researchers of the Centro Nacional Patagónico (CENPAT) that has never been compiled before. The scope of this book extends to any natural science researcher, graduate student or curious reader of the world interested on the Cenozoic history of Península Valdés.

The study region of this book, Península Valdés (Fig. 1), is located on the coast of the Argentine Patagonia, between parallels 42°05′ and 42°53′S and meridians 63°05′ and 64°37′W (Fig. 1). It covers an area of 3600 km^2, and its limits are the Golfo San Matías (San Matías Gulf) to the north, the Atlantic Ocean to the east and south, and the Golfo Nuevo (Nuevo Gulf) and Golfo San José (San José Gulf) to the west. Connected to the continent by an isthmus of only 11 km wide and less than 30 km long (Istmo Carlos Ameghino; Carlos Ameghino Isthmus), Península Valdés is almost an island considered a part of the Patagonian steppe, a cool semidesert. It has a dynamic coastal zone with active sand dunes, numerous cliffs, spits, bays and coastal lagoons. The interior of the land is Patagonian desert steppe with dry climate and strong winds. Because the coast and gulfs of Península Valdes are sites of global significance for the conservation of marine mammals (southern right whale, southern elephant seals, southern sea lions and orcas) it was inscribed by UNESCO as a World Heritage Site in 1999. Península Valdés is an almost unpopulated region, except for the presence of a few isolated sheep farms and the touristic location of Puerto Pirámides (Fig. 1) on the shore of Golfo Nuevo. This small town, with no more than 500 permanent inhabitants, receives around 200,000 tourists each year due to the attractive proximity to whale-sighting spots.

The Península Valdés region

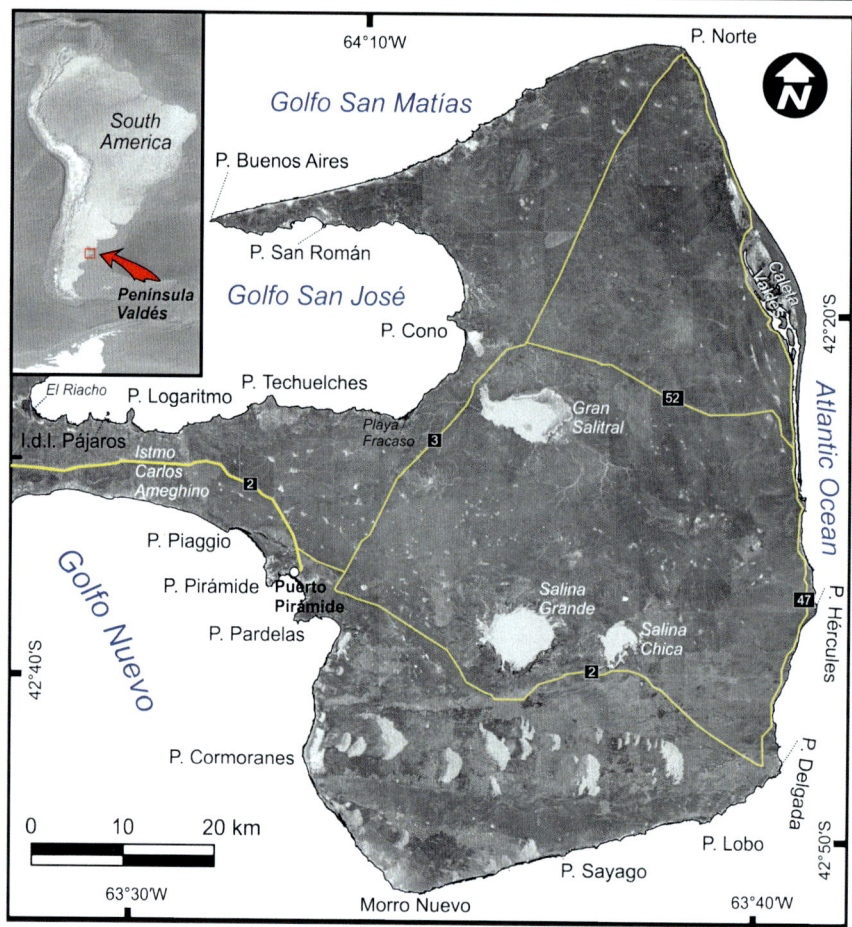

Fig. 1 Location of the study area, Península Valdés, Patagonia, Argentina

This book is designed to be accessible and practical for readers both familiar and unfamiliar with Península Valdés. It is organized in 11 chapters that are strongly interrelated. The sequence of each chapter was designed in other to support the description of the following chapters and to avoid the repetition of information. In addition, chapters were organized in order to be read separately and include references and a glossary of 10–15 key terms within the chapter. Thus it would be possible both to read the 11 chapters of the book as a whole manuscript or as separated papers.

The volume contains many figures and photos, and appendixes of terrestrial living species defined in the area. It also includes seven unpublished thematic maps of Península Valdés, all of them designed at the same scale (e.g., Geomorphological map, Vegetation map, Geohydrological map).

Chapter "Climatic, Tectonic, Eustatic and Volcanic Controls on the Stratigraphic Record of Penísnula Valdés" of the book provide a general framework and geodynamic evolution where the Península Valdés is located from the early Paleozoic to the present. It is analysed at different scales how the interplay between climate, tectonic, sea-level and volcanic processes, set the sedimentary routing system that had governed the final geologic records of this area.

Chapter "Geology of Península Valdés" offers a review of previous geological and paleontological studies in the Península Valdés region and surrounding area since the eighteenth century until present. This section emphasizes the stratigraphic framework (subsurface and surface geology), structural geology, historical geology and the geological resources of Península Valdés.

From a Geo-paleontological approach, the Miocene Marine Transgressions of Península Valdés described in Chapter "Miocene Marine Transgressions: Paleoenvironments and Paleobiodiversity" focus on the sedimentary palaoenvironments and paleobiodiversity (palynomorphs, foraminifers, and marine invertebrates, cetaceans, pinnipeds, marine fishes and birds, as well as continental mammals, birds and fishes). This chapter integrates a vast amount of sedimentological and paleontological information and provides, probably for the first time, an integrated reconstruction of the paleoenvironmental conditions that the Península Valdés region had experienced during the Miocene.

The present climate in Península Valdés is analysed in Chapter "The Climate of Península Valdés Within a Regional Frame" within a regional climate framework to highlight its uniqueness in relation to the rest of north eastern Patagonia. This chapter explains the interplay between large scaled factors and local ones as the almost insularity of the study area. A description is presented through the records of temperature, precipitation and wind and the influence of the oceanic climate. It is shown that although Península Valdés is mostly dominated by the Patagonian arid conditions, its climate has some singularities that make it a less arid and milder climate whit some Mediterranean features.

Chapter "Late Cenozoic Landforms and Landscape Evolution of Penísnula Valdés" is focused on the landforms and on the landscape evolution of Península Valdés. This chapter offers a useful tool to understand the Late Cenozoic landscape dynamic of this area. It also represents an indispensable key to understand topics covered in other chapters as soil genesis and distribution, hydrogeological dynamics, vegetation patterns, geo-ecological functions and processes, and distribution of archaeological material.

Chapter "Vegetation of Península Valdés: Priority Sites for Conservation" describes the main vegetation units of Península Valdés at a meso-scale with emphasis on relevant physiognomic and floristic characteristics. Within this frame, dominant vegetation units were identified at a 1:250,000 scale, reflecting the variety of environmental conditions of Península Valdés.

The soil–geomorphology relationships and pedogenic processes are described in Chapter "Soil–Geomorphology Relationships and Pedogenic Processes in

Península Valdés". In this chapter a general discussion on soil parent materials and soil formation processes is presented taking into account the geomorphological setting offered in Chapter "Late Cenozoic Landforms and Landscape Evolution of Península Valdés" to characterize the soil–geomorphic relationships.

The soil degradation processes are summarized in Chapter "Soil Degradation in Peninsula Valdes: Causes, Factors, Processes and Assessment Methods", where a review on the current knowledge of causes, mechanisms, and factors in the Península Valdés rangelands is analysed. Also, this chapter offers a review of soil degradation assessment methods and several soil degradation studies designed in the region.

Chapter "Groundwater Resources of Península Valdés" analyses the groundwater resources of Península Valdés in order to established the relation between the late Cenozoic stratigraphy and geomorphology with the hydrogeology and hydrodynamic (recharge, flow, and discharge of groundwater) of the region. The groundwater quality and source are well supported by geochemistry and isotopic compositions, respectively.

Chapter "Archaeology of the Península Valdés: Spatial and Temporal Variability in the Human Use of the Landscape and Geological Resources" deals with the archaeology of Península Valdés, focused on two investigative aspects: (1) synthesize the diverse ways in which the native hunting-gatherers used the landscape and its resources through time and (2) the use of mineral resources, mainly rock fragments and soil argillaceous minerals, in the lithic and ceramic technologies, respectively.

Finally, Chapter "Animal Diversity, Distribution and Conservation" describes the patterns of biological diversity of the major taxa inhabiting Península Valdés. This chapter summarizes the ecological information available on the species of arthropods, reptiles, terrestrial birds and mammals known to occur at Península Valdés within the context of the Monte and Patagonia eco-regions.

In Summary, this book is the result of a collective effort of more than 30 specialists and 24 reviewers. We are extremely thankful for their motivation, persistence, critical analysis and patience. This book represents the state of current understanding and hopefully highlights the enormous potential for future study that the Península Valdés has.

Puerto Madryn,   Pablo Bouza and Andrés Bilmes, Editors
August, 2016

List of Reviewers

Cristina Bellelli, INAPL, Ciudad Autónoma de Buenos Aires, Argentina
Gabriel Bernardello, IMBIV (CONICET-UNC), Córdoba, Argentina
Gerardo Bocco Verdinelli, Centro de Investigaciones en Geografía Ambiental (UNAM), Morelia, Méjico
Eleonora Carol, CIG (CONICET-UNLP), La Plata, Argentina
Marta Collantes, FCEN (UBA), Ciudad Autónoma de Buenos Aires, Argentina
Cristina Dapeña, INGEIS (CONICET-UBA), Ciudad Autónoma de Buenos Aires, Argentina
Perla Imbellone, CISAGUA (UNLP), La Plata, Argentina
Federico Kacoliris, FCNyM, La Plata, Argentina
Luciano Lopez, INREMI (UNLP), La Plata, Argentina
Alejandro Monti, IGEOPAT (UNPSJB), Trelew, Argentina
Hector Morrás, INTA, Castelar, Argentina
Daniel E. Martinez, IIMYC (CONICET), Mar del Plata, Argentina
Gustavo Martínez, INCUAPA (CONICET-UNICEN), Tandil, Argentina
Andrés Novaro, INIBIOMA (CONICET-Universidad Nacional del Comahue), Rio Negro, Argentina
Roberto Page, SEGEMAR, San Martín, Argentina
Juan Esteban Panebianco, INCITAP (UNLPAM-CONICET), Santa Rosa, La Pampa
Víctor Ramos, IDEAN (UBA—CONICET), Ciudad Autónoma de Buenos Aires, Argentina
Sebastian Richiano, CIG (CONICET-UNLP), La Plata, Argentina
Juan Rivera, IANIGLA (CONICET-Universidad Nacional de Cuyo-Gobierno de la Provincia de Mendoza), Mendoza, Argentina
Federico Robledo, FCEN (UBA), Ciudad Autónoma de Buenos Aires, Argentina
Augusto Varela, CIG (CONICET-UNLP), La Plata, Argentina
Gonzalo Veiga, CIG (CONICET-UNLP), La Plata, Argentina
Sergio Vizcaino, Museo de La Plata, La Plata, Argentina
Marcelo Zarate, INCITAP (UNLPAM-CONICET), Santa Rosa, La Pampa

Climatic, Tectonic, Eustatic, and Volcanic Controls on the Stratigraphic Record of Península Valdés

Andrés Bilmes, Leandro D'Elia, José Cuitiño, Juan Franzese and Daniel Ariztegui

Abstract The Península Valdés region is situated in an intraplate position of the South American Plate, in the Patagonian foreland close to the Argentine Continental shelf. This region has a complex geotectonic evolution that started more than 400 Ma and involves the conformation of Northern Patagonia as a part of Gondwana during the Paleozoic, the opening of the Atlantic Ocean during the Mesozoic and the configuration of the Andean margin during the Cenozoic. At different scales, the interplay between climate, tectonic, sea-level, and volcanic processes, set the sedimentary routing system that had governed the final geologic records of the Península Valdés region and control the transfer of terrigenous sediments from source to sink. The stratigraphic record of the region was not only influenced by local factors. Processes developed far away from Península Valdés, both in the Southern Andes or in the continental shelf had influenced the late Cenozoic record of this region.

Keywords Sediment routing system · Climate · Tectonics · Eustasy · Volcanism

---

A. Bilmes (✉) · J. Cuitiño
Instituto Patagónico de Geología y Paleontología (IPGP), Consejo Nacional de Investigaciones Científicas y Técnicas (CONICET)—CCT Centro Nacional Patagónico (CENPAT), Boulevard Almirante Brown 2915, U9120ACD Puerto Madryn, Chubut, Argentina
e-mail: abilmes@cenpat-conicet.gob.ar

J. Cuitiño
e-mail: jcuitino@cenpat-conicet.gob.ar

L. D'Elia · J. Franzese
Centro de Investigaciones Geológicas (CIG), Universidad Nacional de La Plata—CONICET, Calle Diagonal 113 N8 275, B1904DPK La Plata, Argentina
e-mail: ldelia@cig.museo.unlp.edu.ar

J. Franzese
e-mail: franzese@cig.museo.unlp.edu.ar

D. Ariztegui
Department of Earth Sciences, University of Geneva, Rue des Maraichers 13, 1205, Geneva, Switzerland
e-mail: Daniel.Ariztegui@unige.ch

1 Geotectonic Setting and Geodynamic Evolution

Península Valdés is located in the Patagonian foreland close to the Argentine Continental shelf, halfway from the Southern Andes and the Continental slope (Fig. 1). It is located in an intraplate position of the South American Plate which is moving to the NW 1.7 cm/a (Schellart et al. 2011; Fig. 1b).

From a geologic perspective, Península Valdés is located in northern Patagonia at the eastern edge of a main tectonic element: the North Patagonian Massif or Somuncurá Massif (Ramos 2008; Fig. 1b). At the same time, it is connected to the east with the marine continental shelf, a feature that dominates the geotectonic configuration of eastern Patagonia since the Late Jurassic (Fig. 1). The North Patagonian Massif limits to the north with two different basement terranes: the Pampia terrane to the northwest and the Río de la Plata craton to the northeast (Fig. 2). To the south it is connected to the Deseado Massif, the other main continental block that conforms the eastern margin of the Patagonia terrane

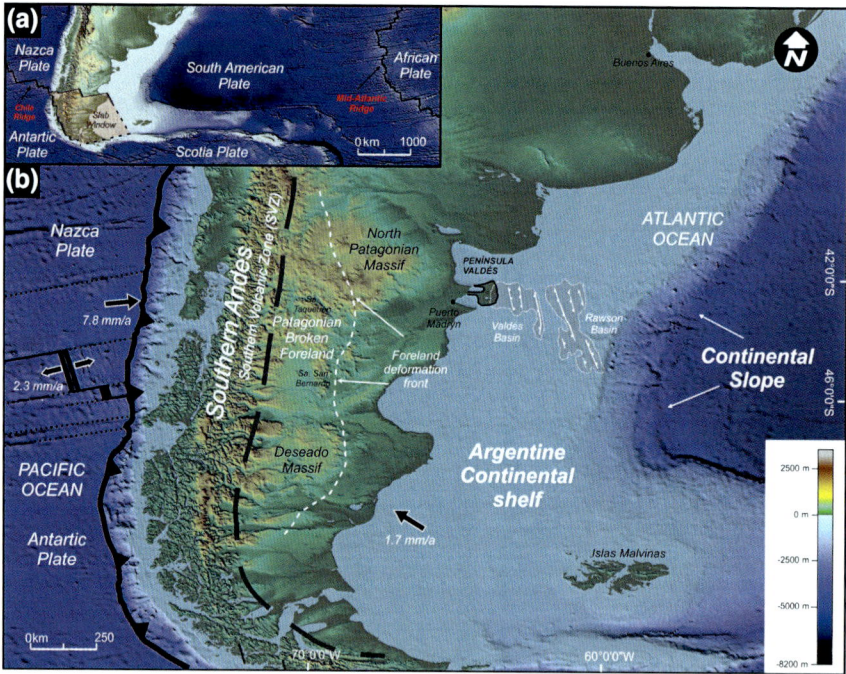

Fig. 1 Present geotectonic settings of Península Valdés, plate configuration and location of the study zone. **a** The South American Plate in the global geotectonic context. **b** Schematic regional map showing main morphotectonic characteristics of the Southern Andes, Southern Andes Foreland and offshore areas that are mention in this chapter. Subsurface normal faults in the Valdes and Rawson basins from Continanzia et al. (2011)

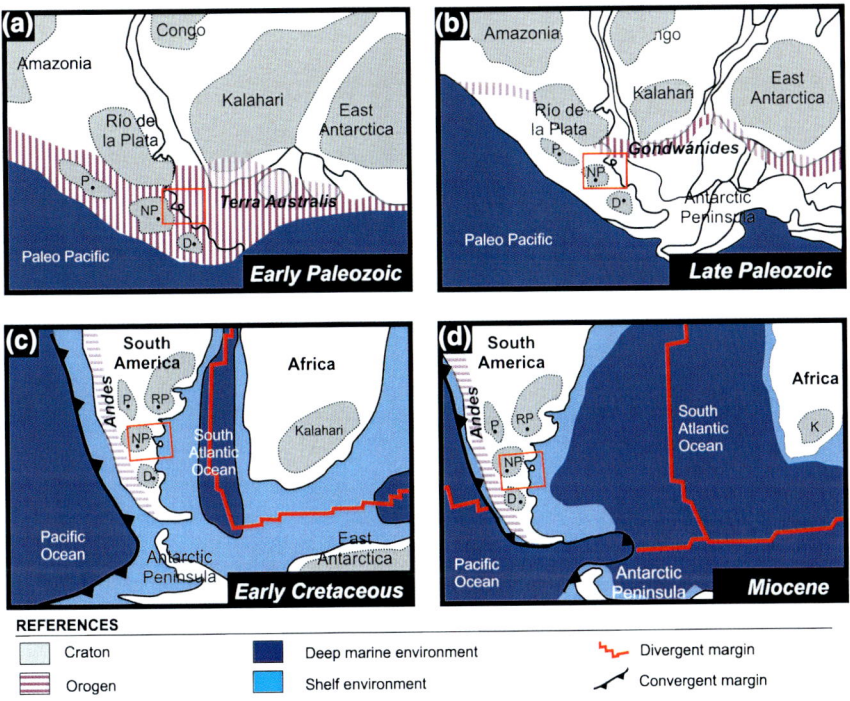

Fig. 2 Geodynamic evolution of the Península Valdés region during the Phanerozoic **a** and **b** Conformation of the Northern Patagonia during the lower Paleozoic and upper Paleozoic, respectively; **c** The *rifting* and opening of the Atlantic Ocean during the Mesozoic; **d** Andean and Andean Foreland configuration margin during the Cenozoic. Location of the Península Valdés region through time is highlighted with a *red line*; North Patagonian massif (NP); Deseado Massif (D); Rio de la Plata (RP); Pampia (P); Kalahari (K). Reconstruction taken from Scotese (2001)

(Ramos 2008; Fig. 2). It is important to highlight that the geotectonic setting of the region where Península Valdés is located has experienced several geotectonic settings in the past 400 Ma. For instance, until the Late Jurassic the landmasses that nowadays form the continents of South America and Africa were connected as part of the Gondwana super-continent, whereas before the Miocene the Andean chain had only a minor topographic expression (Fig. 2).

In summary, the tectonic framework of the region where Península Valdés is located, may be addressed by three "key events": the conformation of Northern Patagonia as a part of Gondwana during the Paleozoic, the opening of the Atlantic Ocean during the Mesozoic and the evolution of the Andean margin during the Cenozoic (Fig. 2).

1.1 The Conformation of Northern Patagonia During the Paleozoic

The evolution of northern Patagonia during the Paleozoic has been a matter of considerable debate during the last decades. While some authors consider Patagonia as an autochthonous part of Gondwana other hypotheses postulate an allochthonous origin (Ramos and Naipauer 2014 and references therein). In the last years, detailed geochronological and petrological studies together with new paleontological findings gathered new evidences in order to understand the tectonic evolution of the region.

The older known stratigraphic elements of Northeast Patagonia (not cropping out in the Península Valdés region) are Cambrian–Ordovician metamorphic rocks intruded by sinorogenic to postorogenic granitoids (Ramos 2008; López de Luchi et al. 2010; Rapalini et al. 2013; Greco et al. 2015). Many evidence support that the North Patagonian Massif was one of the blocks that conformed, together with the Pampia terrane, the Rio de la Plata terrane and the Kalahari terrane of southern Africa, the southern margin of Gondwana in Early Paleozoic times (Ramos 2008; González et al. 2011; Rapalini et al. 2013; Pángaro and Ramos 2012; Pankhurst et al. 2014; Fig. 2a).

During the Late Ordovician and Silurian extensional tectonics led to the origin of an ocean between Patagonia and the Río de la Plata craton. This extensional regime controlled the development of a passive margin represented by the Silurian–Devonian quartzites of the Sierra Grande Formation preserved in eastern Patagonia. It is important to note that sandstones located in the subsurface of Península Valdés can be correlated with the Sierra Grande Formation (Marinelli and Franzin 1996; Gregori et al. 2013; see Chapter "Geology of Península Valdés"), suggesting that the distribution of the Silurian–Devonian passive margin would have included the region of Península Valdés (Fig. 2a). This context of *rifting* and d*rifting* led to the final separation of part of the collided block as present Patagonia (Ramos and Naipauer 2014) and was possibly followed during the Early Carboniferous by the collision and amalgamation of other continental blocks that conformed the entire Patagonia terrane (e.g., The Deseado Massif or the Southern Patagonia terrane; Pankhurst et al. 2006; Chernicoff et al. 2013, respectively; Fig. 2a).

At Carboniferous times, the spreading in the narrow ocean developed between the Río de la Plata craton and northern Patagonia stopped. A major tectonic reconfiguration led to the development of a convergent margin with south-dipping subduction initiated at the northern margin of the Patagonia terrane (Chernicoff et al. 2013).

The final closure between Patagonia and Gondwana occurred during the Carboniferous, even though, collision, deformation and uplift took place in Early Permian times (Ramos 2008; Rapalini et al. 2013; Fig. 2b). The compressive stress regime lasted in this sector of South America until the Late Permian (Ramos 2008).

The Gondwanide Orogenic Belt was created in that period including deformational structures that extended across South America, Africa, and Antarctica (Keidel 1921; Fig. 2b).

1.2 The Rifting and Opening of the Atlantic Ocean During the Mesozoic

The Argentine Continental shelf is 2300 km long and has an average of 400 km wide reaching more than 800,000 km^2 (Fig. 1). This giant tectonic feature originated as a consequence of the thermotectonic processes that led to the breakup of Gondwana and d*rifting* of the South American and African plates since the Mesozoic era (Fig. 2c). *Rifting* and opening of the Atlantic Ocean progressed through different stages along a heterogeneous and highly segmented margin (Blaich et al. 2009). The process of continental extension started at the end of the Triassic and lasted until the opening of the South Atlantic in the Early Cretaceous (Macdonald et al. 2003; Fig. 2c). According to these authors, the early syn-rift phase was accompanied by strike-slip faulting and block rotation, whereas the later extension occurred along with magmatic cycles associated with the impact of the Karoo mantle plume in the Lower Jurassic. During the Middle Jurassic the development of a Large Silicic Igneous Province—the Chon Aike magmatic province—affected the most of northeastern Patagonia (Kay et al. 1989). Is important to note that for other authors initial *rifting* started at this time (c. 183 Ma) and non-earlier (Pángaro and Ramos 2012). Near Península Valdés the rocks of the Chon Aike magmatic province are observed as isolated outcrops and on the subsurface (see Chapter "Geology of Península Valdés"). These rocks are included in the Jurassic Marifil Formation (Cortés 1981; Haller 1982; see Chapter "Geology of Península Valdés").

Rifting prior to seafloor spreading in the southernmost Atlantic is believed to have occurred in the early Jurassic (190 Ma) and involved dextral movement between Patagonia and the northern subplates until the Early Cretaceous (126.7 Ma; Macdonald et al. 2003; Torsvik et al. 2010; Seton et al. 2012). As a result several basins along the Patagonian margin were developed (Autin et al. 2013; Fig. 1b). In the North Patagonian region, the structural trend of the Argentine continental platform abruptly changes from WNW (observed to the north) to NNW (observed in the Península Valdés region) (Fig. 1). In this segment of the continental shelf, in front of Península Valdés, two basins developed between the coast and the slope: the Valdés and the Rawson basins (Keeley and Light 1993). The Valdés Basin has an area of 57,000 km^2 and a maximum thickness of 3000 m (Fig. 1b). The evolution of this basin is relevant for the understanding of the study area because this basin extends to the subsurface of the Península Valdés region reaching a thickness of around 1100 m (borehole YPF.Ch PV.es-1; Marinelli and Franzin 1996; Barredo and Stinco 2010; see Chapter "Geology of Península Valdés"). The main structure is composed of half graben roughly oriented parallel to the coast (Fig. 1). The NNW trending normal faults fits with the amount of displacement of the late Paleozoic features and may have been controlled by the reactivation of the inherited structures (Urien and Zambrano 1996; Max et al. 1999; Macdonald et al. 2003; Ramos 2008; Pángaro and Ramos 2012).

Stabilized seafloor spreading in the southern segment of the South Atlantic rift commenced at around 127 Ma (Fig. 2c), after a prolonged phase of volcanism affecting the southern South Atlantic conjugate margins (Heine et al. 2013; Becker et al. 2012). In the Rawson Basin (Fig. 1b), the breakup unconformity that traces the change from *rifting* to *drifting* is reported to be of Aptian age (c. 118 Ma; Franke 2013). Considering the close similarity in the evolution this age can be assumed as the passage from rift to drift also in the Valdés Basin. Since the Early Cretaceous, the region is involved in the Atlantic passive continental margin (Fig. 2c).

1.3 Configuration of the Andean Margin and Foreland During the Cenozoic

It is generally accepted that the beginning of the Andean margin configuration is determined by the start of the subduction at the western margin of the South American plate, a process synchronic with the *rifting* of the Atlantic Ocean. This process started in the Lower Jurassic with subduction-extension (190–185 Ma; Ramos et al. 2011; D'Elia et al. 2012) followed by a period of compression starting at 100 Ma (Ramos 2010; Tunik et al. 2010). The last period accelerated after 90 Ma, when the South American plate started to drift to the west as a consequence of the opening of the South Atlantic Ocean (Somoza and Ghidella 2012; Folguera and Ramos 2011; Fig. 2c). The configuration and development of the Andes and Andean Foreland was not a steady process, as was inferred in pioneer works (Feruglio 1949; Groeber 1956; among others). It includes diachronic deformational phases from the latest Cretaceous to the late Quaternary that affected simultaneously regions near the orogenic front and regions hundred kilometers from the trench in the Andean Foreland (Cobbold et al. 2007; Guillaume et al. 2009; Folguera and Ramos 2011; Orts et al. 2012; Bilmes et al. 2013; Allard et al. 2015; Gianni et al. 2015; Fig. 2d).

In the Patagonian Andes and foreland, deformational processes in the orogenic front started around 122 Ma and ended by 90 Ma (Baker et al. 1981; Suárez et al. 2010; Folguera and Iannizzotto 2004; Folguera and Ramos 2011). Then, previous to 80 Ma, contractional deformation shifted to the east, uplifting many of the present day mountains of the Andean foreland (e.g., Sierra de San Bernardo, Sierra de Taquetrén; Bilmes et al. 2013; Allard et al. 2015; Gianni et al. 2015; Fig. 1b). It was a period of important mountain uplift not only in the orogenic front, but also in the foreland region, with uplifts of up to 1600 m (Bilmes et al. 2013). From the Late

Cretaceous to the Paleocene magmatic belts near and far from the orogenic front were developed (i.e., Pilcaniyeu and Maiten belts; Rapela et al. 1984; Mazzoni 1994; Madden et al. 2005; Paredes 2008) interpreted as repeated eastward shifts of the arc in the region (Folguera and Ramos 2011). Later during late Oligocene to early Miocene a large mafic igneous province, known as the "Meseta de Somuncura" was developed in the foreland region, about 200 km west from the Península Valdés region (Kay et al. 2007; Fig. 1b).

A renewed episode of orogenic grow started in the lower Miocene around (18 Ma; Folguera and Ramos 2011; Orts et al. 2012; Fig. 2d) and shifted toward the foreland during the lower to middle Miocene (19–14.8 Ma; Guillaume et al. 2009; Giacosa et al. 2010; Bilmes et al. 2013; Gianni et al. 2015). Explosive volcanism coexisted with Miocene mountain-building processes (Mazzoni and Benvenuto 1990) interpreted by some authors as eastward arc expansion (Folguera and Ramos 2011) (Figs. 2d and 3).

In the Patagonian Andes and Andean Patagonian Foreland the late Miocene–Pliocene is characterized by a regional uplift, a process that is still observed today (Guillaume et al. 2009; Pedoja et al. 2011; Folguera et al. 2015a; Fig. 3). As a consequence of this foreland uplift several Late Miocene–Pliocene fluvial terraces were tilted (Guillaume 2009) and all Pleistocene Atlantic shorelines were remarkably elevated, in some cases more than 100 m with respect to present day sea-level (Pedoja et al. 2011). The uplift rate of the Península Valdés region has been estimated based on the age and topographic position of these shorelines in 0.14 ± 0.02 mm/a (Pedoja et al. 2011). This enhanced uplift along the Patagonian coast is linked to dynamic uplift, a particular geotectonic mechanism that is related to the subduction of a mid-ocean ridge (i.e., Chile ridge) and the opening of a *slab window* (Guillaume 2001; Pedoja et al. 2011; Folguera et al. 2015a) (Figs. 1a and 3).

2 Controls on the Stratigraphic Record of the Península Valdés Area

The evolution of the region where nowadays is Península Valdés started more than 400 Ma ago. Thus, it is essential to understand processes that operate on Earth at a time scale people most often not used to. The geological processes operating on Earth's surface produce only subtle changes in the landscape during human lifetime, but over a period of thousands or millions of years, the effect of these processes can be significant. Given enough time, an entire mountain range can be reduced to a featureless lowland, and clastic material derived from the erosion can be transported and deposited hundred kilometers away. The preservation of sedimentary record can form vertical rock successions of thousands of meters thick, as those observed in subsurface in Península Valdés, where there is a thick sedimentary pile of more than 2000 m (see Chapter "Geology of Península Valdés"). A pile of sediments and rocks constitute the stratigraphic record of a region and represent a valuable dataset about the past events in the Earth's history of the area (Fig. 3).

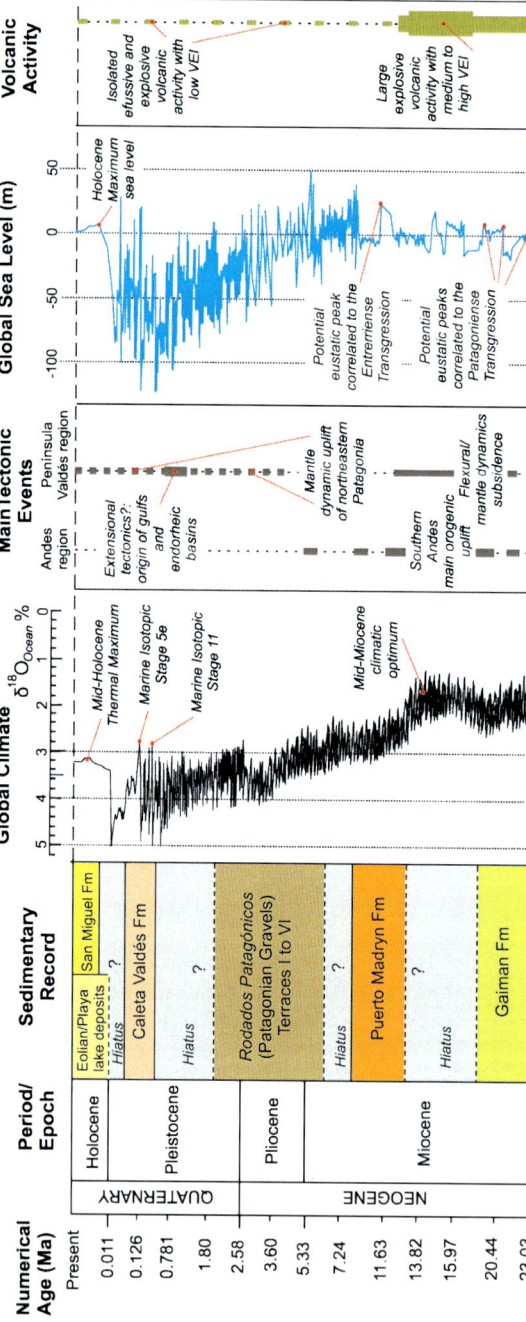

Fig. 3 Controls that influence the stratigraphic record of the Peninsula Valdés region during the late Cenozoic. Numerical age not at scale. Global deep-sea oxygen isotope data from Zachos et al. (2001), Rohling et al. (2014); Deep-sea oxygen isotopes are used as a *proxy* of global average paleo temperature. Global Sea-level data from Miller et al. (2011), Violante et al. (2014). Tectonic and volcanic data of the Southern Andes—Foreland from Mazzoni and Benvenuto (1990), Orts et al. (2012), Bilmes et al. (2013), Morabito and Ramos (2012), Ramos et al. (2015). *Volcanic Explosivity Index* (VEI)

The stratigraphic record of a region is also determined by the dynamic interplay between erosion, sediment transfer, temporary storage, and long-term deposition of the sediments, from source to sink: the sediment routing system (Allen 1997, 2008; Allen and Allen 2013; Fig. 4). From the catchment areas to the sea, the system is controlled by a distinctive set of variables including climate, tectonics, sea-level changes and eventually volcanism. Each of these factors has different influence depending on both the position and the stage within the sediment routing system, from the erosional engine in *proximal areas* to become the infill of the *Sedimentary basin*. The nature of the stratigraphic record may be summarized taking into account the factors that (1) determine the sediment supply to the *Sedimentary basin* and (2) originate the available space for the accumulation of sediments, termed as accommodation space. In coastal and shallow marine areas the accommodation space is determinate by sea-level, whereas in continental environments it is set by base-level, since it determines the geomorphologic *equilibrium profile* of the alluvial-fluvial systems (Muto and Steel 2000; Spalletti and Colombo 2005; Nichols 2009; Allen and Allen 2013).

The interplay between climate, tectonic, sea-level, and volcanic processes, set the sedimentary supply and accommodation space of a sedimentary routing system, which will rule the final geological record of a given region (Figs. 3 and 4). Climate affects the stratigraphic record at different scales. Regional climate, combined with topography and the nature of the vegetation and bedrock, determines the erosional rates (weathering and sediment flux) of a sedimentary routing system, controlling the sediment supply to the *Sedimentary basin*. Climate may also affect the accommodation space. For instance, global climate change may cause variations in the volume of the oceans, affecting sea-level, whereas a regional climate change in a closed basin will adjust the lake level, modifying the available space for the accumulation of sediments. Tectonic processes, such uplift or subsidence, are the most important controls on the stratigraphic record. They involve relative

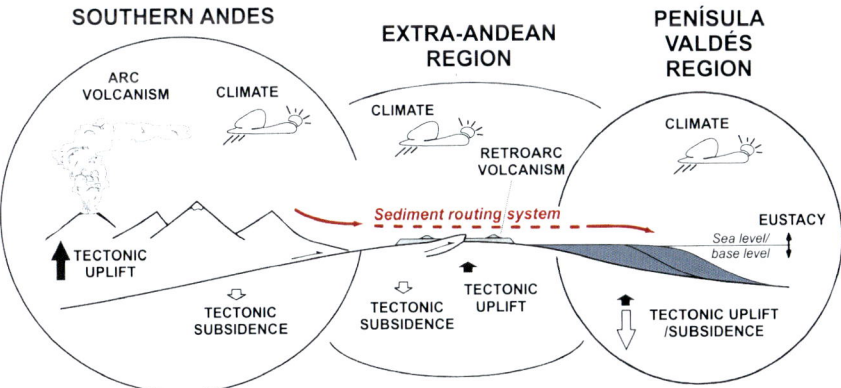

Fig. 4 Conceptual model of the Sediment routing system of the Península Valdés region. Modified from Castelltort and Van Den Driessche (2003)

movements of the substratum at different scales. For example, tectonic uplift affects the topographic relief, modifying the *equilibrium profile* of the alluvial-fluvial systems. Therefore, a change in the *equilibrium profile* has an impact on the erosional rates, as well as in the available space for the accumulation of sediments, adjusting both the sediment supply and accommodation space of the systems. Additionally, tectonic subsidence is the main factor controlling long- or short-term space availability for the accumulation of sediments (i.e., accommodation space). In regions close to the sea, subsidence could produce a relative sea-level rise increasing the accommodation space of sub-aerial depositional areas as well as coastal and shallow marine areas. Finally, volcanism may affect in different ways the sediment routing system. At proximal zones, close to volcanic areas, volcanoes produce as new relief, modifying, or creating new source or catchment areas. During eruptive cycles a volcano or a volcanic chain will deliver different proportions of volcanic particles to the *sedimentary environments*. Depending on the eruption style (effusive or explosive) and magnitude of the eruptions, volcanic materials will affect the sediment flux from *proximal area*s to distal areas (up to hundreds kilometers from the volcanic centers; Fig. 4). Large explosive eruptions introduce huge amounts of sediments in a geologically instantaneous time, producing temporary changes in the mechanism of transport, drainage patters, and depositional features, as well as an increase in sediment supply to the sedimentary systems, until they recover the pre-eruption conditions.

As described in the first part of this chapter, Península Valdés, is located halfway from the Patagonian Andes and the Continental slope (Fig. 1). During the late Cenozoic, climate, sea-level changes, tectonic, and volcanic processes related to the evolution of the Andes and the opening of the Atlantic Ocean are recorded (Fig. 3). In the following sections, each of these stratigraphic controls that affected the geological record of Península Valdés during the late Cenozoic will be considered in detail, introducing key concepts for the reading of the subsequent book chapters.

2.1 Climate

One of the main contributions of paleoenvironmental and paleoclimatic research to the reconstruction of past climate has been the validation that during much of the last 65 million years and beyond, Earth's climate system has been continuously changing. The study of sedimentary archives using a wide variety of well-dated proxies has shown that our planet has been switching from intervals with warm and ice-free poles (*greenhouse-world*), to cold intervals characterized by massive continental ice-sheets and polar ice caps (*ice-world*; e.g., Zachos et al. 2001 and references therein). This should not be surprising since Earth's orbital parameters and plate tectonics—the main forces triggering long-term and global climate—are changing permanently (Fig. 3). Orbital factors vary at different rhythms that remain stable for millions of years. Most of the higher frequency changes result from periodic and quasi-periodic oscillations in Earth's orbital parameters affecting both distribution and amount of solar energy arriving to our planet. Whilst *eccentricity* (~100,000 years) rules the amplitude of *precession* (~23,000 years) and as a result the total annual/seasonal solar energy budget, *obliquity* (~41,000 years) changes the latitudinal distribution of insolation. Hence, orbital factors provide even and mostly predictable long-term fluctuations of climate that are known as Milankovitch cycles. There is good evidence that they regulate the amount of insolation reaching our planet and thus global climate. However, these orbital changes alone are insufficient to explain past climate and hence other features are also important to consider. They include variations in the configuration and distribution of the continents; the opening and closing of oceanic gateways that in turn are ruling ocean circulation; and changes in the atmosphere composition particularly concerning greenhouse gases. These three aspects are closely related and most of all driven by plate tectonics. To the previously described changes in the Earth orbital and tectonic boundary conditions it is necessary to superimpose more regional features that can have a large impact in local climate patterns. One of them is the rain shadow effect associated to emerging mountain chains such as the Himalayan or the Andean cordillera that produce intensification of rainfall on the wind-ward side of the range and an effect of aridity on the leeward side (Blisniuk et al. 2005; Barry 2008).

The late Cenozoic climate in the Península Valdés region has been far from static (Fig. 3) including periods with much warmer or much cooler conditions than today. Fossil tropical fauna from the Puerto Madryn Formation suggest that climate during the Middle Miocene was more humid and warmer than today's (see Chapter "Miocene Marine Transgressions: Paleoenvironments and Paleobiodiversity"). Ice-wedge casts and cryoturbation structures developed in the Rodados Patagonicos (Patagonian gravels) suggest a cooler climate during the Pliocene–lower Pleistocene associated with persistent arid periglacial conditions (see Chapter "Late Cenozoic Landforms and Landscape Evolution of Península Valdés"; Fig. 3). Conversely, the isotopic composition of calcretes embedded in the Rodados Patagónicos (see Chapter "Soil-Geomorphology Relationships and Pedogenic Processes in Península Valdés"; Fig. 3) indicate a warmer and drier interglacial with an average annual temperature of approximately 20 °C during the middle Pleistocene (today is 13.6 °C, see Chapter "The Climate of Península Valdés Within a Regional Frame").

At present there is not a model that could explain alone the causes that have triggered late Cenozoic climate changes in the Península Valdés region. However, some scenarios have been proposed for specific time intervals. Milankovich cycles have been associated with glacial/interglacial changes that are documented in the Rodados Patagónicos (Clapperton 1993; Trombotto 2002; Bouza 2012). On the other hand the absence of an Andean rain shadow effect during the middle Miocene has been proposed to explain the warmer and more humid climate conditions that existed in the Península Valdés region during the accumulation of the Puerto Madryn Formation (Palazzesi et al. 2014; see Chapter "Miocene Marine Transgressions: Paleoenvironments and Paleobiodiversity"; Fig. 3).

2.2 Tectonics

The tectonic uplift and subsidence involve relative movement of the substratum produced by crustal or mantle processes. Both mechanisms may proceed at different scales, such as block uplift or fault subsidence located close to tectonic faults, or uplift/subsidence related to lithospheric processes (e.g., flexural subsidence, thermal uplift) involving regional scales. Fault block uplift or fault subsidence are spatially associated, as its name indicates, with faults. For instance, in contractional tectonic regimes, reverse faults are mainly related to elevated blocks, whereas in extensional tectonic regimes, faults are not only related to fault block uplift, but also are the main inducing factors of mechanical subsidence. At a larger scale, subsidence or uplift may involve lithospheric processes, such as orogeny overload caused by thrust tectonic, which may originate long-wavelength subsidence related to flexure of the lithosphere: isostatic topography. In this case, *orogenic wedge* in *Retroarc foreland basins* or isolated load in Retroarc broken foreland basins (i.e., a foreland basin that is segmented by isolated or partially connected basement uplifts) may originate the flexural subsidence of the lithosphere. In addition, mass anomaly and heat transfers in the upper part of the mantle may trigger other kind of long-wavelength tectonic processes which are referred to as dynamic topography. This large-scale tectonic process may be developed in convergent margins, passive margins, as well as in intraplate settings. For example, slab pull associated to a subduction system in convergent margins may drive dynamic topography subsidence by mantle flow, whereas a thermal anomaly driven by intracontinental hot spot or *slab window* creates a regional uplift.

During the late Cenozoic the Península Valdés area was closely related to the tectonic processes associated with the evolution of the Andean Orogeny (Fig. 3). The two principal tectonic processes occurring at the Andean margin at this latitude are the variations in the angle of the subducted plate (Folguera and Ramos 2011; Orts et al. 2012; Folguera et al. 2015b) and the subduction of the Chile Ridge (Guillaume et al. 2009; Pedoja et al. 2011). The dynamic of the subduction controlled the tectonic evolution of the Andean orogeny as well as of the retroarc areas. During the Neogene, the retroarc of the North Patagonian Andes was configured as a broken foreland basin (Bilmes et al. 2013), whereas the influence of the subducted Chile Ridge become to act, at this latitude, from late Miocene times (Guillaume et al. 2009; Pedoja et al. 2011). In this tectonic setting, several fault tectonics, and lithospheric processes of uplift and subsidence were mentioned. The Miocene contractional tectonics related to the configuration of the Patagonian Broken Foreland involved fault block uplifts as far as 700 km east of the Andean Trench (Bilmes et al. 2013). Thus, the source and catchment areas of the fluvial systems that flow to the Península Valdés area were affected, controlling the depositional scenario of the sediments of the Puerto Madryn Formation (see Chapter "Miocene Marine Transgressions: Paleoenvironments and Paleobiodiversity"). In turn, contractional tectonic would have driven major tectonic effects in the extra Andean areas. Recently, some authors have suggested that the tectonic subsidence associated with either flexural subsidence or dynamic effect of the mantle (dynamic topography), could have caused accommodation space of the continental and marine Miocene–Pliocene succession preserved in extraandean regions of northern Patagonia, near the Atlantic coast (Folguera et al. 2015b). At the Península Valdés area, this lapse corresponds to deposition of the Gaiman and Puerto Madryn formations (see Chapters "Geology of Península Valdés" and "Miocene Marine Transgressions: Paleoenvironments and Paleobiodiversity"), for which the accommodation space would be also associated with the same process of tectonic subsidence (Fig. 3). For the Quaternary other kind of tectonic processes have been recorded in the Península Valdés area. To this latitude, the subduction of the Chile Ridge associated to other coupling mechanisms (see Folguera et al. 2015b) resulted in a long-wavelength (>1000 km) dynamic uplift of northern Patagonia (Guillaume et al. 2009; Pedoja et al. 2011). Thus, the tectonic uplift would be mainly responsible of the exposure of the Pleistocene marine terraces in the Península Valdés area (see Chapter "Late Cenozoic Landforms and Landscape Evolution of Península Valdés"). In addition, a tectonic control associated to extensional faults that affected the pre-Quaternary successions of the area has been proposed to explain the origin of the sea gulfs and Salinas of the Península Valdés region (see Chapter "Late Cenozoic Landforms and Landscape Evolution of Península Valdés"; Fig. 3).

2.3 Eustasy and Relative Sea-Level

The original idea of a varying sea-level appeared during the late nineteenth century, when geologists realized that some ancient marine sediments and fossils indicate water depths different from present day. The term eustasy refers to changes of sea-level—i.e., the sea-level can raise or fall referenced to center of the Earth—occurring at a global scale (Fig. 3). The concept implies long-term (more than 100 years) fluctuations not related to geologically instantaneous meteorological or tidal sea-level variations. Recent eustatic variations (1–10 ka; 1 ka = 1000 years) can be estimated by measuring and dating shoreline markers and tropical reefs and atolls in quiet tectonic areas of the world. In the ancient geological record (1–100 Ma), the correlation of erosion surfaces caused by sea-level falls, and marine flooding deposits caused by sea-level rises, are estimates of the eustatic worldwide variations.

The eustatic sea-level fluctuations are produced by several mechanisms. One of the most relevant processes in the variation of the water volume in the ocean is related to continental ice-sheets expansion or decay (*glacioeustasy*). Another geologically significant process is the variations in the volume of the oceanic basins worldwide, triggered by tectonic forces such as the rate of spreading of the mid-oceanic ridges (*tectonoeustasy*). Minor processes that affect water volume in the oceans include: desiccation and inundation of marginal seas; thermal expansion and contraction of seawater; and variations in groundwater and lake storage. Eustatic changes produced by thermal expansion/contraction of seawater and desiccation/inundation of marginal seas occur at rates of 1 cm/a but with low amplitudes (5–10 m; Miller et al. 2011), and are hardly recognized in the ancient stratigraphic record. Thermal expansion coupled to glacier melting is responsible for milimetric sea-level rise measured during the last century caused by global warming (IPCC 2014). Eustatic changes produced by glaciation/deglaciation of the poles (i.e., glacioeustasy) occur in the order of 1–10 ka and the magnitude of sea-level oscillation is up to 200 m with higher rates of 2 cm/a (Miller et al. 2011). These time lapses are strongly related to the astronomically forced climate changes, referred as Milankovitch cycles. During the Pleistocene, when glaciation events are well constrained, several large eustatic sea-level variations were recorded. During this period, a large area of the Patagonian continental shelf experienced sub-aerial exposure due to a eustatic sea-level fall (see Chapter "Late Cenozoic Landforms and Landscape Evolution of Península Valdés"). Variations in the volume of the oceanic basins are long-term and slow processes which are in the order of 1–100 Ma and the magnitude of the sea-level variation for such processes is in the order of several tens of meters, up to 200 m. This process cause large-scale stratigraphic features such as thick marine sediment accumulation (sea-level rise) or large erosion surfaces (sea-level fall). The sedimentation of the early Miocene Gaiman Formation, that is part of the continental-scale Patagoniense Transgression (see Chapter "Miocene Marine Transgressions: Paleoenvironments and Paleobiodiversity"), can be explained by means of this long-term eustatic mechanisms (Fig. 3).

Regarding relative sea-level change, modern geology recognizes the influence of tectonic processes in uplifting and subsiding continental areas. It is easy to understand that the presence of marine fossils at the top of the Southern Andes (2000–3000 m a.s.l.) was produced by strong tectonic forces that moved those marine rocks upward (Fig. 1b). However, in continental margins close to sea-level the distinction between eustatic versus tectonic controls on sea-level is not straightforward. Subtle tectonic subsidence in these areas will produce a marine flood, and then the record of the flood itself is not a "proof" of a eustatic variation, but a relative sea-level change. As an example, Pleistocene marine terraces lying above the present day sea-level coast of the Península Valdés region are linked to tectonic uplift and not to a eustatic sea-level change (see Chapter "Late Cenozoic Landforms and Landscape Evolution of Península Valdés").

2.4 Volcanism

Volcanic activity has an effect of great magnitude on the stratigraphic record (Fig. 4). The type of the volcanic edifices—composite volcano or calderas—is superimposed over the tectonic relief (topography) and tectonic subsidence (overload or volcanic subsidence). On the other hand, the influence of volcanic activity on *Sedimentary basins* is substantial due to the volume and rate of supply which is generally orders of magnitude larger than nonvolcanic sedimentary systems (Fisher and Smith 1991; Manville et al. 2009). For instance, the average present day discharge of the Brahmaputra River is ~ 0.67 km^3/a (Goodbred and Kuehl 2000), whereas the 1991 eruption of Mount Pinatubo or the 2008 eruption of the Chaitén volcano, delivered around 5–8 km^3 in less than 6 months (cf. Scott et al. 1996a, b; Lara 2009) modifying completely the *sedimentary environment* (Hayes et al. 2002; Umazano et al. 2014). Volcanic products have a strong aggradational tendency (Smith 1987; Smith and Lowe 1991; Haughton 1993), either as topographic positive features or by filling preexisting depressions. During volcanic eruptions, the depositional landscape may be modified in only a matter of hours or days *in proximal areas* (e.g., Cas and Wright 1987; Thouret 1999; Davidson and De Silva 2000; Németh and Ulrike 2007). Depending on the type and magnitude of the volcanic eruption, the material delivered to the *Sedimentary basin*s may affect the stratigraphic record up to hundreds of kilometers from the volcanic centers (Cuitiño and Scasso 2013; Fig. 4).

The Península Valdés is located in a distal position with respect to the volcanic centers of the Andean Southern Volcanic Zone (SVZ) or the volcanic centers associated with the evolution of the basaltic retroact plateau (Fig. 1). However, the volcanic influence on the geological record is important and evidence of its direct and indirect impact on the stratigraphic units of the area (Figs. 3 and 4). During the late Oligocene-early Miocene the large back-arc mafic volcanic field of Somuncurá took place (Kay et al. 2007). The volcanic activity was mainly related to alkali and tholeitic lava flows with, to a lesser extent, the development of monogenetic eruptive cones. By its low value in the *Volcanic Explosivity Index*, this type of volcanic activity did not have a direct influence on the surrounding *sedimentary environment*s. However, the basaltic volcanic activity caused a delayed effect in the source regions and catchment areas of the fluvial systems, controlling the provenance of the sediments of the Puerto Madryn Formation and the Rodados Patagónicos deposits (see Chapters "Geology of Península Valdés", "Miocene Marine Transgressions: Paleoenvironments and Paleobiodiversity" and "Late Cenozoic Landforms and Landscape Evolution of Península Valdés"). During the Miocene, an important period of silicic magmatic activity was recorded in the North Patagonian Andes, evidenced by exhumed plutonic stocks in the Main Cordillera and the occurrence of widespread ignimbrite units in retroarc regions (Pankhurst and Rapela 1998; Mazzoni and Benvenuto 1990; Fig. 3). A huge volume of silicic volcaniclastic materials was delivered to the *Sedimentary basin*s. It not only affected the sedimentary record in the retroarc foreland basins (Franzese et al. 2011; Orts et al. 2012), but also affect the provenance of the geological record up to the continental shelf sedimentary system. This is the case of the de deposits of the Miocene Gaiman Formation of the Península Valdés region (see Chapter "Miocene Marine Transgressions: Paleoenvironments and Paleobiodiversity"), which even though were deposited at ~ 600 km from the magmatic arc (Fig. 1), they show an extremely important influence of pyroclastic material associated with the Miocene silicic arc volcanism.

3 Perspectives and Future Work

The interplay between climate, tectonic, sea-level, and volcanic processes, set the sedimentary routing system that had governed the final geologic records of the Península Valdés region (Fig. 4). Whereas climate and volcanic processes predominantly control the sedimentary supply to the sedimentary environments of Península Valdés, tectonics, and eustasy mainly control the available space for the accumulation of sediments. This stratigraphic record was not only influenced by local controls. Processes developed far away from Península Valdés in the North Patagonian Andes or in the continental shelf had also influenced (Figs. 3 and 4). Despite the significant volume of previous work, including the detail list of this volume, studies related to the late Cenozoic sedimentary routing system of Península Valdés are still in their infancy. One of the key problems in this region are the uncertainties associated with the ages of the stratigraphic record. New geochronological calibrations in the Península Valdés region are necessary. Further studies would probably tackle some of the standing questions dealing with the interplay between subsidence and eustasy of the Puerto Madryn and Gaiman formations; the interrelation between eustasy and uplift during the deposition of Quaternary shorelines; or the influence of climate and tectonics in the deposition of the Rodados Patagónicos. Understanding the interrelation and effect of climate, tectonics, esutasy, and volcanism in the stratigraphic record of Península Valdés are far from being clear and will offer significant potential for further work that has global implications.

Acknowledgements The authors would like to thank the helpful reviews of Professors Gonzalo Veiga and Victor Ramos which improved the final version of this manuscript. This research has been funded by the CONICET (PIP 0632) and Agencia Nacional de Promoción Científica y tecnológica (PICT 2167).

Glossary

Drifting	This term refers to the passage from the rift to passive continental margin as a result of rapidly attenuated lithosphere and the creation of a proto-oceanic trough
Eccentricity (orbital)	Parameter that determines the shape (and variations) of the ellipsoidal orbit of the Earth around the Sun. The shape of the Earth's orbit has a variation period of about 100,000 years
Equilibrium profile	Is a theoretical surface relative to a local base-level or sea-level that control erosion or deposition of the fluvial/alluvial system. In response to a landscape change the fluvial/alluvial system would react to reestablish equilibrium conditions

Greenhouse-world	Long-term climatic stage of the Earth characterized by global warm temperatures and lack of continental glaciers in the poles, caused by the accumulation of certain gases in the atmosphere
Ice-world	Long-term climatic stage of the Earth's characterized by global cold temperatures and continental ice-sheets in the poles. Glacial-interglacial periods occur during this stage
Obliquity	Is the inclination of the Earth's axis in relation to its plane of orbit around the Sun. Oscillations in the degree of Earth's axial tilt occur on a periodicity of 41,000 years
Orogenic wedge	Part of the foreland basin system with wedge-shape originated by contractional tectonics, including the orogenic belt with the wedge top basins (piggy back and thrust top basins)
Precession (axial)	Is the trend in the direction of the Earth's axis of rotation. It has a period of 23,000 years
Proximal areas	In the sedimentary routing system concept, it is the region where the sediments are created or the depositional zone close to these sources (e.g., mountain ranges, volcanoes, or hillslopes)
Proxy	Environmental parameters of the past that are preserved in the stratigraphic record and can be measured for reconstructing conditions that prevailed during the Earth's history (e.g., oxygen isotopes)
Retroarc foreland basins	A *Sedimentary basin* developed on continental crust along the length of compressional destructive margin (i.e., Andean-type orogen) behind the arc, in which subsidence is caused flexure-induced by thrust loading
Rifting	The process by which the continental lithosphere stretches by extension. It is produced by a system of normal faults
Sedimentary basin	Regions of the earth in which the place to the preservation of sediments occurs as a result of long-term subsidence, creating accommodation space for the infill
Sedimentary environment	A place where sediment is deposited and the physical, chemical, and biological conditions that exist there. Examples: lakes, rivers, marine shelves, deltas

Sedimentary flux	The rate of sediment supply to any depositional basin, governed by the complex interaction of several parameters, such as: bedrock-type, uplift, weathering, climate, erosion, and transportation through drainage systems
Slab window	It is a gap in the subducted oceanic plate through which asthenospheric mantle can flow directly in contact with the overriding plate. It is formed when an oceanic spreading ridge reaches a subduction trench and is subsequently subducted
Volcanic explosivity index (VEI)	A relative measure of the explosiveness of volcanic eruptions, determined by the total volume of volcanic products, the eruption cloud height, and qualitative observations. The scale is open-ended from 0 (nonexplosive eruptions) to 8 (largest volcanoes in history given).

References

Allard JO et al (2015) Conexión cretácica entre las cuencas del Golfo San Jorge y cañadón Asfalto (Patagonia): paleogeografía, implicancias. Rev Asoc Geol Argentina 72(1):21–37

Allen PA (1997) Earth surface processes. Blackwell Publishing Ltd., Oxford

Allen PA (2008) From landscapes into geological history. Nature 451:274–276

Allen PA, Allen JR (2013) Basin analysis, 3rd edn. Wiley-Blackwell, Oxford

Autin J et al (2013) Colorado Basin 3D structure and evolution, Argentine passive margin. Tectonophysics 604:264–279. Available at http://dx.doi.org/10.1016/j.tecto.2013.05.019

Baker PE et al (1981) Igneous history of the Andean Cordillera and Patagonian plateau around latitude 46 degrees S. Philos Trans R Soc Lond A: Math Phys Eng Sci 303(1474):105–149. Available at http://rsta.royalsocietypublishing.org/content/303/1474/105.abstract

Barredo SP, Stinco LP (2010) Geodinámica de las cuencas sedimentarias: su importancia en la localización de sistemas petroleros en la Argentina. Petrotecnia 48–68

Barry RG (2008) Mountain weather and climate 3rd edn. Cambridge University Press, Cambridge

Becker K et al (2012) The crustal structure of the southern Argentine margin. Geophys J Int 189 (3):1483–1504. Available at http://gji.oxfordjournals.org/cgi/doi/10.1111/j.1365-246X.2012.05445.x. Accessed 24 June 2014

Bilmes A et al (2013) Miocene block uplift and basin formation in the Patagonian foreland: the Gastre Basin, Argentina. Tectonophysics 601:98–111. Available at http://linkinghub.elsevier.com/retrieve/pii/S0040195113002953. Accessed 26 Nov 2013

Blaich OA et al (2009) Crustal-scale architecture and segmentation of the Argentine margin and its conjugate off South Africa. Geophys J Int 178(1):85–105

Blisniuk PM et al (2005) Climatic and ecologic changes during Miocene surface uplift in the Southern Patagonian Andes. Earth Planet Sci Lett 230(1–2):125–142. Available at http://www.sciencedirect.com/science/article/B6V61-4F6F69G-1/2/d7caec06cc05bfd270c1be2a6ba746f5

Bouza PJ (2012) Génesis de las acumulaciones de carbonatos en Aridisoles Nordpatagónicos: Su significado paleopedológico. Revista de la Asociacion Geologica Argentina 69(2):300–315

Cas RAF, Wright JW (1987) Volcanic successions: modern and ancient. Chapman and Hall, London

Castelltort S, Van Den Driessche J (2003) How plausible are high-frequency sediment supply-driven cycles in the stratigraphic record? Sediment Geol 157:3–13

Chernicoff CJ et al (2013) Combined U-Pb SHRIMP and Hf isotope study of the Late Paleozoic Yaminué Complex, Rio Negro Province, Argentina: implications for the origin and evolution of the Patagonia composite terrane. Geosci Front 4(1):37–56

Clapperton CM (1993) Nature of environmental changes in South America at the Last Glacial Maximum. Palaeogeogr Palaeoclimatol Palaeoecol 101(3–4):189–208

Cobbold PR et al (2007) Distribution, timing, and causes of Andean deformation across South America. Geol Soc Lond Spec Publ 272(1):321–343. Available at http://sp.lyellcollection.org/content/272/1/321.abstract

Continanzia J, Manceda R, Covellone GM, Gavarrino AS (2011) Cuencas de Rawson y Valdés: Síntesis del conocimiento exploratorio—Visión Actual. VIII Congreso de Exploración Y Desarrollo de Hidrocarburos 47–64

Cortés JM (1981) Estratigrafía Cenozóica y estructura al Oeste de la Península Valdés, Chubut. Consideraciones Tectónicas y Paleogeográficas. Rev Asoc Geol Argentina 37(4):424–445

Cuitiño JI, Scasso RA (2013) Reworked pyroclastic beds in the early Miocene of Patagonia: reaction in response to high sediment supply during explosive volcanic events. Sediment Geol 289:194–209. Available at http://dx.doi.org/10.1016/j.sedgeo.2013.03.004

D'Elia L et al (2012) Volcanismo de sin-rift de la Cuenca Neuquina, Argentina: Relación con la evolución Triásico Tardía-Jurásico Temprano del margen Andino. Andean Geol 39(1):106–132

Davidson J, De Silva S (2000) Composite volcanoes. In: Sigurdsson H et al (eds) Encyclopedia of volcanoes. Academic Press, San Diego, pp 663–682

Feruglio E (1949) Descripción geológica de la Patagonia. Dirección General de Yacimientos Petrolíferos Fiscales, Buenos Aires

Fisher RV, Smith GA (1991) Sedimentation in Volcanic Settings. SEPM (Society for Sedimentary Geology). Special Publication, USA

Folguera A, Iannizzotto NF (2004) The lagos La Plata and Fontana fold-and-thrust belt: Long-lived orogenesis at the edge of western Patagonia. J S Am Earth Sci 16(7):541–566

Folguera A, Ramos V (2011) Repeated eastward shifts of arc magmatism in the Southern Andes: a revision to the long-term pattern of Andean uplift and magmatism. J S Am Earth Sci 32:1–16

Folguera A et al (2015a) A review about the mechanisms associated with active deformation, regional uplift and subsidence in southern South America. J S Am Earth Sci 64:511–529

Folguera A et al (2015b) Evolution of the Neogene Andean foreland basins of the Southern Pampas and Northern Patagonia (34°–41°S), Argentina. J S Am Earth Sci 64:452–466

Franzese JR, D'Elia L, Bilmes A, Muravchik M, Hernández M (2011) Superposición de cuencas extensionales y contraccionales oligo-miocenas en el retroarco andino norpatagónico: la Cuenca de Aluminé, Neuquén, Argentina. Andean Geol 38:319–334

Franke D (2013) *Rifting*, lithosphere breakup and volcanism: comparison of magma-poor and volcanic rifted margins. Mar Pet Geol 43:63–87. Available at http://dx.doi.org/10.1016/j.marpetgeo.2012.11.003

Giacosa R et al (2010) Meso-Cenozoic tectonics of the southern Patagonian foreland: Structural evolution and implications for Au–Ag veins in the eastern Deseado Region (Santa Cruz, Argentina). J S Am Earth Sci 30(3–4):134–150. Available at http://www.sciencedirect.com/science/article/pii/S0895981110000465

Gianni G et al (2015) Patagonian broken foreland and related synorogenic *rifting*: the origin of the Chubut Group Basin. Tectonophysics 649:81–99. Available at http://linkinghub.elsevier.com/retrieve/pii/S0040195115001729

González PD, Tortello MF, Damborenea SE (2011) Early Cambrian archaeocyathan limestone blocks in low-grade meta-conglomerate from El Jagüelito Formation (Sierra Grande, Río Negro, Argentina). Geologica Acta 9(2):159–173

Goodbred SL, Kuehl SA (2000) Enormous Ganges-Brahmaputra sediment discharge during strengthened early Holocene monsoon. Geology 28(12):1083–1086. Available at http://geology.gsapubs.org/content/28/12/1083.abstract

Greco GA et al (2015) Geology, structure and age of the Nahuel Niyeu Formation in the Aguada Cecilio area, North Patagonian Massif, Argentina. J S Am Earth Sci 62(MAY):12–32

Gregori DA et al (2013) Preandean geological configuration of the eastern North Patagonian Massif, Argentina. Geosci Front 4(6):693–708. Available at http://dx.doi.org/10.1016/j.gsf.2013.01.001

Groeber P (1956) La Serie Andesítica patagónica. Sus relaciones, posición y edad. Rev Asoc Geol Argentina 9:39–42

Guillaume B et al (2009) Neogene uplift of central eastern Patagonia: dynamic response to active spreading ridge subduction? Tectonics 28(2):TC2009. Available at http://dx.doi.org/10.1029/2008TC002324

Haller MJ (1982) Descripción geológica de la Hoja 43 h, Puerto Madryn, provincia del Chubut. Servicio Geológico Nacional, Boletín, Buenos Aires, p 184

Haughton PDW (1993) Simultaneous dispersal of volcaniclastic and non-volcanic sediment in fluvial basins: examples from the lower old red sandstone, East-Central Scotland. In: Alluvial sedimentation. Blackwell Publishing Ltd., Hoboken, pp 451–471. Available at http://dx.doi.org/10.1002/9781444303995.ch29

Hayes SK, Montgomery DR, Newhall CG (2002) Fluvial sediment transport and deposition following the 1991 eruption of Mount Pinatubo. Geomorphology 45(3–4):211–224

Heine C, Zoethout J, Müller RD (2013) Kinematics of the South Atlantic rift. Solid Earth 4(2):215–253

IPCC (2014) Climate change 2014: synthesis report. Contribution of working groups I, II and III to the fifth assessment report of the intergovernmental panel on climate change, Geneva, Switzerland

Kay SM et al (1989) Late Paleozoic to Jurassic silicic magmatism at the Gondwana margin: analogy to the middle proterozoic in North America? Geology 17(4):324–328

Kay SM et al (2007) The somuncura large igneous province in Patagonia: interaction of a transient mantle thermal anomaly with a subducting slab. J Petrol 48(1):43–77. Available at http://www.scopus.com/inward/record.url?eid=2-s2.0-33845989892&partnerID=40&md5=edd63d86f4f8d8bbed6e04d04d2a0b53

Keeley M, Light M (1993) Basin evolution and prospectivity of the Argentine continental margin. J Pet Geol 16(4):451–464. Available at http://onlinelibrary.wiley.com/doi/10.1111/j.1747-5457.1993.tb00352.x/abstract

Keidel J (1921) Sobre la distribución de los depósitos glaciares del Pérmico conocidos en la Argentina y su significación para la estratigrafía de la serie del Gondwana y la paleogeografía del Hemisferio Austral. Academia Nacional de Ciencias, Boletin, Córdoba, p 25

Lara LE (2009) The 2008 eruption of the Chaitén Volcano, Chile: a preliminary report. Andean Geol 36(1):125–129. Doi:10.1007/s00445-010-0428-x

López de Luchi MG, Rapalini AE, Tomezzoli RN (2010) Magnetic fabric and microstructures of Late Paleozoic granitoids from the North Patagonian Massif: evidence of a collision between Patagonia and Gondwana? Tectonophysics 494(1–2):118–137

Macdonald D et al (2003) Mesozoic break-up of SW Gondwana: implications for regional hydrocarbon potential of the southern South Atlantic. Mar Pet Geol 20(3–4):287–308

Madden R et al (2005) Geochronology of the Sarmiento Formation at Gran Barranca and elsewhere in Patagonia: calibrating middle Cenozoic mammal evolution in South America. 16 Congreso Geológico Argentino 4:411–412

Manville V, Németh K, Kano K (2009) Source to sink: a review of three decades of progress in the understanding of volcaniclastic processes, deposits, and hazards. Sediment Geol 220(3–4):136–161. Available at http://dx.doi.org/10.1016/j.sedgeo.2009.04.022

Marinelli RV, Franzin HJ (1996) Cuencas de Rawson y Península Valdés. In: Ramos VA, Turic MA (eds) Geología y Recursos Naturales de la Plataforma Continental Argentina, 13° Congreso Geológico Argentino y 3° Congreso de Exploración de Hidrocarburos (Buenos Aires). Relatorio, pp 159–169

Max MD et al (1999) Geology of the Argentine continental shelf and margin from aeromagnetic survey. Mar Pet Geol 16(1):41–64

Mazzoni MM (1994) Conos de cinder y facies volcaniclásticas miocenas en la meseta del Canquel (Scarrit Pocket), provincia del Chubut, Argentina. Rev Asoc Argentina Sedimentología 1(1):15–31

Mazzoni MM, Benvenuto A (1990) Radiometic ages of Tertiary ignimbrites and the Collón Cura Formation, Northwestern Patagonia. XI Congreso Geológico Argentino, San Juan, 1, pp 87–90

Miller KG et al (2011) The Phanerozoic record of global sea-level change. Science 1293 (2005):1293–1298

Muto T, Steel RJ (2000) The accommodation concept in sequence stratigraphy: some dimensional problems and possible redefinition. Sed Geol 130(1–2):1–10

Németh K, Ulrike M (2007) Practical volcanology. lecture notes for understanding volcanic rocks from field-based studies. Geological Institute of Hungary, Occasional Papers, 2

Nichols G (2009) Sedimentology and stratigraphy. Wiley, New York. Available at http://www.lavoisier.fr/livre/notice.asp?id=RKOWARA62L6OWN\nhttp://books.google.com/books?hl=en&lr=&id=zl4L7WqXvogC&oi=fnd&pg=PP10&dq=Sedimentology+and+Stratigraphy&ots=l1HnLL8qt7&sig=BFyQdgY_OCHGMZdh-wpy5h4teM4

Orts DL et al (2012) Tectonic development of the North Patagonian Andes and their related Miocene foreland basin (41°30′–43°S). Tectonics, 31(3):TC3012. Available at http://dx.doi.org/10.1029/2011TC003084

Palazzesi L et al (2014) Fossil pollen records indicate that Patagonian desertification was not solely a consequence of Andean uplift. Nat Commun 5:3558. Available at http://www.ncbi.nlm.nih.gov/pubmed/24675482

Pángaro F, Ramos VA (2012) Paleozoic crustal blocks of onshore and offshore central Argentina: new pieces of the southwestern Gondwana collage and their role in the accretion of Patagonia and the evolution of Mesozoic south Atlantic *Sedimentary basin*s. Mar Pet Geol 37(1):162–183

Pankhurst RJ, Rapela CW (1998) The proto-Andean margin of Gondwana. Geol Soc Spec Publ Available at http://www.scopus.com/inward/record.url?eid=2-s2.0-0032300073&partnerID=40&md5=7aecba4628f4d7be963b8e98e2208a18

Pankhurst RJ et al (2006) Gondwanide continental collision and the origin of Patagonia. Earth Sci Rev 76(3–4):235–257. Available at http://www.scopus.com/inward/record.url?eid=2-s2.0-33744815226&partnerID=40&md5=d8d40b7d9e5cdcc052cc45a373a64ff3

Pankhurst RJ et al (2014) The Gondwana connections of northern Patagonia. J Geol Soc 171 (3):313–328. Available at http://jgs.lyellcollection.org/cgi/doi/10.1144/jgs2013-081

Paredes JM (2008) Basaltic explosive volcanism in a tuff-dominated intraplate setting, Sarmiento Formation (Middle Eocene-lower Miocene), Patagonia Argentina. Latin Am J Sedimentol Basin Anal 15(2):77–92

Pedoja K et al (2011) Uplift of quaternary shorelines in eastern Patagonia: darwin revisited. Geomorphology 127(3–4):121–142

Ramos ME, Tobal JE, Sagripanti L, Folguera A, Orts DL, Giménez M, Ramos VA (2015) The North Patagonian orogenic front and related foreland evolution during the Miocene, analyzed from synorogenic sedimentation and U/Pb dating (∼42°S). J S Am Earth Sci 64:467–485

Ramos VA (2008) Patagonia: a paleozoic continent adrift? J S Am Earth 26:235–251

Ramos VA (2010) The tectonic regime along the Andes: present-day and Mesozoic regimes. Geol J 45(1):2–25. http://www.scopus.com/inward/record.url?eid=2-s2.0-73249141417&partnerID=40&md5=2ef247c674bec29baa8e163ed3e8f17d

Ramos VA, Naipauer M (2014) Patagonia: where does it come from? J Iberian Geol 40(2): 367–379

Ramos VA et al (2011) Evolución Tectónica De Los Andes Y Del Engolfamiento Neuquino Adyacente. In: Leanza H et al (eds) Relatorio del XVIII Congreso Geológico Argentino pp 335–348

Rapalini AE et al (2013) The South American ancestry of the North Patagonian Massif: geochronological evidence for an autochthonous origin? Terra Nova 25(4):337–342

Rapela CW et al (1984) El vulcanismo paleoceno-eoceno de la Provincia Volcánica Andino-Patagónica. IX Congreso Geológico Argentino, Relatorio, 8: 180–213 (Bariloche)

Rohling EJ et al (2014) Sea-level and deep-sea-temperature variability over the past 5.3 million years. Nature 508(7497):477–82. Doi:10.1038/nature13230

Schellart WP et al (2011) Influence of lateral slab edge distance on plate velocity, trench velocity, and subduction partitioning. J Geophys Res Solid Earth 116(10):1–15

Scotese CR (2001) Atlas of Earth History, vol 1. Paleogeography, PALEOMAP Project, Arlington, Texas

Scott KM et al (1996a) Channel and sedimentation responses to large volumes of 1991 volcanic deposits on the east flank of Mount Pinatubo. In: Newhall CG, Punongbayan RS (eds) Fire and mud, eruptions and lahars of Mount Pinatubo. PHIVOLCS Press; University of Washington Press, Philippines; Quezon City, Seattle, pp 971–988

Scott KM et al (1996b) Pyroclastic flows of the June 15, 1991, climactic eruption of Mount Pinatubo. In: Newhall CG, Punongbayan RS (eds) Fire and mud, eruptions and lahars of Mount Pinatubo. PHIVOLCS Press; University of Washington Press, Philippines; Quezon City, Seattle, pp 545–570

Seton M et al (2012) Global continental and ocean basin reconstructions since 200Ma. Earth Sci Rev 113(3–4):212–270. Available at http://dx.doi.org/10.1016/j.earscirev.2012.03.002

Smith GA (1987) The influence of explosive volcanism on fluvial sedimentation: the Deschutes Formation (Neogene) in Central Oregon. J Sediment Petrol 57:613–629

Smith GA, Lowe DR (1991) Lahars: volcano-hydrologic events and deposition in the debris flow-hyperconcentrated flow continuum. Fisher RV, Smith GA (eds) Sedimentation in volcanic settings. Special Publications SEPM (Society Economic Paleontologists and Mineralogists). SEPM, Tulsa, OK, pp 59–70

Somoza R, Ghidella ME (2012) Late Cretaceous to recent plate motions in western South America revisited. Earth Planet Sci Lett 331–332:152–163. Available at http://www.sciencedirect.com/science/article/pii/S0012821X12001173

Spalletti LA, Colombo F (2005) From alluvial fan to playa: an Upper Jurassic ephemeral fluvial system, Neuquen Basin, Argentina. Gondwana Res 8(3):363–383. Available at http://linkinghub.elsevier.com/retrieve/pii/S1342937X05711412

Suárez M et al (2010) 40Ar/39Ar and U–Pb SHRIMP dating of Aptian tuff cones in the Aisén Basin, Central Patagonian Cordillera. J S Am Earth Sci 29(3):731–737. Available at http://linkinghub.elsevier.com/retrieve/pii/S0895981109001667 [Accessed September 28, 2014]

Thouret JC (1999) Volcanic geomorphology-an overview. Earth Sci Rev 47(1–2):95–131

Torsvik TH, Rousse S, Smethurst MA (2010) Reply to comment by D. Aslanian and M. Moulin on "A new scheme for the opening of the South Atlantic Ocean and the dissection of an Aptian salt basin". Geophys J Int 183(1):29–34

Trombotto D (2002) Inventory of fossil cryogenic forms and structures in Patagonia and the mountains of Argentina beyond the Andes. S Afr J Sci 98(3–4):171–180

Tunik M et al (2010) Early uplift and orogenic deformation in the Neuquén Basin: constraints on the Andean uplift from U-Pb and Hf isotopic data of detrital zircons. Tectonophysics 489 (1–4):258–273. Doi:10.1016/j.tecto.2010.04.017

Umazano AM et al (2014) Fluvial response to sudden input of pyroclastic sediments during the 2008–2009 eruption of the Chaitén Volcano (Chile): the role of logjams. J S Am Earth Sci 54:140–157. Available at http://dx.doi.org/10.1016/j.jsames.2014.04.007

Urien CM, Zambrano JJ (1996) Estructura de la plataforma continental Argentina. In: Ramos VA, Turic MA (eds) Geología y Recursos Naturales de la Plataforma Continental Argentina, 13° Congreso Geológico Argentino y 3°. Congreso de Exploración de Hidrocarburos (Buenos Aires), Relatorio, pp 29–66

Violante RA et al (2014) Chapter 6 The Argentine continental shelf: morphology, sediments, processes and evolution since the Last Glacial Maximum. Geol Soc Lond Mem 41(1):55–68. Doi:10.1144/M41.6

Zachos J et al (2001) Trends, rhythms, and aberrations in global climate 65 ma to present. Science 292(5517):686–693. Available at http://www.sciencemag.org/content/292/5517/686.abstract

Geology of Península Valdés

Miguel J. Haller

Abstract The surface of Península Valdés is gently cut, characterized by plateaus, sea cliffs, and spits as well as depressions caused by a complex process of tectonic, deflation, fluvial erosion, and mass wasting. Subsurface geology comprises early Paleozoic rocks, Jurassic volcanics, Cretaceous continental deposits, and Paleogene and Neogene deposits. The oldest rock outcrops are the marine sediments deposits with pyroclastic contribution of early Miocene age. These deposits are covered by sandstones and mudstones of coastal environment and late Miocene age. On these sediments rest gravel beds of the late Pliocene—early Pleistocene. Other deposits, aeolian, coastal, and coastal marine origin are assigned to the Pleistocene and Holocene. The most important mineral economic resource of the region of geological origin is halite (NaCl) accumulated in the floor of the Salina Grande and Salina Chica endorheic basins. Salt has been exploited during the first quarter of this century. On the other hand, sand and gravel quarries are sporadically exploited for local use.

Keywords Paleogene · Neogene · Subsurface information · Geological resources

1 Previous Geological Work: Early Geological and Paleontological Studies

The first geological description of the Tertiary Neogene beds from the Península Valdés region was made by Malaspina in 1789 during his exploration trips around the world. Malaspina (1885, p. 62) wrote in his travel diary: "… the inner coast of

M.J. Haller (✉)
Universidad Nacional de la Patagonia San Juan Bosco, Puerto Madryn, Argentina
e-mail: haller@cenpat-conicet.gob.ar

M.J. Haller
Instituto Patagónico de Geología y Paleontología (IPGP), Consejo Nacional de Investigaciones Científicas y Técnicas (CONICET) - CCT Centro Nacional Patagónico (CENPAT), Boulevard Almirante Brown 2915, ZC: U9120ACD Puerto Madryn, Chubut, Argentina

© Springer International Publishing AG 2017
P. Bouza and A. Bilmes (eds.), *Late Cenozoic of Península Valdés, Patagonia, Argentina*, Springer Earth System Sciences,
DOI 10.1007/978-3-319-48508-9_2

San José seem to be formed by some horizontal layer of sandy blackish earth, others whitish, reddish, overlapping each other in number of twenty, and all probably composed by sand, marls, clay, etc., presenting a rather sterile soil and clear not only from large trees and also any kind of bushes."

Although Alcide d'Orbigny has not visited himself the Patagonian coast south of Carmen de Patagones, he observed the Tertiary beds in the cliffs of the mouth of Río Negro and he considered the Cenozoic marine sediments extending from Entre Ríos to the Strait of Magellan as a unit which he named Terrain Tertiarie Patagonien (d'Orbigny 1842) and assigned them an Eocene age similar to the Tertiary Paleogene beds in the Paris Basin.

Darwin made a short examination in the Golfo San José (North of the Península Valdés region Fig. 1) during his journey on board of the HCS Beagle in 1833. He describes (Darwin 1845) that "… the cliffs are about a hundred feet high; the lower third consists of yellowish-brown, soft, slightly calcareous, muddy sandstone …". Darwin (1845) found several fossils like *Ostrea*, *Pecten*, *Terebratula*, and *Turritella*,

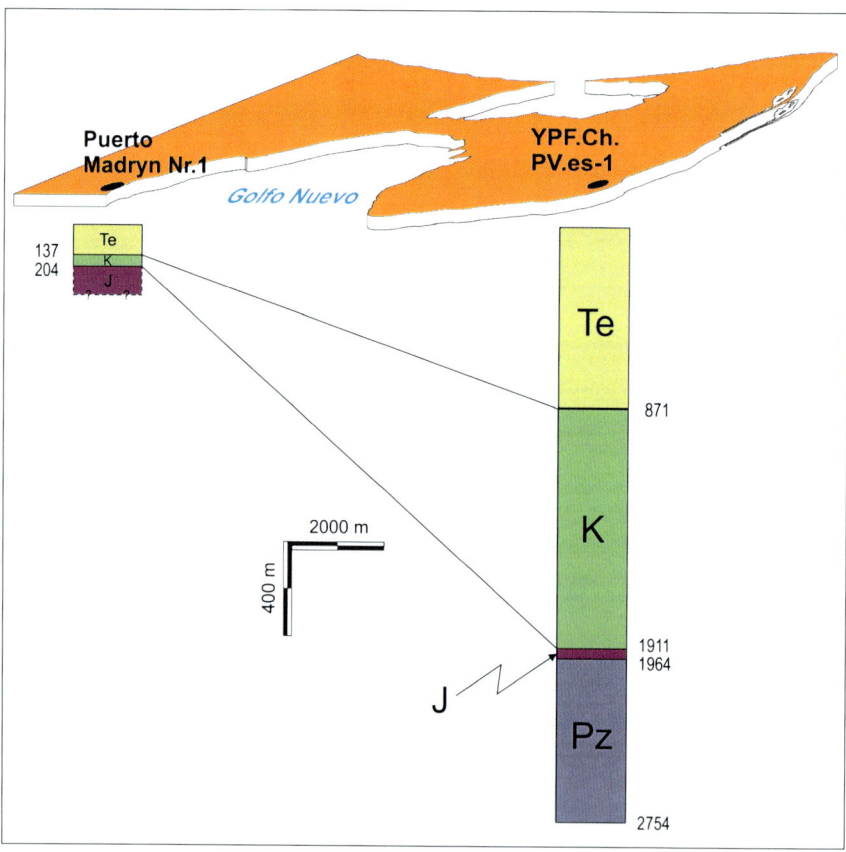

Fig. 1 Schematic chronological columns for YPF.Ch.PV.es-1 and Puerto Madryn Nr. 1 boreholes

and compared them with founding in Entre Ríos, Río Negro, San Julian and Puerto Deseado. He also mentions: "… al the strata appear horizontal, but when followed by eye for a long distance, they are seen to have a small easterly deep." He also saw the gravel on the surface (i.e., Rodados Patagónicos—Patagonian Gravels—), which he later recognized in other parts of Patagonia and named Patagonian Shingle Formation. In Golfo Nuevo (Fig. 1), Darwin described two strata of pale brown mudstone separated by a darker colored argillaceous variety. Darwin mentions that those strata extend along the Patagonian coast until Puerto Deseado, and then to the south they are found at intervals until Santa Cruz River. Darwin included all those beds cropping out on the coast in the Patagonian Tertiary Formation, and stated that they may belong to the same epoch, of "considerable antiquity." He was impressed by the outcrops extension, which he estimated in 27 degrees of latitude, extension he compares to the distance "from the Straits of Gibraltar to the south of Iceland."

With the support of his brother Carlos, Florentino Ameghino made a great contribution to the paleontology and stratigraphy of Patagonia. Ameghino (1890) considered that the Patagonian Tertiary beds were older than those of Entre Ríos.

Other studies at the beginning of the twentieth century, other studies which made minor corrections in the stratigraphy of the Península Valdés region (were made by Roveretto 1921; Windhausen 1921 and Frenguelli 1927).

During the second half of the twentieth century several authors dealt with the marine Tertiary beds from Patagonia, like Bertels (1970), Camacho (1974, 1979a, b, 1980), Riggi (1979a, b, 1980), Di Paola and Marchese (1973).

In the last two decades of the twentieth century, Scasso and del Río (Scasso and del Río 1987; Scasso et al. 2001; del Río 1988, 1990; 1991, 1992) began systematical sedimentological and paleontological studies with modern criteria that allowed deepen understanding the origin, age, and deposition environment of these Tertiary Neogene formations.

2 Stratigraphy

The geology of the region is simple, above a basement of Paleozoic, Cretaceous and Paleogene rocks, only sedimentary rocks are outcropping. The late comprise a depositional record from late early Miocene to Holocene times (see Chapters "Miocene Marine Transgressions: Paleoenvironments and Paleobiodiversity" and "Late Cenozoic Landforms and Landscape Evolution of Península Valdés"). On a relief developed in these Neogene units, recent sediments were deposited. The different units that make up the geology of the region are summarized in Table 1.

Table 1 Stratigraphy of the Península Valdés area

Era	Period	Epoch	Geological unit	Lithology	Maximum thickness (m)
Cenozoic	Quaternary	Holocene	Alluvium and colluvium deposits	Sand, gravel and silt	2–3
			Eolian deposits	Sand and silt	14
			Playa lake sediments and evaporites	Silt, clay, evaporites (halite, glauberite and gypsum)	1
			San Miguel formation	Gravel and sand	6
		Pleistocene	Caleta Valdés Formation	Gravel	25
	Neogene	Pliocene–Pleistocene	Patagonian gravels	Gravel	4
		Late Miocene	Puerto Madryn Formation	Sandstones, siltstones, mudstones and coquinas	80 (350)
	Paleogene	Early Miocene	Gaiman formation	Claystones, siltstones, tuffaceous mudstones, tuffs, and sandstones	20 (280)
		Oligocene		Mudstones, sandstones, calcareous siltstones, marls, claystones	(140)
		Danian	Marine	Claystones and mudstones	(60)
			Continental	Conglomeradic sandstones and claystones	(6)
Mesozoic	Cretaceous	Barremian–Maastrichtian	Continental/marine	Claystones, calcareous sandstones, conglomeradic sandstones, sandstones, conglomerates, siltstones, tuffaceous sandstones, tuffaceous claystones and mudstones	(1134)
	Jurassic			Volcanic agglomerate Rhyolitic volcanics	(53)
Paleozoic	Early?			Mudstones, tuffs, quartzites, quartzitic sandstones and calcareous siltstones	(737)

Shading indicates subsurface units. In brackets: thickness from subsurface data

2.1 Subsurface Information

The state oil company YPF started in 1975 an exploration drilling on the Península Valdés region with the objective of the acquisition of subsurface data on the stratigraphy, lithology, and thickness of the sedimentary rocks of the western flank of the sedimentary Valdés Basin (Mainardi et al. 1980) which extends offshore eastward in the Atlantic Ocean. The results provide a first glimpse of the subsurface stratigraphy of the Península Valdés region.

The YPF.Ch.PV.es-1 well (Fig. 1) is located in the southeast of Península Valdés region. The borehole wellhead elevation is 51 m and reached a depth of 2754 m. According to YPF (1976), the stratigraphic column comprises Paleozoic, Cretaceous, Paleogene and Neogene and Tertiary rocks (Fig. 1).

Paleozoic rocks drilled section is 737 m thick without reaching the base of the unit. These rocks are dominated by slightly micaceous gray siltstones and silty argillites, interbedded, with fine to very fine grained and very consolidated quarzitic sandstones. Based on the lithology and the stratigraphic position, these rocks were correlated with marine quartzites of the Sierra Grande Formation (Harrington 1962), that are outcropping 175 km northwest in the southwestern province of Río Negro, whose age was assigned to the Silurian-Devonian (Spalletti et al. 1991).

A 53 m-thick igneous body of rhyolitic composition covered by a volcanic agglomerate separates the Paleozoic rocks from the Cretaceous deposits. Following Continanzia et al. (2011), those igneous rocks may correlate with a very important igneous event which took place during the Jurassic in Patagonia and has been dated in 183 ± 2 Ma (Rb/Sr) by Rapela and Pankhurst (1993), and 186.2 ± 1.5 Ma (Ar/Ar) by Alric et al. (1996). In the Atlantic coast where the Península Valdés is located, these rocks are included in the Marifil Formation (Malvicini and Llambías 1974).

The Cretaceous rocks thickness is 1089 m thick. The Cretaceous deposits constitute the initial filling of the Valdés basin. The section consists of reddish brown and light greenish gray claystone, partly slightly sandy, with quartz pebbles. Toward the base, fine green and whitish pyroclastic beds are dominant with participation of mud and sand. The pyroclastic rocks are interbedded with fine to very fine grained reddish brown sandstones with clay matrix. These rocks are correlated with continental sediments of the Chubut Group (Lesta 1968) outcropping west of Península Valdés. The age of the Chubut Group is assigned to Barremian—Maastrichtian (Chebli et al. 1976; Casal et al. 2015). Caramés et al. (2004) postulate that some Maastrichtian arenaceous foraminifera found in the uppermost beds of the Cretaceous section of well YPF.Ch.PV.es-1 indicate a shallow marine environment.

The upper section of borehole YPF.Ch.PV.es-1, of 830 m thickness, consists of Tertiary Paleogene and Neogene sediments represented by yellowish gray and greenish gray siltstones; gray and green slightly marly and partially sandy mudstones and claystones. They are interbedded with fine to medium grained green and gray sandstones, sometimes conglomeratic. The Tertiary beds contain abundant fossil remains of shells, coral, and spicules. Microfossils allowed Masiuk et al. (1976) to differentiating differentiate sections of lower Paleocene, Eocene, and Miocene age deposited in a medium shelf depositional environment. Later, Caramés et al. (2004) based on foraminifera content established that the Cenozoic deposits of the well YPF.Ch.PV.es-1 is represented by four sections: (1) 60 m deposits of late Danian age; (2) 140 m deposits of Oligocene age; (3) 280 m deposits of late Oligocene—Lower Miocene age and; (4) 295 m deposits of middle to late Miocene age. The lower, Danian section, is composed by pale brown silty sandstones and fine to coarse gray sandstones containing fossil remains of shells, corals and equinoderms. These deposits are partially correlated with Arroyo Verde Formation cropping out in coastal Patagonia. The following section, of Oligocene age, is formed by gray mudstones, limy siltstones, marls, and claystones with fossil fish remains and abundant pyrite. This section is correlated by Caramés et al. (2004) with the San Julian Formation from southern Patagonia. The late Oligocene—early Miocene section is composed by an assemblage of interbedded claystones, siltstones, mudstones, tuffaceous mudstones, tuffs, and sandstones with disseminated pyrite and equinoderm spicules. This section is correlated by Caramés et al. (2004) with Gaiman Formation and should be the subsurface continuation of the strata cropping out west of Península Valdés. The upper section—middle to late Miocene—is formed by interbedded siltstones, fine to medium sandstones, silty sandstones, and mudstones with shell remains. This upper section is correlated by Caramés et al. (2004) with the Puerto Madryn Formation cropping out in several places of the Península Valdés region.

It is interesting to point out that the Large Igneous Province event of Jurassic age represented in northern Patagonia by the extended Marifil Formation (Malvicini and Llambías 1974), appears in the hydrogeological exploration borehole Puerto Madryn Nr. 1 at a depth of 189 m (INGyM 1965, Fig. 1) in contrast to the well YPF.Ch.PV.es-1 where it appears at a depth of 1964 m. In the Drilling Puerto Madryn Nr. 1, Cretaceous sediments reach 52 m in thickness, while the Tertiary is 137 m thick. Figure 1 shows schematic chronological columns for YPF.Ch.PV.es-1 and Puerto Madryn Nr. 1 boreholes, showing the thickness increase of Cretaceous and Tertiary strata toward the East, consistent with the development of the sedimentary Valdés Basin.

2.2 Surface Geology

Surface outcrops comprise Tertiary Neogene and Quaternary rocks (Fig. 2). On a relief carved in these units, recent sediments were deposited. Figure 2 shows the geological map of the Península Valdés region.

Geological map of Península Valdés

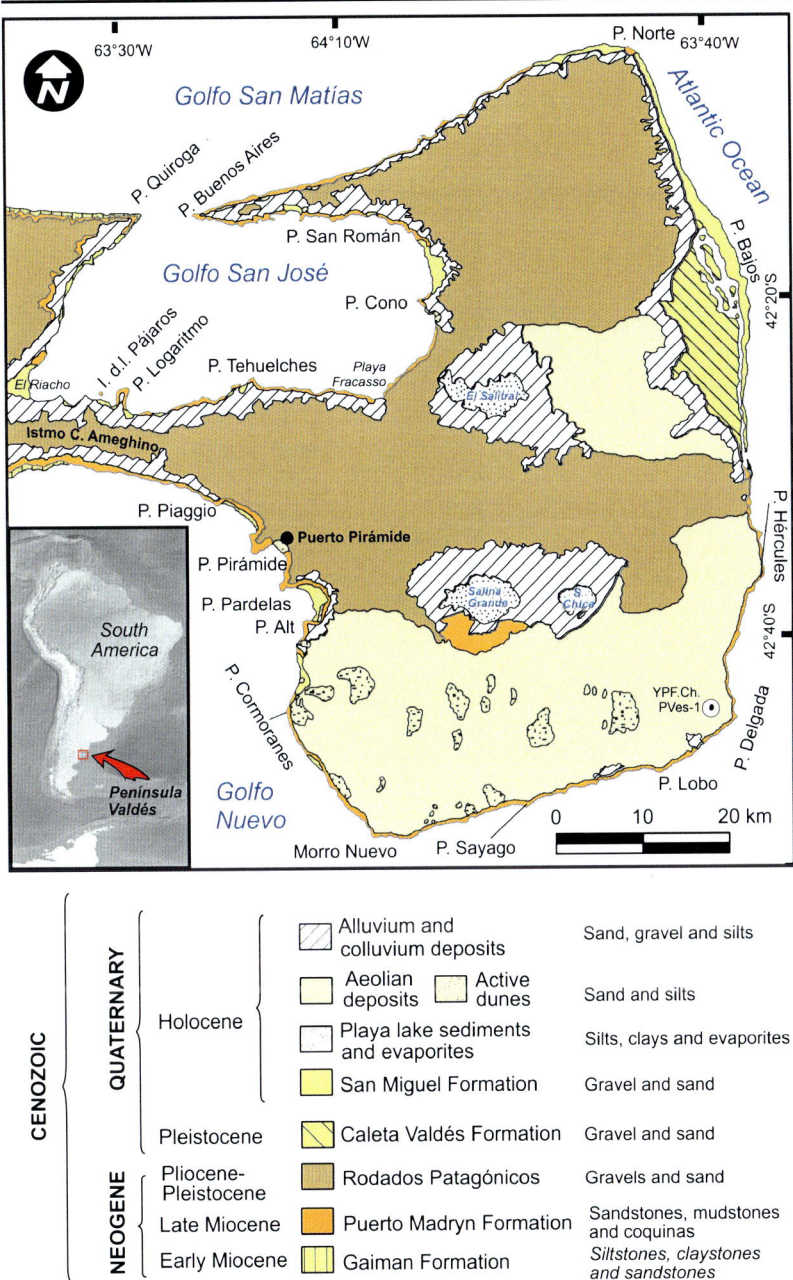

Fig. 2 Geological map of Península Valdés. Modified from Haller et al. (2001)

2.2.1 Gaiman Formation (Haller and Mendía 1980; Early Miocene)

Background

The Gaiman Formation denomination is used here to name the mudstones of tuffaceous volcanoclastic nature deposited during the Patagonian Marine Cycle and cropping out on the Atlantic coast of north Patagonia.

Marine sediments from the Atlantic coast of Patagonia are part of the *Tertiarie Patagonien* from d'Orbigny (1842), who named this way to all the strata of the Patagonian Atlantic margin. These sediments have also been observed in the Golfo Nuevo and elsewhere in Patagonia by Darwin (1846), who grouped them in the Patagonian Tertiary Formation.

The ages, correlation, and nomenclature of the Patagonia Paleogene and Neogene deposits of the Atlantic coastal cliffs have been the subject of long-lived discussion (Hatcher 1900; Ameghino 1906; Feruglio 1949; Camacho 1967; Riggi 1979a; Chiesa and Camacho 1995; Frassineti and Covacevich 1999; Casadío et al. 2000; Parras et al. 2008 among others). Haller and Mendía (1980) recommend restricting the use of the term Patagonia Group or Patagonia Formation to its type area in southern Patagonia and use another lithostratigraphic name for the sediments cropping out in other basins. In that sense, they proposed the name Gaiman Formation for these deposits for the area between Camarones and Sierra Grande. Later, for the Atlantic coastal region of Comodoro Rivadavia (46° S.L.) Bellosi (1990) defined as Chenque Formation the strata previously named as Patagoniano or Patagonia Formation.

Areal Distribution and Lithology

In the Península Valdés area, only reduced outcrops are found in the cliffs bordering the Golfo San Matías to the south, extending to Punta Quiroga. From here, outcrops extend along the west bank of the Golfo San José 3000 m to the southwest. Other reduced outcrops are found in the north western bordering cliffs of Golfo Nuevo.

The Gaiman Formation consists of tuffaceous mudstones with a few interspersed sandy levels. They are generally light-colored and have a massive structure, but in some banks normal or lenticular lamination can observed. Bioturbations—similar to those described by Lech et al. (2000) in the Río Chubut valley outcrops—are common, with many tracks of annelids and marks of perforating organisms. Trace fossils are the main source of paleoenvironmental information within this unit (see Chapter "Miocene Marine Transgressions: Paleoenvironments and Paleobiodiversity").

Stratigraphic Relationships

The base of this unit is unknown in the field of Península Valdés. To the west it is based on the limestones of the Arroyo Verde Formation (Haller 1982) of late early Paleocene age. It is covered by erosional unconformity by the Puerto Madryn Formation.

Age

As mentioned above, the age of the Gaiman Formation has been the subject of numerous discussions. In the Puerto Madryn area, the Gaiman Formation strata lie on the Sarmiento Group, carrying Colpodon fossils of late Colhuehuapian age. According to Marshall (1985), the Colhuehuapian beds cover the early Miocene. On the basis of their stratigraphic relationships, the correlation with the sea-level curves and fossil content, Scasso and Castro (1999) assigned an early Miocene age for the Gaiman Formation. More recently, Cuitiño et al. (2015) obtained very precise $^{86}Sr/^{87}Sr$ ages for the Chenque Formation in Golfo San Jorge Basin which indicate early to middle Miocene age fort those deposits.

2.2.2 Puerto Madryn Formation (Haller 1979, Late Miocene)

Background

Puerto Madryn Formation includes the sandstones and mudstones outcropping in the cliffs of Golfo San José and Golfo Nuevo and in the cliffs bordering the Península Valdés (Fig. 2). As mentioned previously, these layers were first observed by Malaspina in 1789 during his exploration trips around the world. Malaspina (1885) wrote in his travel diary: "... the cliffs seemed to be formed by horizontal beds of ... probably sands, marls, clays, etc. ...". Darwin (1846), who considered the strata from Península Valdés contemporaries to the Patagonian Tertiary Formation, as he named the marine Tertiary strata outcropping along the eastern coast of Patagonia. Carlos Ameghino was the first scholar to describe these strata after Malaspina and Darwin. Ameghino (1890) distinguished three chronostratigraphic units: the lower, called Paranaense with fossil content similar to that of Parana, Entre Rios; the Mesopotamiense similar to the banks of Rio Negro in Carmen de Patagones; and the Patagoniense of Golfo Nuevo composed by sandstones and volcanic marlstones which corresponds to the Gaiman Formation, but its age was misinterpreted by C. Ameghino because the latter occupies sometimes higher topographic levels.

The famed argentine paleontologist Florentino Ameghino, Carlos' brother, correctly locate stratigraphically this unit, by separating them with the name "Entrerriana Formation" in a later work (Ameghino 1897). That "Entrerriana Formation" is characterized by a fauna composed of *Ostrea patagonica*, *O. alvarezi*, *Pecten paranaensis*, etc. Subsequently, Wilckens (1905) considered these layers as the product of a transgression occurred in the late Tertiary, contemporary to the one that occurred in the Paraná basin, transgression which he called "Parana Stuffe".

In his important contribution, Ihering (1907) establishes the criteria followed by later paleontologists, assigning the Entrerriana Formation marine deposits containing *Ostrea patagonica* and *Ostrea alvarezi* and the layers with *O. madryna* to Rionegrense Formation. Roveretto (1921) specified the boundaries between the two units and found the existence of a Rionegrense characterized by the presence of *O. madryna* and *O. ferrarisi*.

Windhausen (1921) separated an Entrerriana Formation of marine origin, from the "Rio Negro Sandstones" of continental origin. The author admits a transitional passage between the two units. Frenguelli (1927) made a detailed stratigraphic and paleontological study in the region, considering that there are two chronostratigraphic units: Patagoniano and Entrerriano. In the latter, he distinguished three units: Entrerriense, Continental Rionegrense, and Marine Rionegrense. Between the last two mentioned units, there would have been a period of erosion. Meanwhile, according to Feruglio (1949), there would not be a definite boundary between Entrerriense and Rionegrense but a gradual passage. Feruglio (1949) considered that the disconformity between the two units are local phenomena and do not necessarily indicate a significant hiatus.

Distribution and Lithology

The Puerto Madryn Formation extends on the cliffs surrounding San José and Nuevo gulfs as well on the cliffs of the north, west, and south coast of Península Valdés (Figs. 2 and 3a). Some minor exposures are found on the southern border on the depression of Salinas Grande and Chica.

The Puerto Madryn Formation includes mudstones, sandstones, siltstones, bioclastic sandstones, and intraformational conglomerates. According with Scasso et al. (2012), the sedimentary assemblage represents a transgressive-regressive cycle (Scasso et al. 2012; and Chapter "Miocene Marine Transgressions: Paleoenvironments and Paleobiodiversity"). The lower section of the Puerto Madryn Formation represents the transgressive phase while the upper section shows a regressive phase.

The profile in Punta Cono in the eastern coast of the Golfo San José (Fig. 2) begins with yellowish gray mudstone and massive structure. Continue white siltstones with pyroclastic contribution. Above there is a coquina of *Ostrea patagonica* which in turn is covered by a succession of mudstones and sandstones. A bank with diatoms and volcanic glass stands out. On the top lies friable mudstone with abundant gypsum. Total thickness is 80 m.

In Punta San Román in the northern coast of Golfo San José, the profile of Puerto Madryn Formation shows white mudstone with pyroclastic components, covered by pale gray fine grained polymictic sandstones. The sandstones contain several coquina levels with Ostrea. Upwards crop out tuffaceous fine grained friable volcaniclastic friable fine grained sandstones. They have a bluish gray color and show normal lamination and calcareous interbedding. On the top are oyster banks with abundant *O. madryna* and less common *O. alvarezi*.

On the northern coast of Península Valdés, between Punta Buenos Aires and Punta Norte, the Puerto Madryn Formation crops out continuously on the coastal cliffs. In Punta Norte, the thickness of Puerto Madryn Formation reaches 45 m.

On the eastern coast of the Península, this unit crops out in the cliffs close to Punta Cero and extends to the south until Punta Delgada and continues toward the West on the southern margin of Península Valdés.

Fig. 3 a Puerto Madryn Formation. Cliff at Punta Pirámide, Golfo Nuevo; **b** Fault affecting the Puerto Madryn Formation in the northern shore of Golfo Nuevo (fault throw is approximately 2 m); c Structural sketch based on geophysical data according to Kostadinoff (1993). The *dotted lines* indicate basement faults. Letters inside squares indicate location of pictures of Fig. 3a, b

In Punta Hércules, the profile starts with very fine grained coquina sandstone with calcite cement. It shows cross-bedding where coarse-grained sandstones alternate with more bioclastic beds. Upwards, continues an alternation in normal bedding of fine-grained sandstones and siltstones. The set has a pale yellowish brown color. They are covered by yellowish gray sandy mudstones. Follow mudstones of pinkish gray color with levels with ostrea and casts of pectinidae. To the top, alternate sandstone and mudstone banks. The total thickness is 44 m.

The outcrops of Puerto Madryn Formation extend almost continuously on the northern coast of Golfo Nuevo from Morro Nuevo to the west.

Stratigraphic Relations

The Puerto Madryn Formation rests is uncomfortably upon the Gaiman Formation, as can be seen in Punta Quiroga. On the other hand, it is uncomfortably covered, by the Caleta Valdés Formation and Patagonian Gravels, of Pleistocene age.

Age

Field relations indicate an age between the lower Miocene and the Quaternary. Ostracods present suggest, according to García (1970) a late Miocene age. $^{40}K/^{39}Ar$ dating on three glass concentrates from a tuff in the upper part of the marine upper beds of Puerto Madryn Formation yielded an average age of around 9.4 Ma (Zinsmeister et al. 1980). $^{87}Sr/^{86}Sr$ dating from pectinidae and oyster shells from beds of the lower section of Puerto Madryn Formation yielded a mean age of 10 ± 0.3 Ma (Scasso et al. 2001). Malacofauna studied by del Río (1988), Martínez and del Río (2002) and del Río (2004) in turn indicate a late Miocene age for this unit. A palynological study (Palazzesi and Barreda 2004) supports a late Miocene age as well the Mammalian fossil content (Dozo et al. 2010) supports the late Miocene age for Puerto Madryn Formation. Given the consistency of the above results, a late Miocene age for the Puerto Madryn Formation is accepted.

2.2.3 Rodados Patagónicos/Patagonian Gravels (Pliocene–Early Pleistocene)

Background

According to the criteria set by Fidalgo and Riggi (1970), the Rodados Patagónicos are here named for the sandy gravel deposits that cover the highest mesa-like landforms. Despite some authors (Panza 2002; Martínez and Coronato 2008) suggest abandoning the name Rodados Patagónicos arguing that this denomination applies to different genetic processes and different formation periods. However Rodados Patagónicos is here used considering that the name is self-descriptive and has a historical root in the geological literature of Patagonia.

Distribution and Lithology

The Rodados Patagónicos are distributed on the remains of the oldest aggradation plain, whose remnants are in the central sector of the isthmus Carlos Ameghino and on the western region of the Península Valdés.

These deposits are made up of banks polymictic conglomerates with sandy-silty clay matrix, cemented partly by limy material. The gravels banks show an extended lateral continuity. The upper section of the gravel banks does not show a definite fabric, with the major axes of the clasts randomly arranged. However, the lower section shows a noticeable orientation of the elongated clast forms. The clasts are well rounded and are predominantly subspherical to slightly elongated and their composition are of siliceous, andesitic, and basaltic volcanics. Sometimes, loessial brown fine sands are intercalated at the base of the Patagonian Gravels. The thickness of this unit reaches 3 m. Clast size decreases progressively toward the east; in the isthmus Carlos Ameghino the clasts reach a diameter larger than 3.5 cm, while in the east coast of the Península Valdés, the maximum larger diameter size is 2 cm.

The presence of cryogenic disturbances in the Patagonian Gravels like fossil ice wedges cryodisturbances is known since the description of Liss (1969) of fossil ice wedges is present on this unit (see Chapter "Late Cenozoic Landforms and Landscape Evolution of Península Valdés"). Other observable features are calcareous carbonate crusts and disturbance of the primary sedimentary fabric (Vogt and del Valle 1994; Trombotto and Ahumada 1995 and Trombotto 1996; see Chapter "Soil–Geomorphology Relationships and Pedogenic Processes in Península Valdés").

Sedimentary Environment

According to Cortelezzi et al. (1965, 1968), the Patagonian Gravels were deposited by a braided fluvial system. Beltramone and Meister (1993) stated that the dispersant river environment would have high energy, with variations of the flow system during the sedimentation cycle.

Stratigraphic Relations

The Patagonian Gravels rely on erosional unconformity above the Miocene sediments of the Puerto Madryn Formation separated by an unconformity. On the other hand, they constitute the highest level of aggradation in the Península Valdés region, currently being destroyed by erosion by surface water runoff and partially by the wind. They are only covered by a skeletal soil with sparse vegetation and thin aeolian sand dunes.

Age

Field relations indicate a post-Miocene age for this unit. Considering the degree of evolution of the landscape, the tabular deposits of the Patagonian Gravels are in an advanced state of erosion, so that in the region of the isthmus Carlos Ameghino connecting the Península Valdés to the mainland, the mantle of gravel reaches only a few hundred-meter width. Furthermore, the deposition of such volumes of gravel requires the availability of large amounts of water on the continent, as during interglacial periods of deglaciation. The presence of cryodisturbances implies a periglacial environment during or short after the deposition of the gravels, also coherent with a deglaciation stage. For these reasons the Patagonian Gravels are given a Pliocene—early Pleistocene age.

2.2.4 Caleta Valdés Formation (Haller et al. 2001, Middle—Late Pleistocene)

Background

The first description of the beach ridges on the eastern margin of Península Valdés was provided by Roveretto (1921). He distinguishes two older levels and two younger levels formed by recent marine action (see Chapter "Late Cenozoic Landforms and Landscape Evolution of Península Valdés"). The name Caleta

Valdés Formation is given to the polymictic conglomerates with sandy matrix that make the ancient shorelines at the eastern end of the Península Valdés, presently at an altitude of 15–20 m (Kokot et al. 2005).

Distribution, Lithology, and Fossil Content
These deposits are spread throughout the west bank of the creek Valdés. Its thickness reaches 25 m. They are composed of well-rounded medium to coarse gravel. They have coarse sand matrix. Among the pebbles are found remains of *Adelomedon ancilla, Ameghinomya antigua, Aulacomya magallánica, Brachidontes rodriguezi, Crepidula dilatata, Lucapinella henseli, Mytilus edulis, Panopea abreviatta, Patinigeria magallanica, Pitaria lahillei, P. rostrata, Protothaca antiqua*, and *Saramangia exalbida* (Codignotto 1983; Fasano et al. 1983; and Rutter et al. 1989).

Depositional Environment
Gravels of the Caleta Valdés Formation were deposited by the action of longshore currents and therefore represent accretion marine deposits.

Stratigraphic relations
The Caleta Valdés Formation rests uncomfortably on the late Miocene sediments of the Puerto Madryn Formation. It is located on a lower topographic level than the Patagonian Gravels but above the present level of beach ridges. Its surface has a limited developed sandy soil and has an herbaceous cover with scarce bushes (see Chapter "Soil–Geomorphology Relationships and Pedogenic Processes in Península Valdés").

Age
Stratigraphic relations indicate a post-Miocene and pre-Holocene age for this unit. The geological literature presents several radiometric dating for Caleta Valdés Formation in its type locality. Codignotto (1983) reports a ^{14}C age of 38,700 years old done on samples of Chione sp. in life position. Fasano et al. (1983) ^{14}C yielded ages of 41,000 ± 4000; 39,000 ± 3200 and 34,000 ± 1700 years on articulated mollusk shells. Meanwhile, Rostami et al. (2000) obtained 115,000 ± 5000 and 126,000 ± 10,000 years Th/U ages for the topographically higher levels of Caleta Valdés Formation. Other ESR ages have been given by Schellmann (1998) comprised between 136,000 and 46,000 years for the uppermost level of Caleta Valdés Formation. Based on the cited radiometric ages and the topographic position above the Holocene beach ridges, Caleta Valdés Formation is assigned to the Pleistocene.

2.2.5 San Miguel Formation (Haller 1982, Holocene)

Background
Haller (1982) applied this term to gravels and sands with abundant mollusk shell fragments that are cropping out in the vicinity of Puerto Madryn at higher elevation than present beach deposits. Following these criteria in this chapter all deposits of beach and beach ridges of the high margins of the Península Valdés, and the Golfo Nuevo, Golfo San José, and Golfo San Matías gulfs are named in this way.

Distribution and Lithology

The San Miguel Formation is topographically eight meters above the current line of high tide. In the Golfo San José, there are reduced outcrops to the east and west of the Punta San Roman and north and south of Punta Cono (Fig. 2). In the Golfo Nuevo the outcrops are located east of Punta Pardela (Fig. 2). On the east coast of Península Valdés, the San Miguel Formation constitutes the Holocene beach ridges that limit the Caleta Valdés lagoon (Kokot et al. 2005).

The San Miguel Formation is mainly composed of medium to coarse gravel, with a matrix of fine gravel, coarse sand, and fragments of bivalves. The lithology of the gravels corresponds to mesosiliceous volcanics rocks with varying degrees of alteration and very subordinately to plutonic rocks and flint.

There is a relationship between the sizes of the boulders and distribution of deposits. Those deposits that are surrounded by paleo-cliffs such as those to the east of Punta Pardelas, west of Punta Ninfas in the Golfo Nuevo, and at the north of Punta Cono in Golfo San José, are made up of rounded and subspherical gravel pebble size. Meanwhile the other side, deposits facing the open sea and unprotected from cliffs are formed by boulders, cobbles, and pebbles showing low sphericity and flattened shapes. The San Miguel Formation contains numerous fragmented remains of bivalves and gastropods.

Depositional Environment

The textural characteristics of the San Miguel Formation and its relationship to the location on the coast, suggest that this unit was deposited in two coasts sub-environments. Those located within the gulfs formed by medium-sized gravels, protected by paleo-cliffs were deposited in a beach sub-environment subject to wave action of high to medium energy. On the other hand, the thicker deposits of flattened shapes were accumulated by the action of littoral drift currents in the form of beach ridges.

Stratigraphic Relations

The San Miguel Formation is sparsely vegetated on its surface. It passes laterally to present-day beach deposits. On the east coast of the Península Valdés region, the accumulation process continues currently (Kokot et al. 2005).

Age

Characteristics mentioned above indicate that the San Miguel Formation was deposited during the Holocene. Several ^{14}C ages for these deposits are found in the geological literature: Codignotto (1987) and Codignotto and Kokot (1988) yielded ^{14}C ages between 5725 ± 105 and 1330 ± 80 years AP. Weiler and Meister obtained coherent ^{14}C ages in El Riacho area of 6250 ± 90; 6090 ± 110; 5990 ± 60; and 1140 ± 50 years AP. More recently, Brückner et al. (2007) published ^{14}C ages comprised between 334 ± 50 and 6518 ± 61 years BP. Those data indicate, as pointed out by the mentioned authors, that a sea level highstand occurred about 6500 years ago with the deposition of elevated beaches at an altitude of ≈ 8 m above actual sea level.

2.2.6 Unconsolidated Deposits not Grouped Within a Formal Lithostratigraphic Unit

Playa Lake Sediments and Evaporites

The endorheic depressions in the region have their bottoms covered by very fine sediments, like silt, clayed silt, and clay, clear brown to light gray colored. In many cases, fine sediments may bear associated with evaporites, predominantly halite. In the Salitral closed depression, these mineral are accompanied by sulfates as glauberite and gypsum. The thickness of the evaporites ranges between 1 and 3 mm.

In the Salinas Grande and Salina Chica there are accumulations of evaporite crusts. There is an alternance of muds and evaporites whose thickness is larger than one meter.

Aeolian Deposits

The southern surface of Península Valdés shows an accumulation of sediments deposited by wind action (Fig. 2). Among them, it is possible to distinguish active sand dunes and slightly also older aeolian deposits, colonized by vegetation and subjected to erosion by subaerial agents.

Lithologically, it is fine-to-medium sand with a minor pelitic fraction and very low fraction of gravel size. The composition is feldspathic quartz, with little involvement of volcanic glass and fragments of organic origin, such as remains of shells. The heavy fraction that concentrates on levels distinguishable by their color is formed of tourmaline, epidote, amphibole, and pyroxene.

Alluvium and Colluvium Deposits

Alluvium and colluvium deposits cover areas scattered throughout the region. They consist of unconsolidated deposits of light gray to light brown color, whose grain size corresponds to medium-fine sand, mixed with varying proportions of silt, clay, and scattered pebbles.

The accumulations are relatively thin and they have their origin in the material eroded from the different geological units cropping out in the península.

3 Structural Geology

The region is characterized by a series of gravitational type faults (Fig. 3b, c). The Quaternary cover makes the structural features observable only in the cliffs bordering the sea. Subvertical faults are observed and the maximum throw are about four meters. Most fractures are observed submeridional direction. A series of joints with the same direction complete the structural features of the region. The mentioned tectonic disturbances affect the Puerto Madryn Formation of Neogene age.

Other structural features are interpretable from the geomorphological features. Thus, the straight path of the coast bordering Golfo San Matias in the south and the east coast of the Península Valdés region, suggesting a structural control on the morphology.

Strata of the Gaiman and Puerto Madryn formations have soft warping with slopes not exceeding 5°, so it is not possible to establish whether they are of syngenetic or tectonic nature.

Geophysical studies by Kostadinoff (1993) allowed postulating the existence of a structural high on the Península Valdés, which is separated westward from the mainland by a structural low (Fig. 3c). The structural low location is approximately coincident with the gulfs Golfo San José and Golfo Nuevo (Fig. 3c). This structure is derived from former crustal heterogeneities and possibly defines a horst-graben morphology.

4 Historical Geology

Considering only the surface geology, the following historical geology can be proposed:

During the early Miocene the area was covered by a continental shelf sea, which had an important pyroclastic contribution as a result of volcanic eruptions that occurred in the Andean chain (see Chapter "Climatic, Tectonic, Eustatic and Volcanic Controls on the Stratigraphic Record of Península Valdés" and this chapter). This sea deposits remained registered in the Gaiman Formation.

Subsequently, during the middle Miocene, the region was raised accompanying the Andes uplift, producing the tectonic readjustment of the Patagonian foreland. Later, in the middle-late Miocene, there was a new marine ingression in the region with the deposition of Puerto Madryn Formation. The depositional environment of the base of the Puerto Madryn Formation was a temperate sea near the coast. The marine sedimentation was gradually replaced due to a larger continental contribution of sediments.

During Pliocene-Pleistocene times, fluvial systems deposited the Patagonian Gravels which formed an extended aggradation surface.

In the area located to the west of what is now the Caleta Valdés an accretionary regime deposited beach ridges during the Pleistocene, registered in the Caleta Valdés Formation. Probably in this period, the intense winds existing in the periglacial system of the region, caused deflation of Nuevo and San José gulfs favored by old tectonic discontinuities, and began the formation of Salinas Grande and Chica depressions.

During the Holocene postglacial, the waters reached approximately its present level and about 6500 years ago there was another relative sea level rise, leaving the San Miguel Formation record, formed by gravel and shell fragments in a beach environment.

5 Geological Resources

Mineral resources within the area of Península Valdés are limited to salt deposits, located in Salina Grande and Salina Chica. Also, sporadically, sand and gravels are extracted.

The mining history of the region has its beginnings in the late nineteenth century, with the installation of a company dedicated to the exploitation of the salt flats. The extracted salt was transported by railroad to the natural port of Pirámide being shipped to Buenos Aires.

Mining continued until 1920; after a period of inactivity around 1950 it resumes operation in the southern area of Salina Grande. At this stage the production supplied the fishing industry based in the city of Puerto Madryn.

During the decade of 1990 only one mine operated in Salina Chica providing salt for an aluminum slag recovery factory in Puerto Madryn. During 1995 the mine produced 1520 tons of salt with 98% of Na Cl content.

Presently, there are no operating salt mines in Península Valdés although several mining claims are still valid.

5.1 Salt Deposits

Several authors have contributed to the knowledge of the salt deposits in Península Valdés. Brodtkorb and Re (1962) and Brodtkorb (1980) described in their works the geological features and the possible origin of the depressions containing the salt; the latter author also indicates economic parameters and chemical characteristics of these deposits.

The following geological features were obtained from the above-mentioned works and complemented by own field observations. The deposits are composed of salt layers of variable outlines and thickness which rest on mud.

The salt bodies present in the base several glauberite horizons whose thicknesses vary from 0.20 to 1 m; especially in certain parts of the central area. Glauberite is manifested as isolated individuals or as clustered gypsum pseudomorph rosettes.

The typical profile (top-down) of the salt is according to Brodtkorb (1980), 1–3 mm white salt which overlies a crust of silt and salts 1–3 mm thick. Underneath, about 1 m foul-smelling blackish green mud. Thin horizons of 1–2 mm in thickness of halite crystals were also observed. Continues a layer of volcanic ash from about 50 mm thick, followed by a layer of silt with 2 m glauberite and gypsum crystals.

Mud plasticity silt gradually decreases downwards increasing the sand percentage in the mud.

According to Brodtkorb (1980), indicated reserves for Salina Grande are 52,222,300 tons; while for Salina Chica measured reserves are 6,013,600 tons. Chemical composition is indicated in Table 2.

Table 2 Chemical composition of the salt in Salina Grande and Salina Chica, after Brodtkorb (1980)

	Salina Grande	Salina Chica
	% weight	% weight
NaCl	80.25	82.50
$MgCl_2$	0.83	0.63
$CaCl_2$	0.19	1.72
Na_2SO_4	0.97	2.17
$CaSO_4$	6.99	5.88
Insoluble	10.65	7.09

5.2 Industrial Rocks (Sand and Gravels)

This material is circumstantially worked out for its use in the scarce building activity in the region.

The sector of dunes located in the vicinity of Puerto Pirámide and in the southern part of the Península is exploited in a very primitive form using shovels and small trucks.

The material—called by the villagers "blasted sand"—is a loose aggregate of particles whose particle size varies from fine to medium sand.

Gravels are used for the construction and improvement of the road network. It is common to see the gravel quarries on the road sides that are operated at certain times of the year.

The mined material comes from the Patagonian Gravels covering part of the surface of the Península.

6 Perspectives and Future Work

While the simple stratigraphy of the region is well understood, further attention should be given to framework of facies and sequence stratigraphy. Further attention must be paid to precise the chronostratigraphy and biostratigraphy of the Miocene deposits. There is considerable scope to precise the dating of the Pliocene-Quaternary coastal sediments. There is also considerable scope for structural geology of Península Valdés and surrounding gulfs as well with its relation with the Valdés sedimentary basin.

Acknowledgements The author would like to acknowledge the valuable critical review made by Roberto Page and Luciano Lopez. Also he is in debt for comments and advice made by the editors of this book, Pablo Bouza and Andrés Bilmes, which substantially improved the manuscript.

Glossary

Boulder	Large rock fragment formed by detachment from its parent consolidated rock by weathering and erosion, which has one linear dimension of at least 25.4 cm
Cenozoic	The current geologic era, which began 66.4 million years ago and continues to the present
Cobbles	Clast of rock having a particle size of 6,4–25,6 cm
Formation	Primary rock unit used in the subdivision of a sequence and may vary in scale from tens of centimeters to kilometers
Glauberite	Sodium calcium sulfate mineral with the formula $Na_2Ca(SO_4)2$
Gravel	Unconsolidated rock fragments that have a general particle size range of 2–64 cm
Lithostratigraphy	Part of stratigraphy that deals with the description and nomenclature of the rocks of the Earth based on their type and their stratigraphic relations
Neogene	A geologic period starting 23 million years ago and ending 2.6 million years ago with the beginning of the Quaternary
Paleogene	A geologic period and system that began 66 and ended 23 million years ago
Pebbles	Rock fragments (often rounded) of around 1–10 cm in size
Polymictic	Composed of several minerals or rock types
Pyroclastic	Fragments of rock and volcanic ash thrown out of a volcano
Quaternary	Is the current and most recent geological period
Sedimentary basins	Are regions of Earth of long-term subsidence creating accommodation space for infilling by sediments
Subsidence	Is the motion of the Earth's surface as it shifts downward relative the sea level
Volcanics	Volcanic rocks formed from magma erupted from a volcano

References

Alric V et al (1996) Geocronología $^{40}Ar/^{39}Ar$ del volcanismo jurásico de la Patagonia Extraandina. 13° Congreso Geológico Argentino y 3° Congreso de Exploración de Hidrocarburos, Actas 5:243–250 (Buenos Aires)

Ameghino C (1890) Exploraciones geológicas en Patagonia. Boletín del Instituto Geográfico Argentino 9(1):3–46 (Buenos Aires)

Ameghino F (1897) Mammiféres crétacés de l' Argentine. Deuxime contribution a la connassaince de la faune mammologique des couches a Pyrotherium. Boletín del Instituto Geográfico Argentino 18 (Buenos Aires)

Ameghino F (1906) Les formations sedimentaires du Cretacé superieur et du Tertiaire de Patagonie. Anales del Museo Nacional de Buenos Aires 8 (Buenos Aires)

Bellosi ES (1990) Formación Chenque: Registro de la transgresión patagoniana (Terciario medio) de la cuenca de San Jorge, Argentina. XI Congreso Geológico Argentino Actas 2:57–60 (San Juan)

Beltramone C, Meister C (1993) Paleocorrientes de los Rodados Patagónicos. Tramo Comodoro-Trelew. Asociación Geológica Argentina, Revista 47(2):147–152 (Buenos Aires)

Bertels A (1970) Sobre el "Piso Patagoniano" y la representación de la época del Oligoceno en Patagonia Austral, República Argentina. Asociación Geológica Argentina, Revista 25:495–501 (Buenos Aires)

Brodtkorb A (1980) Some Sodium Chloride Deposits from Patagonia, Argentina. In: Coogan AH, Haube L (eds) V Symposium on Salt. Northern Ohio Geological Society, Hamburg, vol 1, pp 31–39

Brodtkorb A, Ré N (1962) Los depósitos salinos del Bajo del Gualicho y de la península de Valdés, Provincias de Río Negro y Chubut. I Jornadas Geológicas Argentinas. Anales 3 (Buenos Aires)

Brückner H et al (2007) Erste Befunde zu Veränderungen des holozänen Meeresspiegels und zur Größenordung holozäner 14C-Reservoireffekte im Bereich des Golfo San José (Península Valdés, Argentinien). Bamberger Geographische Schriften 22:93–111

Camacho HH (1967) Las transgresiones del Cretácico superior y Terciario de la Argentina. Asociación Geológica Argentina, Revista 22(4):253–280 (Buenos Aires)

Camacho H (1974) Bioestratigrafía de las formaciones marinas del Eoceno y Oligoceno de la Patagonia. Academia de Ciencias Exactas, Físicas y Naturales, Anales 26 (Buenos Aires)

Camacho H (1979a) Descripción Geológica de las Hojas 47b y 48b (Bahía Camarones), Provincia del Chubut. Servicio Geológico Nacional, Boletín N° 153 (Buenos Aires)

Camacho H (1979b) Significados y usos de "Patagoniano", "Patagoniense", "Formación Patagónica", "Formación Patagonia" y otros términos de la estratigrafía del Terciario marino argentino. Asociación Geológica Argentina, Revista 34(3):235–242 (Buenos Aires)

Camacho H (1980) La Formación Patagonia, su nuevo esquema estratigráfico y otros temas polémicos. Asociación Geológica Argentina, Revista 35(2):276–281 (Buenos Aires)

Caramés A, Malumián N, Náñez C (2004) Foraminíferos del Paleógeno del Pozo Península Valdés (PV.es-1), Patagonia septentrional, Argentina. Ameghiniana 41(3):461–474 (Buenos Aires)

Casadío S, Feldmann R, Foland K (2000) $^{40}Ar/^{39}Ar$ age and oxygen isotope temperature of the Centinela Formation, southwestern Argentina: An Eocene age for crustacean-rich "Patagonian" beds. J S Am Earth Sci 13:123–132

Casal GA, Allard JO, Foix N (2015) Análisis estratigráfico y paleontológico de afloramientos del Cretácico superior en la cuenca del Golfo San Jorge: propuesta de nueva unidad litoestrati-gráfica para el Grupo Chubut. Asociación Geológica Argentina, Revista 72(1) (Buenos Aires)

Chebli G et al (1976) Estratigrafía del Grupo Chubut en la región central de la Provincia homónima. In 6° Congreso Geológico Argentino, Actas 1:375–392 (Buenos Aires)

Chiesa JO, Camacho HH (1995) Litoestratigrafía del Paleógeno marino en el noroeste de la provincia de Santa Cruz, Argentina. Monografías de la Academia Nacional de Ciencias Exactas, Físicas y Naturales de Buenos Aires 11, Parte 1: 9–15 (Buenos Aires)

Codignotto JO (1983) Depósitos elevados y/o de acreción Pleistoceno-Holoceno en la costa Fueguino-Patagónica. Simposio Oscilaciones del Nivel del Mar durante el Último Hemiciclo Deglacial en la Argentina, Actas 12–26 (Mar del Plata)

Codignotto JO (1987) Cuaternario marino entre Tierra del Fuego y Buenos Aires. Asociación Geológica Argentina, Revista 42(1–2):208–212 (Buenos Aires)

Codignotto JO, Kokott RR (1988) Evolución geomorfológica holocena en caleta Valdés, Chubut. Asociación Geológica Argentina, Revista 43(4):474–481 (Buenos Aires)

Continanzia J et al (2011) Cuencas de Rawson y Valdés: Síntesis del conocimiento exploratorio - Visión Actual. VIII Congreso de Exploración y Desarrollo de Hidrocarburos 47–64 (Mar del Plata)

Cortelezzi CR, De Salvo O, De Francesco F (1965) Estudio de las gravas Tehuelches de la región comprendida entre el río Colorado y el Río Negro, desde la costa de la provincia de Buenos Aires hasta Choele-Choel. Acta Geológica Lilloana 6:65–85 (S. M. de Tucumán)

Cortelezzi CR, De Salvo O, De Francesco F (1968) Estudio de las gravas Tehuelches en la región comprendida entre el río Colorado y el río Negro desde la costa atlántica hasta la cordillera. III Jornadas Geológicas Argentinas, Actas 3:123–145 (Buenos Aires)

Cuitiño JI et al (2015) Sr-stratigraphy and sedimentary evolution of early Miocene marine foreland deposits in the northern Austral (Magallanes) Basin, Argentina. Andean Geol 42(3):364–385 (Santiago)

Darwin Ch (1845) Journal of researches into the natural history and geology of the countries visited during the voyage of H.M.S. Beagle around the world, under the command of Capt. Fitz Roy, R.A. Second Edition, corrected with additions. London. Delphi Classics 2015

Darwin C (1846) Geological observations on South America. Being the third part of the geology of the voyage of the Beagle, under the command of Capt. Fitzroy, R.N. during the years 1832 to 1836. Smith Elder and Co. London, p 280

del Río C (1988) Bioestratigrafía y cronoestratigrafía de la Formación Puerto Madryn (Mioceno medio) - provincia del Chubut - Argentina. Academia Nacional de Ciencias Exactas, Físicas y Naturales, Anales, 40:231–254 (Buenos Aires)

del Río C (1990) Composición, origen y significado paleoclimático de la malacofauna "Entrerriense" (Mioceno medio) de la Argentina. Academia Nacional de Ciencias Exactas, Físicas y Naturales, Anales 42:205–224 (Buenos Aires)

del Río C (1991) Revisión sistemática de los bivalvos de la Formación Paraná (Mioceno medio) provincia de Entre Ríos - Argentina. Academia Nacional de Ciencias Exactas, Físicas y Naturales, Monografía 7:11–26 (Buenos Aires)

del Río C (1992) Middle Miocene bivalves of the Puerto Madryn Formation, Valdés Península, Chubut Province, Argentina (Nuculidae-Pectinidae) Part I. Palaeontographica Abt A 225(1–3):1–58 (Stuttgart)

del Río CJ (2004) Tertiary marine Molluscan Assemblages of Eastern Patagonia (Argentina): a biostratigraphic analysis. J Paleontol 78(6):1097–1112

Di Paola EC, Marchese HG (1973) Litoestratigrafía de la Formación Patagonia en el área tipo (Bajo de San Julián-desembocadura del río Santa Cruz). Provincia de Santa Cruz. República Argentina. 5° Congreso Geológico Argentino. Actas 3:207–222

d'Orbigny A (1842) Voyage dans l'Amerique méridionale, exécuté pendant les années 1826–33, III, et 4. Paris

Dozo MT et al (2010) Late Miocene continental biota in Northeastern Patagonia (Península Valdés, Chubut, Argentina). Palaeogeogr Palaeoclimatol Palaeoecol 297:100–106

Fasano JL, Isla FI, Schnack EJ (1983) Un análisis comparativo sobre la evolución de ambientes litorales durante el Pleistoceno tardío - Holoceno: Laguna Mar Chiquita (Buenos Aires) - Caleta Valdés (Chubut). Simposio Oscilaciones del Nivel del Mar durante el último Hemiciclo Deglacial en la Argentina, Actas 27 (Mar del Plata)

Feruglio E (1949–1950) Descripción geológica de la Patagonia. Dirección General Yacimientos Petrolíferos Fiscales, Buenos Aires, vols 1, 2 and 3

Fidalgo F, Riggi JC (1970) Consideraciones geomórficas y sedimentológicas sobre los Rodados Patagónicos. Asociación Geológica Argentina, Revista 25(4):430–443 (Buenos Aires)

Frenguelli J (1927) El Entrerriense del Golfo Nuevo en el Chubut. Boletín de la Academia Nacional de Ciencias 39 (Córdoba)

Frassinetti D, Covacevich V (1999) Invertebrados fósiles marinos de la Formación Guadal (Oligoceno superior- Mioceno inferior) en Pampa Castillo, Región de Aisén, Chile. Boletín del Servicio Nacional de Geología y Minería de Chile 51:1– 96 (Santiago)

García ER (1970) Ostracodes du Miocene de la République Argentine ("Entrerriense") de la Peninsule Valdés. IV Colloque Africain de Micropaleontologie 391–417 (Abidjan)

Haller MJ (1979) Estratigrafía de la región al poniente de Puerto Madryn, provincia del Chubut, República Argentina. 7° Congreso Geológico Argentino, Actas. Tomo 1:285–297 (Buenos Aires)

Haller MJ (1982) Descripción geológica de la Hoja 43 h, Puerto Madryn, provincia del Chubut. Servicio Geológico Nacional, Boletín 184, 5 figs., 6 lám., 8 cuad., 1 mapa. Buenos Aires

Haller MJ, Mendía JE (1980) Las sedimentitas del ciclo Patagoniano en el litoral atlántico norpatagónico. Coloquio "R. Wichmann". Asociación Geológica Argentina. In: Mendía JR, Bayarsky A (1981) (eds) Estratigrafía del Terciario en el valle inferior del río Chubut. VIII Congreso Geológico Argentino, Actas 3:593–606 (Buenos Aires)

Haller MJ, Monti AJ, Meister CM (2001) Hoja Geológica 4363-I Península Valdés, Provincia del Chubut. Programa Nacional de Cartas Geológicas de la República Argentina, 1:250.000. Boletín N° 266, pp 1–34; 1 mapa. Servicio Geológico Minero Argentino. Buenos Aires

Harrington HJ (1962) Paleogeographic development of South America. Bull Am Assoc Pet Geol 46(10):1773–1813 (Tulsa)

Hatcher JB (1900) Sedimentary rocks of Southern Patagonia. American J Sci Serie 4, 9(50): 85–108

Ihering HV (1907) Les mollusques fossiles du Tertiaire et du Cretacé superieur de l'Argentine. Anales del Museo Nacional de Buenos Aires 3:7 (Buenos Aires)

INGyM (1965) Perfiles de perforaciones. Período 1916–1925. Instituto Nacional de Geología y Minería. Publicación N° 152. Buenos Aires

Kokot RR, Monti AAJ, Codignotto JO (2005) Morphology and short-term changes of the Caleta Valdés Barrier Spit, Argentina. J Coast Res 21(5):1021–1030 (West Palm Beach, Florida)

Kostadinoff J (1993) Estudio geofísico de la península de Valdés y los golfos nordpatagónicos. Asociación Geológica Argentina, Revista 47(2): 229–236 (Buenos Aires)

Lech RR, Aceñolaza FG, Grizinik MM (2000) Icnofacies Skolithos-Ophiomorpha en el Neógeno del Valle Inferior del Río Chubut, provincia de Chubut, Argentina. Serie Correlación Geológica 14:147–161 (S.M. de Tucumán)

Lesta P (1968) Estratigrafía de la cuenca del golfo San Jorge. III Jornadas Geológicas Argentinas 1:251–289 (Buenos Aires)

Liss CC (1969) Fossile Eiskeile (?) an der Patagonischen Atlantikküste. Zeitschrift für Geomorphologie, N.F., 3d.13, Heft 1. Berlin-Stuttgart

Mainardi EC, Turic MA, Stubelj R (1980) La exploración petrolífera en la Plataforma Continental Argentina. ARPEL, 35 Reunión a nivel de expertos. México

Malaspina A (1885) La vuelta al mundo por las corbetas Descubierta y Atrevida. Madrid

Malvicini L, Llambías E (1974) Geología y génesis del depósito de manganeso Arroyo Verde, provincia del Chubut, República Argentina. 5° Congreso Geológico Argentino, Actas 2: 185–202 (Buenos Aires)

Marshall LG (1985) Geochronology and land-mammal biochronology of the transamerican faunal interchange. In: Stehli F, Webb SD (eds) The great American biotic interchange. Plenum Press, New York, pp 49–85

Martínez OA, Coronato AMJ (2008) The late Cenozoic fluvial deposits of Argentine Patagonia, In: Rabassa J (eds) The late Cenozoic of Patagonia and Tierra del Fuego. Dev Quat Sci 11: 205–226 (Elsevier)

Martínez S, del Río C (2002) Las provincias malacológicas miocenas y recientes del Atlántico sudoccidental. Anales de Biología 24:121–130

Masiuk V, Decker D, García Espiasse A (1976) Micropaleontología y sedimentología del Pozo YPF (Ch.P.V.es-1), Península de Valdés, provincia del Chubut, República Argentina. Importancia y correlaciones. ARPEL, 24, YPF, 22 p (Buenos Aires)

Palazzesi L, Barreda V (2004) Primer registro palinológico de la Formación Puerto Madryn, Mioceno de la provincia del Chubut, Argentina. Ameghiniana 41:355–362 (Buenos Aires)

Panza JL (2002) La cubierta detrítica del Cenozoico superior. In: Haller MJ (eds) Geología y Recursos Naturales de Santa Cruz, 15° Congreso Geológico Argentino, Relatorio, pp 259–284

Parras A et al (2008) Correlation of marine beds based on Sr- and Ar-date determinations and faunal affinities across the Paleogene/Neogene boundary in southern Patagonia, Argentina. J S Am Earth Sci 26:204–216

Rapela C, Pankhurst R (1993) El Volcanismo riolítico del noreste de la Patagonia: Un evento meso-jurásico de corta duración y origen profundo. 12° Congreso Geológico Argentino y 2° Congreso de Explotación de Hidrocarburos, Actas 4:179–188 (Buenos Aires)

Riggi JC (1979a) Nuevo esquema estratigráfico de la Formación Patagonia. Asociación Geológica Argentina, Revista 34(1):1–11 (Buenos Aires)

Riggi JC (1979b) Nomenclatura, categoría litoestratigráfica y correlación de la Formación Patagonia en la costa atlántica. Asociación Geológica Argentina, Revista 34(3):243–248 (Buenos Aires)

Riggi JC (1980) Aclaración y ampliación de conceptos sobre el nuevo esquema estratigráfico de la Formación Patagonia. Asociación Geológica Argentina, Revista 35(2):282–189 (Buenos Aires)

Rostami K, Peltier WR, Mangini A (2000) Quaternary marine terraces, sea-level changes and uplift history of Patagonia, Argentina: comparisons with predictions of the ICE-4G (VM2) model of the global process of glacial isostatic adjustment. Quatern Sci Rev 19:1495–1525

Roveretto G (1921) Studi di geomorfologia argentina. V. La Penisola Valdez

Rutter N, Schnack E, del Río CJ, Fasano J, Isla F, Radtke U (1989) Correlation and dating of Quaternary littoral zones along the Patagonian coast, Argentina. Quatern Sci Rev 8:213–234

Scasso RA, del Río C (1987) Ambientes de sedimentación, estratigrafía y proveniencia de la secuencia marina del Terciario superior de la región de Península Valdés, Chubut. Asociación Geológica Argentina, Revista 42(3–4):291-321 (Buenos Aires)

Scasso RA, Castro LN (1999) Cenozoic phosphatic deposits in North Patagonia, Argentina: phosphogenesis, sequencestrati-graphy and paleoceanography. J S Am Earth Sci 12:471–487

Scasso RA, McArthur JM, del Río CJ, Martínez S, Thirwall MF (2001) $^{87}Sr/^{86}Sr$ late Miocene age of fossil molluscs in the "Entrerriense" of the Valdés Península (Chubut, Argentina). J S Am Earth Sci 14:319–329

Scasso RA, Dozo MT, Cuitiño JI, Bouza P (2012) Meandering tidal-fluvial channels and lag concentration of terrestrial vertebrates in the fluvial-tidal transition of an ancient estuary in Patagonia. Latin Am J Sedimentology Basin Anal 19, 27–45 (Buenos Aires

Schellmann G (1998) Jungkänozoische Landschaftsgeschichte Patagoniens (Argentinien). Andine Vorlandvergletscherungen, Talentwicklung und marine Terrassen. Essener Geographische Arbeiten 29:216 S (Essen)

Spalletti L, Cingolani C, Merodio J (1991) Ambientes y procesos generadores de las sedimentitas portadoras de hierro en la plataforma silúrico-eo-devónica de la Patagonia, República Argentina. Revista del Museo de La Plata (nueva serie) 10:305–318

Trombotto D (1996) The old cryogenic structures of Northern Patagonia: the Cryomere Penfordd. Z. Geomorph.N.F. 40:385–399 (Berlin-Stuttgart)

Trombotto D, Ahumada AL (1995) Die Auswirkung alter Kryomere auf die "Rodados Patagónicos" in Nordpatagonien, Argentinien. Eiszeitalter u. Gegenwart 45: 93–108 (Hannover)

Vogt T, del Valle HF (1994) Calcrites and cryogenic structures in the area of Puerto Madryn (Chubut, Patagonia, Argentina). Geogr Ann 76A(1–2):57–75

Wilckens O (1905) Die Meeresablagerungen der Kreide und Tertiärformation in Patagonien. Neues Jahrb. f. Min., G. u. P., 21. Stuttgart

Windhausen A (1921) Informe sobre un viaje de reconocimiento geológico en la parte noreste del Territorio del Chubut, con referencia especial a la provisión de agua de Puerto Madryn. Con un estudio petrográfico de algunas rocas por R. Beder. Dirección General de Minas, Boletín 24B. Buenos Aires

YPF (1976) Legajo del Pozo YPF.Ch.PV.es-1. Inédito. Buenos Aires, 122 pp

Zinsmeister WJ, Marshall IG, Drake RE, Curtis G (1980) First radioisotope (Potassium-Argon) Age of Marine Neogene Rionegro Beds in Northeastern Patagonia Argentina. Science 212:440

Miocene Marine Transgressions: Paleoenvironments and Paleobiodiversity

José I. Cuitiño, María T. Dozo, Claudia J. del Río, Mónica R. Buono, Luis Palazzesi, Sabrina Fuentes and Roberto A. Scasso

Abstract Two major marine transgressions covered part of Patagonia during the Miocene and both are recorded in the Península Valdés region. The older (early Miocene) is represented by the volumetrically scarce outcrops of the Gaiman Formation, composed by shelf mudstones and fine sandstones. The late Miocene transgression is represented by the Puerto Madryn Formation, widely distributed in Península Valdés and composed of mudstones, sandstones and shell beds, being the focus of this work. Sediments of this unit were deposited in inner shelf, nearshore, tidal channel and tidal flat environments. Fossil content is abundant and diverse, including palynomorphs, foraminifers, marine invertebrates (dominated by molluscs), cetaceans, pinnipeds, marine fishes and birds, as well as continental mam-

J.I. Cuitiño (✉) · M.T. Dozo · M.R. Buono
Instituto Patagónico de Geología y Paleontología (IPGP), Consejo Nacional de Investigaciones Científicas y Técnicas (CONICET)—CCT Centro Nacional Patagónico (CENPAT), Boulevard Brown 2915, U9120ACD Puerto Madryn, Chubut, Argentina
e-mail: jcuitino@cenpat-conicet.gob.ar

M.T. Dozo
e-mail: dozo@cenpat-conicet.gob.ar

M.R. Buono
e-mail: buono@cenpat-conicet.gob.ar

C.J. del Río · L. Palazzesi · S. Fuentes
Museo Argentino de Ciencias Naturales Bernardino Rivadavia—CONICET, Av. Ángel Gallardo 460, C1405DJQ Ciudad Autónoma de Buenos Aires, Argentina
e-mail: cdelrio@macn.gov.ar

L. Palazzesi
e-mail: lpalazzesi@macn.gov.ar

S. Fuentes
e-mail: sfuentes@macn.gov.ar

R.A. Scasso
Departamento de Ciencias Geológicas, FCEyN, Universidad de Buenos Aires, Instituto de Geociencias Básicas, Ambientales y Aplicadas de Buenos Aires CONICET, Intendente Güiraldes 2160, C1428EHA Ciudad Autónoma de Buenos Aires, Argentina
e-mail: rscasso@gl.fcen.uba.ar

mals, birds, and fishes. Isotopic and biostratigraphical data suggest a late Miocene age for the Puerto Madryn Formation, although some middle Miocene biostratigraphical indicators are present. Paleoenvironmental information suggests oceanic and continental temperatures warmer than present day, evidenced by the Caribbean molluscan association and the continental vertebrate and palinological associations, respectively. Instead, cetaceans, dinoflagellates, and some marine fishes, suggest colder oceanic temperatures. Precipitations were also higher than present, evidenced by the presence of freshwater mammals, birds, fishes, and plants. This work highlighted some gaps in the geological and paleontological knowledge including geochronology, stratigraphic control of paleontological studies and the knowledge of poorly known fossil groups, which should be the focus of future investigations.

Keywords Península Valdés · Stratigraphy · Fossils · Puerto Madryn Formation · Paleontology · Biostratigraphy

1 Introduction

The aim of this chapter is to bring together the abundant sedimentological and paleontological information recorded in the Miocene marine strata preserved in the Península Valdés region. A large number of paleontological studies made in the last decades for fossil plants, marine and continental vertebrates and marine invertebrates are summarized and analyzed in their stratigraphic and sedimentologic context. Their chronological, biostratigraphical, and paleobiogeographical implications are discussed as well as their paleodiversity and paleoecology. The latter is compared with the sedimentological evidence in order to bring together the *paleoenvironmental* information preserved in the Miocene marine rocks, to reconstruct the landscape and to highlight the major *paleoenvironmental* changes that occurred during the late Cenozoic in northeastern Patagonia.

2 The Miocene "Patagoniense" and "Entrerriense" Marine Transgressions

Two major continental-scale Miocene marine flooding events are recorded in Patagonia: the early Miocene *Patagoniense Transgression* and the late Miocene *Entrerriense Transgression* (Fig. 1). The term "*transgression*" used here is informal and bears chronostratigraphic significance. It refers to a marine sedimentation event represented by a succession of strata deposited on continental areas in a certain period of time (e.g., the early Miocene). These transgressions comprise the entire marine sedimentation cycle, including both their transgressive and regressive

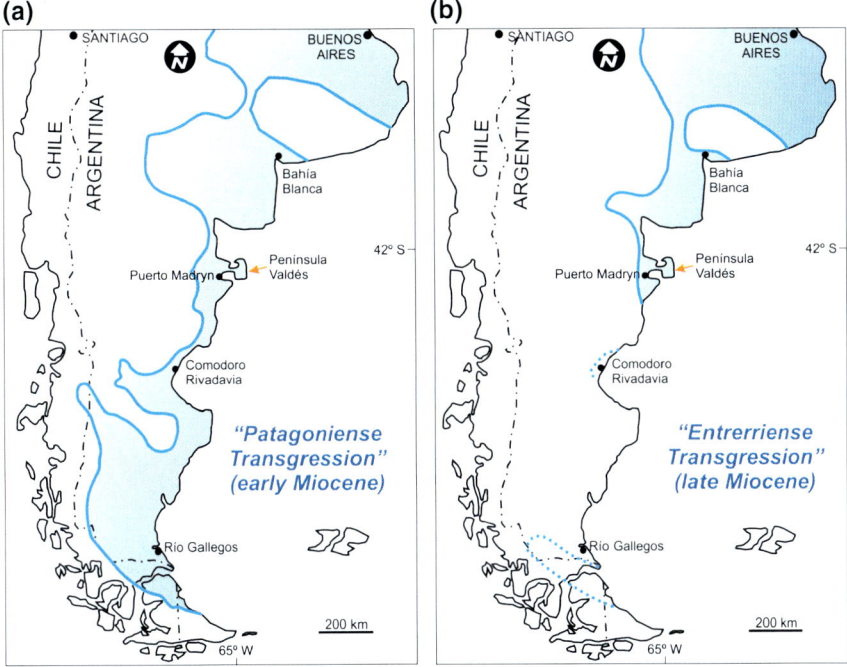

Fig. 1 Regional maps showing the distribution of sedimentary deposits assigned to the Miocene transgressions in southern South America, including both outcrops and subsurface data. **a** early Miocene Patagoniense Transgression, **b** late Miocene Entrerriense Transgression. The *dotted line* indicates possible areas where the Entrerriense Transgression can be found. Modified after Malumián and Náñez (2011)

parts. Several lithostratigraphic units can be assigned to each single transgression and minor age differences are expected from one basin to another.

The *Patagoniense Transgression* is one of the major marine flooding events recorded during the Cenozoic in South America. Marine deposits related to this event are recognized in Patagonia from Tierra del Fuego in the South to Río Negro provinces in the North (Fig. 1a) (Malumián and Náñez 2011), in the subsurface of the northern Chacoparanaense Basin (Marengo 2015), as well as in some Atlantic basins in Brazil (Rossetti et al. 2013). The middle-late Miocene *Entrerriense Transgression* includes shallow marine and estuarine deposits widely distributed in South America, especially in the Chaco–Paranaense Basin (e.g., Marengo 2015) and in some subandean basins from northwest Argentina to western Brazil (e.g., Hernández et al. 2005). In Patagonia, the *Entrerriense* marine deposits are restricted to the northeast of Chubut and east of Río Negro provinces (Fig. 1b) (Scasso and del Río 1987; Malumián and Náñez 2011).

These two major marine flooding events are recorded in the sedimentary succession of the Península Valdés. The early Miocene *Patagoniense Transgression* is poorly exposed in this area (Fig. 2a), being represented by the Gaiman Formation

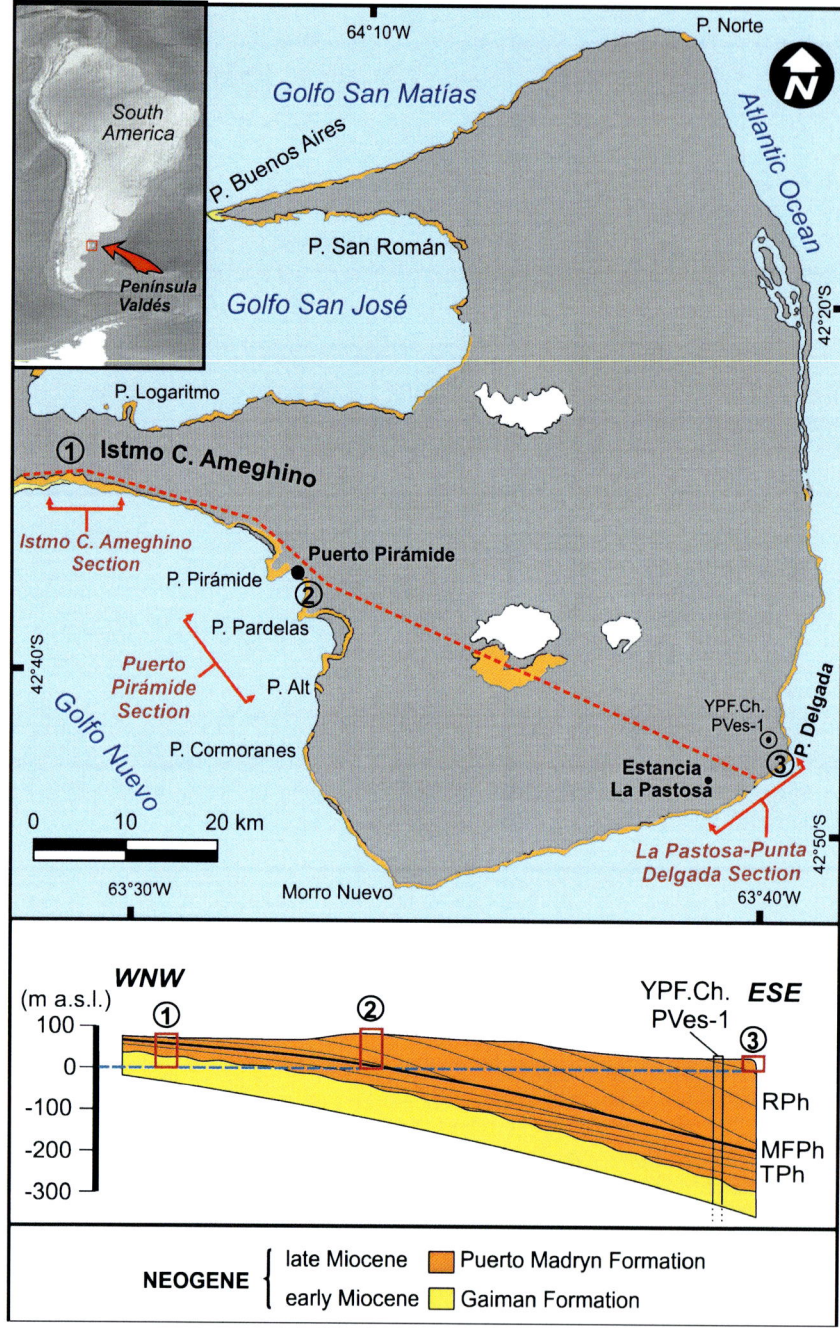

◀ **Fig. 2** Areal and stratigraphical distribution of the Miocene marine deposits in the Península Valdés region. **a** Simplified geologic map of Península Valdés highlighting the outcrop distribution of the Gaiman (*yellow*) and Puerto Madryn (*orange*) formations (modified from Chapter "Geology of Península Valdés"). Key localities, the location of the studied sections and the orientation of the cross section are also indicated, **b** Cross section showing the stratigraphic configuration of the Miocene strata in the Península Valdés. *TPh*: *Transgressive Phase*; *MFPh*: *Maximum Flooding Phase*; *RPh*: *Regressive Phase*. *Red boxes* indicate the position and thickness of the measured sections in outcrops

(Haller and Mendía 1980), with outcrops restricted to the small cliffs of the Istmo Carlos Ameghino (Carlos Ameghino Isthmus; see Chapter "Geology of Península Valdés") and to the intertidal wave-cut platform (informally named restinga) of Punta Buenos Aires (Fig. 2a). The sedimentary deposits of the Gaiman Formation are monotonous and poorly exposed. The *Entrerriense Transgression* is recorded in the sedimentary deposits that make up the Puerto Madryn Formation (Haller 1979), which are widely distributed in Península Valdés, nicely exposed at the coastal cliffs, as well as in some creeks and lowlands in the continental interior of the peninsula (Fig. 2a).

The early Miocene Gaiman Formation shows a poor quality of preservation of the calcareous marine fossils and only marine vertebrate remains (e.g., penguins, cetaceans) are well preserved. On the contrary, the Puerto Madryn Formation, deposited during Serravalian to Tortonian times (~ 14–7 Ma; see Sect. 5), bears a remarkably rich and well-preserved assemblage of marine invertebrates, as well as other fossil groups such as palynomorphs and marine and continental vertebrates which are summarized in this chapter (see Sect. 4). It also shows superb exposures of shallow marine to estuarine deposits that, in combination to the paleontological record, allow reconstructing marine and estuarine paleocommunities and *paleoenvironments* in great detail (see Sect. 6). Given the strong differences concerning the outcropping area, variety of sedimentary features, fossil diversity and quality of preservation of fossils in each lithostratigraphic unit in Península Valdés, the *paleoenvironmental* reconstruction performed in this chapter is strongly biased toward the Puerto Madryn Formation.

3 Sedimentary Paleoenvironments Through Time

3.1 *Paleoenvironments of Deposition of the Gaiman Formation*

The Gaiman Formation in Península Valdés crop out at the Istmo Carlos Ameghino below the lower beds of the Puerto Madryn Formation (Fig. 2a). This unit is composed mainly of whitish mudstones and fine sandstones, with a high proportion of pyroclastic material (see Chapter "Geology of Península Valdés"). It shows a homogeneous appearance because physical *sedimentary structures* or stratification

Fig. 3 Photographs of representative sedimentary deposits described for the Miocene marine strata in the Penínusla Valdés. **a** Plan view of the thoroughly bioturbated, structureless mudstones of the Gaiman Formation at Istmo Carlos Ameghino, **b** Nearshore, open marine deposits of the *Facies Association* 1. This bed represents a coarsening upward succession from thoroughly *bioturbated* (high *bioturbation index*) *shelf* mudstones (FA2) at the base, to upper shoreface, highly packed bioclastic deposits (FA1) at the top; lower beds of La Pastosa-Punta Delgada Section, **c** Laminated, *shelf* mudstones of the *Facies Association* 1, representing part of the *Maximum Flooding Phase*. They are covered by structureless, *bioturbated* fine sandstones of lower shoreface (FA2); lower beds of the Puerto Pirámide Section, **d** Strongly abraded bioclastic tidal channel deposits (FA3) showing large-scale cross-stratification. The base of the channel is pointed by *red arrows*; middle beds of the Puerto Pirámide Section, **e** Sandy tidal channel deposits (FA3) with erosive, concave-up base (*yellow arrows*). Note dark gray coloration provoked by the volcanic composition of the sand particles. *White arrows* point to thick mud drapes and *orange arrows* point to the base of an heterolithic tidal channel (FA3) that erodes part of the underlying sandy tidal channel; middle beds of the Puerto Pirámide Section, **f** Thick (~5 m) heterolithic tidal channel deposits (FA3) displaying inclined heterolithic stratification (IHS), upper beds of the Puerto Pirámide Section. Hammer (*circled*) for scale is 30 cm long

patterns are obscured by an intensive and penetrative marine *bioturbation* (Fig. 3a). Interbedded with these structureless deposits, centimeter thick layers of lighter color can be recognized. They have sharp horizontal bases and *bioturbated* tops, and are interpreted as ash fallout deposits. The sparse fossil macroinvertebrates, including oysters, turritellids gastropods, decapods, echinoids and other bivalves, are dissolved or replaced by silica or gypsum, and systematic identification is not possible (Scasso and del Río 1987). On the other hand, marine vertebrates are relatively well preserved, including cetaceans, penguins and fishes (see Sect. 4.3). Trace fossils are the main source of *paleoenvironmental* information within this unit. Feeding and dwelling structures are common (Fig. 3a), including *Teichichnus, Rossrlia, Thalassinoides, Planolites, Chondrites, Asterosoma* and *Helicodromites* ichnogenera, and are interpreted as representatives of the *Cruziana* ichnofacies (Lech et al. 2000; Scasso and Bellosi 2004; Scasso et al. 2010). These ichnofaunas are comparable to that described for the open-marine deposits of the age-equivalent Chenque Formation (e.g., Carmona et al. 2008), characterized by intense *bioturbation*, complex tiering structures and a very high diversity.

The Gaiman Formation is the result of sedimentation in a low-energy, *shelf* marine environment coeval with a strong and continuous explosive volcanic activity in the continent evidenced by the elevated pyroclastic content of the unit. Fine-grained deposits and the trace fossil suite indicate an open, low-energy offshore marine environment, perhaps a wide bay, whereas the thoroughly *bioturbated* ichnofabric and the high phosphate content suggest low sedimentation rates (Scasso and Castro 1999; Fazio et al. 2007). Phosphate concretions are associated to microenvironments developed within burrows during the early diagenesis (Fazio et al. 2007).

3.2 Paleoenvironments of Deposition of the Puerto Madryn Formation

The Puerto Madryn Formation shows greater sedimentary variability in comparison with the Gaiman Formation. Physical *sedimentary structures* are usually well preserved and only in certain layers thoroughly *bioturbated* deposits can be observed as well (Scasso and del Río 1987). Fossils are abundant and well preserved in this unit and many taxonomic groups are represented including marine invertebrates, marine vertebrates, continental vertebrates, and palynomorphs. Thick and abundant shell beds are a conspicuous feature of this unit (e.g., del Río et al. 2001).

Sedimentologic analysis of the Puerto Madryn Formation in the Península Valdés region allows recognizing four main *paleoenvironments* of deposition, related to four *Facies Associations*. These were defined observing grain size, *sedimentary structures*, bioturbation, fossil content, and geometry of the beds.

3.2.1 Facies Association 1. Open Marine, Nearshore Deposits

Nearshore deposits of the Puerto Madryn Formation are common in the lower part of the unit, which is mostly exposed in the western area of the Península Valdés and were the focus of a detailed study by del Río et al. (2001). These deposits are composed of fine- to coarse-grained sandstones, shell beds, and in some occasion few conglomerates. Common physical *sedimentary structures* are hummocky cross-stratification, parallel stratification, and medium scale through cross-stratification. Shell beds may show a poorly defined horizontal stratification, although most are massive (Fig. 3b). Fragmentation of hard parts is variable and usually shell beds show mixing of fragmented and entire shells. Barren fine- to medium-grained sandstones are usually thoroughly *bioturbated* and display elements of the *Skolithos* and *Cruziana* ichnofacies, and may show thin shell beds. The nearshore deposits show meter-scale coarsening upward cycles culminating in highly packed shell beds (Fig. 3b).

Facies Association 1 suggests that the nearshore deposits of the Puerto Madryn Formation accumulated in a marine shoreface environment. Wave action, especially during storm events, was the main transport and depositional agent, evidenced as hummocky cross-stratification structures in lower shoreface deposits. The higher energetic environment, the upper shoreface, is represented by shell beds with coarse-grained sandy matrix (Fig. 3b). Lower shoreface deposits are characterized by thoroughly *bioturbated* fine-grained sandstones. The coarsening upward cycles reflect shallowing upward successions from lower to upper shoreface environments (Fig. 3b), typical of wave dominated shorelines in open shallow marine systems (del Río et al. 2001).

3.2.2 Facies Association 2. Shelf Deposits

Facies Association 2 is characterized by fine-grained sediments that grade from mudstones to fine-grained sandstones (*Facies* 4a from Scasso and del Río 1987). Most of these deposits are thoroughly *bioturbated* displaying the *Cruziana* ichnofacies key ichnogenera, including *Teichichnus, Rosselia, Thalassinoides, Planolites, Chondrites and Asterosoma*. Thin shell beds intercalate within the *bioturbated* deposits showing low abrasion and fragmentation degree (*Facies* 3 from Scasso and del Río 1987). Some tabular, meter scale laminated dark mudstone beds can be recognized within the *shelf* successions (Fig. 3c) showing low degree of *bioturbation* and in some cases a monospecific suite of *Chondrites* isp.

These sediments were deposited in a shallow marine basin, below the fair-weather wave base. The high trace fossils diversity and the elevated *bioturbation index* observed, suggests low sedimentation rate in a well-oxygenated marine bottom, under the influence of the sunlight. Occasionally, storm events reworked the marine bottom creating thin shell beds or thin laminated sandstones layers. The laminated mudstones lack body fossil invertebrates and display low-diverse trace fossil suites (e.g., *Chondrites* isp.). They probably represent episodes of low-energy and biologic stress in the bottom of the water column on the *shelf*. Noteworthy, these laminated mudstones are the only palynologically fertile layers of the Puerto Madryn Formation (see Sect. 4.1), suggesting proper conditions (oxygen depletion?) for organic material preservation. These conditions are interpreted as the result of a deepening event, which constitutes the maximum depth that the sea achieved during deposition of the unit (del Río et al. 2001).

3.2.3 Facies Association 3. Tidal Channel Deposits

Tidal channel deposits show the highest variety of sedimentologic attributes. They grade from conglomeratic shell beds to mudstone deposits. Despite the great grain size variation, some common features can be recognized, including an erosive base, lenticular shape, finning upward trend, and dominance of tide-generated *sedimentary structures* such as mud drapes, bipolar paleocurrent orientations, or rhythmic alternations of sand–mud couplets. According to the nature of the sedimentary infill, three types of tidal channel deposits are recognized in the Puerto Madryn Formation: (1) cross-bedded bioclastic channels; (2) cross-bedded sandstone channels, and (3) mudstone–heterolithic channels. The first type is composed essentially by highly abraded bioclastic material (shell hash) and show medium- to large-scale cross-bedding (*Facies* 1 from Scasso and del Río 1987) (Fig. 3d). The second type is dominated by gray sandstones with cross-bedding, ripple lamination, or planar lamination (*Facies* 5 from Scasso and del Río 1987) (Fig. 3e). Gray coloration is provoked by the basic to intermediate volcanic composition of the sand particles. Mud drapes are ubiquitous in these deposits, whereas bioclastic material is only present as reworked concentrations at the lower part of the sedimentary bodies. *Bioturbation* shows low diversity and variable intensity, from high to low *bioturbation indexes*. The third type of tidal channel deposits is composed of mudstones or muddy heterolithic deposits (*Facies* 4b from Scasso and del Río 1987). They often show inclined layers forming inclined heterolithic stratification (IHS) deposits (Fig. 3f). These channels show thin, rippled sand partitions, although some are entirely composed of pyroclastic mud. Fossils are scarce, usually dominated by reworked, non-abraded oysters, which lie on the basal erosion surface in association with mud intraclasts. Some continental vertebrate remains are found within these channel lags (see Sect. 4.4). *Bioturbation indexes* are low or *bioturbation* is absent and usually shows low diversity.

Tidal channels of the Puerto Madryn Formation were formed in a shallow marine, tide-influenced environment (Scasso and del Río 1987; Scasso et al. 2012). The main difference among the three types of tidal channels here defined is the grain size and the scale of the cross-bedding sets. Both elements reflect variations in depth and tidal current energy, from an outer, high-energy estuary to an inner estuary, grading to tidal-fluvial channels in the innermost part. Coarse bioclastic tidal channel deposits, with meter scale cross-bedded sets suggest deep subtidal channels formed after strongly erosive processes that promoted an intensive bioclastic reworking from the underlying strata. The sandy tidal channels, with sparse body and trace fossils, suggest sudden inputs of continental volcanic sand into the coastal zone, which subsequently underwent tidal reworking. These tidal channels show moderate-to-high tidal energy and represent a landward position in relation to the bioclastic tidal channels. The low *bioturbation indexes* observed in these deposits suggest stressing bottom conditions, probably caused by freshwater input and the strong sedimentary dynamics of this environment. Muddy heterolithic tidal channels represent the most landward sedimentary *paleoenvironment* recognized in the Puerto Madryn Formation. They display tidal as well as fluvial features. The low diversity and low *bioturbation index*, and the lack of autochthonous marine fossils, suggest elevated freshwater discharge that caused stressing conditions for marine life. Abundant centimeter-thick single mudstones drapes are ubiquitous in tidal channels, suggesting elevated concentration of mud in suspension which became trapped in the system due to the convergence of tidal and fluvial currents.

3.2.4 Facies Association 4. Tidal Flat Deposits

These deposits are composed of horizontally bedded sandstones, mudstones, and heterolithic beds (*Facies* 4b of Scasso and del Río 1987) (Fig. 3e). Sandstone layers show parallel lamination, ripple lamination, or small scale cross-bedding. Heterolithic deposits dominate this *facies* and range from sand-dominated to mud-dominated. Rhythmic thickness alternation of sand–mud couplets and bipolar paleocurrent directions of alternating current ripples are frequently observed. Wave ripples in fine sandstone layers can be seen intercalated within the heterolithic strata. Mudstone beds without sand partitions are less common, and typically are composed of pyroclasite material (ash). Fining upward trends are common within this *facies* and internal erosive surfaces, blanketed by mud intraclasts, are frequently found separating packages of different heterolithic strata. Tidal flats deposits may grade laterally or show erosive contacts with adjacent tidal channels (e.g., Figs. 3e and 4b). Marine invertebrates are scarce and usually appear associated to erosion surfaces together with mud intraclasts. Oysters in life position, forming nests or small biostromes can be observed intercalated within tidal flat deposits (*Facies* 2 of Scasso and del Río 1987). Trace fossil diversity (ichnodiversity) and *bioturbation indexes* are usually low. Nevertheless, higher *bioturbation indexes* produced by only one ichnogenera can be recognized in certain layers.

The sedimentary features of the *Facies Association* 4 suggest deposition in a flat (no channelized) environment under the dominance of tidal depositional processes or in tidal flats laterally attached to tidal channels (Fig. 4b). Rhythmic bedding and bipolar paleocurrents are a clear indicator of tidal action. Sand-dominated layers may be assigned to a subtidal to lower intertidal flats, whereas heterolithic beds may be assigned to intertidal flats. Thick beds composed of pyroclastic mud suggest outburst of sediments caused by explosive volcanic activity. Low diversity of marine invertebrates and trace fossils indicates a stressing environment for marine life, probably caused by subaerial exposure and freshwater input from rivers.

Oysters, which have shown tolerance to such estuarine conditions, are the only marine organism preserved in these deposits. Wave generated *sedimentary structures* suggest wave action on the tidal flats, probably caused by sporadic storms.

3.3 Spatial Distribution of the Facies Associations of the Puerto Madryn Formation

The contact between the Gaiman and Puerto Madryn formations dips gently (less than 1°) to the east (Scasso and del Río 1987) and controls the general stratigraphic conformation of the Neogene beds in Península Valdés (Fig. 2b). This reflect, at least in part, the original seaward sedimentation slope of the system (Scasso and del Río 1987), since no significant deformational forces acted in this area after deposition, excepting few normal faults with no more than 4 m of vertical throw (see Chapter "Geology of Península Valdés") and gentle local tilts. Because the outcrops are analyzed along the coastal cliffs of the Península Valdés (i.e., along a horizontal NNW-ESE transect, Fig. 2b) the intersection of this transect with the strata gives a general trend with the older (lower) beds exposed to the west and the younger (upper) beds exposed to the east. Figure 2b shows the distribution of three sections representative of the stratigraphic succession. These are ordered stratigraphically from base to top as follows: (1) the lower (western) Istmo Carlos Ameghino Section (Fig. 4a); (2) the middle Puerto Pirámide Section (Fig. 4b), and (3) the upper La Pastosa-Punta Delgada Section (Fig. 4c). The lower Istmo Carlos Ameghino Section is exposed in the west of the Península Valdés along the cliffs around the Istmo Carlos Ameghino (Fig. 2). The middle Puerto Pirámide Section is exposed in the cliffs of the northeast of Golfo Nuevo (Nuevo Gulf), between the Puerto Pirámide sea lion colony and Punta Alt (Fig. 2). The upper La Pastosa-Punta Delgada Section is exposed to the southeastern extreme of the Península Valdés, around the cliffs of the Estancia La Pastosa and Punta Delgada lighthouse (Fig. 2).

The Istmo Carlos Ameghino Section shows the base of the Puerto Madryn Formation overlying the Gaiman Formation (Fig. 4a). A sharp erosive surface, in some cases bearing *Glossifungites* trace fossils, separates both units (Fig. 4a) (Scasso et al. 1999). The most common deposits in this section are *shelf* mudstones (*Facies Association* 2) and nearshore bioclastic deposits (*Facies Association* 1) (Figs. 4a and 5). An abundant and high diversity assemblage of marine invertebrates, especially of molluscs, was described for this section (e.g., del Río et al. 2001; see Sect. 4.2). Three main shell beds are described for this section, namely the Lower Shell Bed (LSB), Middle Shell Bed (MSB), and the Upper Shell Bed (USB) (Fig. 5).These conspicuous shell beds are laterally continuous for several kilometers and are useful elements for correlation among localities (del Río et al. 2001).

The Puerto Pirámide Section, lying stratigraphically in between the other two sections, shows a minor proportion of *shelf* deposits of the *Facies Association* 2 to

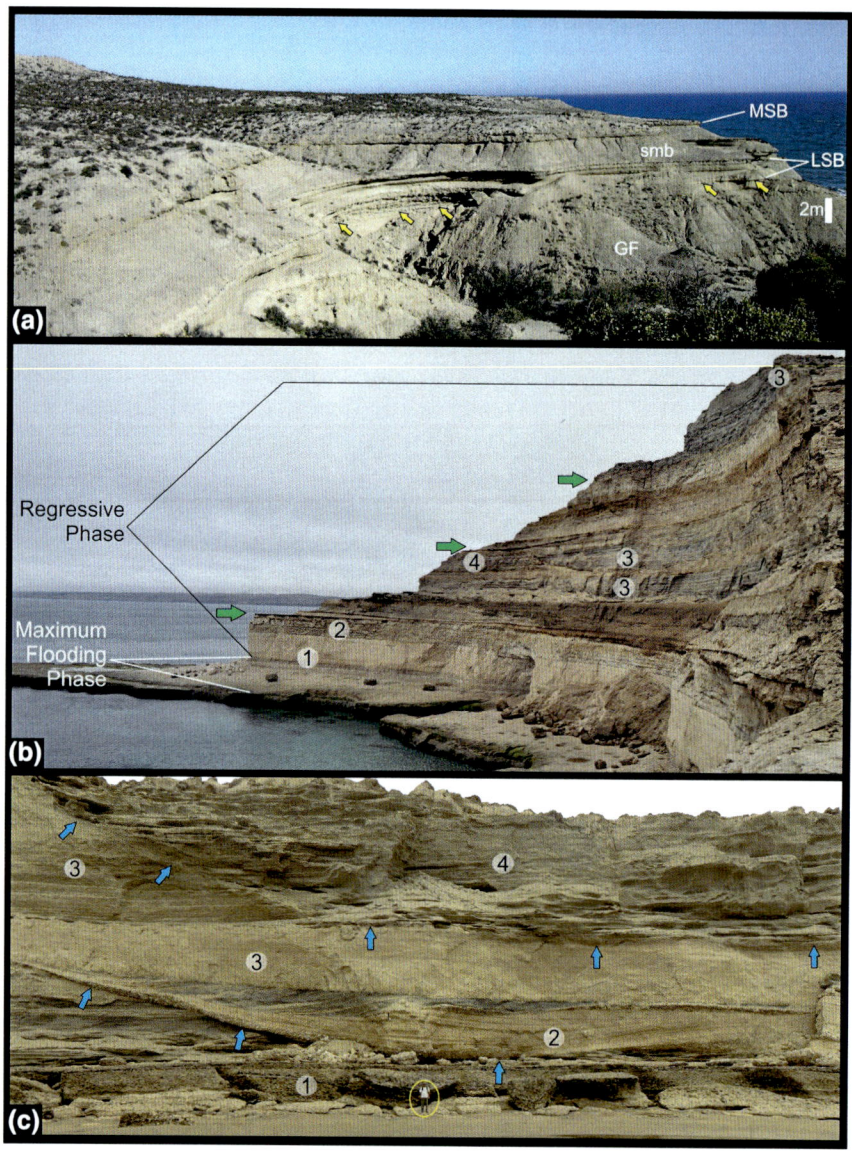

◀ **Fig. 4** Field photographs of representative sections (see Fig. 2 for sections location) of the Península Valdés Neogene marine successions. **a** The Istmo Carlos Ameghino Section showing the Gaiman Formation (*GF*) below and the *Transgressive Phase* of the Puerto Madryn Formation above. The *yellow arrows* point to the erosion surface separating both units, which is covered by the Lower Shell Bed (*LSB*, del Río et al. 2001). The Middle Shell Bed (*MSB*) is separated from the underlying *LSB* by a *shelf* mudstone bed (*smb*), **b** The Puerto Pirámide Section (50 m *thick*). The *Maximum Flooding Phase* and the lower part of the *Regressive Phase* are recognized here. *Green arrows* indicate erosive flooding surfaces corresponding to the base of low-hierarchy cycles. *Numbers* indicate *1* shelf mudstone deposits, *2* open marine, nearshore deposits, *3* tidal channel deposits and *4* tidal flat deposits, **c** The La Pastosa-Punta Delgada Section. Muddy heterolithic tidal channels and tidal flats dominate this section. *Blue arrows* indicate basal tidal channel erosion surfaces. Numbers indicate *1* nearshore, open marine deposits, *2* muddy heterolithic tidal channel deposits, *3* muddy tidal flat deposits and *4* sandy tidal channel deposits. Person (*circled*) as scale

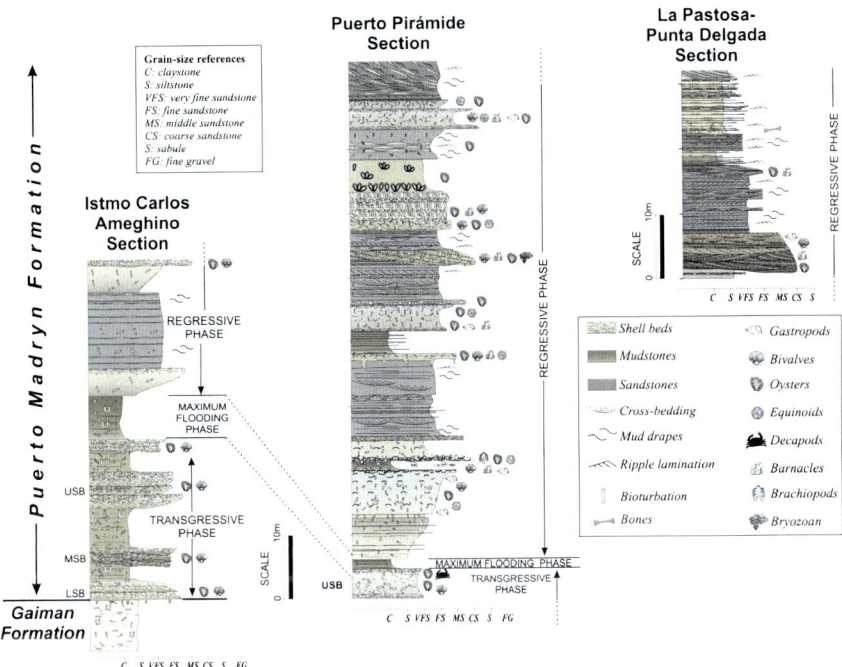

Fig. 5 Representative sedimentary columns of the three sections defined for the Miocene marine deposits in the Península Valdés region (see Fig. 2 for section locations). *LSB* Lower Shell Bed, *MSB* Middle Shell Bed, *USB* Upper Shell Bed. Modified after del Río et al. (2001) and Scasso et al. (2015)

the lower part, being dominated upward by thin open nearshore marine deposits of the *Facies Association* 1 and a thick complex of tidal channel (*Facies Association* 3) and tidal flat (*Facies Association* 4) deposits to the upper part (Scasso et al. 2010, 2015) (Fig. 4b). To the lower part at sea level, the USB can be recognized (del Río et al. 2001). Bioclastic, outer estuary tidal channels are common in this section, as well as sandy tidal channel deposits and oyster biostromes to the upper part.

The La Pastosa-Punta Delgada Section shows dominance of tidal channels and tidal flat deposits (Fig. 4c) of the *Facies Associations* 3 and 4, respectively (e.g., Scasso et al. 2012, 2015). Nearshore open marine deposits of the *Facies Association* 1 are represented by sparse thin shell beds. Muddy tidal channels and muddy tidal flats are the most remarkable feature of this part of the Puerto Madryn Formation. Correlation to the western sections is not straightforward since no key stratigraphic markers are found among them. Given the gentle dip of the strata to the east (Fig. 5), an upper position in relation to the Puerto Pirámide Section is proposed for La Pastosa-Punta Delgada Section.

3.4 Stratigraphic Evolution of the Miocene Marine Deposits

The early Miocene Gaiman Formation, representative of the *Patagoniense Transgression*, shows a general base to top trend from *shelf* to nearshore deposits along the Chubut River valley, about 100 km southwest of Península Valdés (Scasso and Castro 1999). However, because of the low vertical extension of the exposures in the Istmo Carlos Ameghino Section, it is not possible to recognize a trend there. The fact that the deposits of this unit in Península Valdés were deposited in a *shelf* environment, suggest that the overlying regressive nearshore deposits should had been removed by subsequent erosion after deposition (Scasso and del Río 1987). Accordingly, the Gaiman Formation is composed of *shelf* deposits in Península Valdés. The contact between the Gaiman and the Puerto Madryn formations (Fig. 4a) was interpreted as main surface of nondeposition, involving little erosion (Scasso and del Río 1987). However, Palazzesi et al. (2006) suggested a significative hiatus between the top of the Gaiman Formation and the base of the Puerto Madryn Formation in the Istmo Carlos Ameghino, and therefore the upper part of the unit may have been eroded before or during the early stages of accumulation of the Puerto Madryn Formation. Currently, this contact is interpreted as a major ravinement and coplanar surface which represents a hiatus of few millions of years (Scasso et al. 1999) in agreement with the earlier interpretation of an erosional unconformity given by Feruglio (1949).

Within the Puerto Madryn Formation, a high hierarchy transgressive–regressive (T–R) stratigraphic cycle can be defined, comprising the entire stratigraphic thickness of the unit. The cycle presented here is slightly modified from that proposed by del Río et al. (2001). It begins with a *Transgressive Phase*, and is followed by a thin *Maximum Flooding Phase* [originally defined as the Maximum Highstand Phase by del Río et al. (2001)], which in turn is capped by the *Regressive* (highstand) *Phase* (Fig. 5). The *Transgressive Phase* initiates above the basal erosive ravinement surface, and represents a continuous upward deepening of the marine environment. This phase is best represented in the Istmo Carlos Ameghino Section in the west of the Península Valdés, and consists of up to 30 m of nearshore to *shelf* sediments (Fig. 5). This phase culminates with the *Maximum Flooding Phase* (Fig. 5), which is a 10–15 m thick, mud-dominated bed composed of laminated or thoroughly *bioturbated shelf* deposits

(Fig. 3c), representing the maximum depth that the sea achieved during the deposition of the Puerto Madryn Formation. Overlying this phase is the *Regressive Phase* (Fig. 5), which represents a gradual shallowing of the marine depositional system, culminating with tidal-fluvial transition deposits. This is the thickest phase, with about 100 m in the Puerto Pirámide Section (Fig. 5), although probably a larger thickness can be measured integrating the upper part of the Puerto Pirámide Section and the La Pastosa-Punta Delgada Section. The oil well YPF.Ch.P Ves-1, located close to the area of La Pastosa-Punta Delgada Section (Fig. 2a), penetrated about 300 m of sediments assignable to the Puerto Madryn Formation (Caramés et al. 2004, see Chapter "Geology of Península Valdés"). Most of this thickness is interpreted to represent the *Regressive Phase* (Figs. 2b and 5).

Within the major transgressive–regressive (T–R) stratigraphic cycle of the Puerto Madryn Formation, several lower hierarchy cycles can be recognized. These are punctuated by transgressive surfaces displaying firm or hard substrate trace fossils, overlain by reworked shell beds or strongly *bioturbated* bioclastic sandstones (e.g., Fig. 3d). In the *Transgressive Phase*, the flooding surfaces are followed by *shelf* deposits which in turn grade upward to nearshore deposits. In the *Regressive Phase*, the flooding surfaces are covered by tidal channel deposits.

As expected in a changing *paleoenvironmental* scenario, the fossil content of the Puerto Madryn Formation is not evenly distributed. The lower *Transgressive Phase* bears most of the marine invertebrate fauna yet described (see Sect. 4.2), representing mostly shallow, open marine environments. Also, most of the marine vertebrate remains (cetaceans, fishes, penguins; see Sect. 4.3) were recovered from these beds. The *Maximum Flooding Phase*, although a thin layer, bears a special significance because this is the only bed that yielded palynologic material (see Sect. 4.1). Unique *paleoenvironmental* conditions during the *Maximum Flooding Phase* favored the preservation of organic material. The *Regressive Phase* bears a mix of marine and terrestrial fossil assemblages. Marine invertebrate assemblages are restricted to thin nearshore, open marine deposits, and in some cases the assemblages are poorly diverse or strongly reworked. Continental vertebrate assemblages (see Sect. 4.4) are exclusively preserved in the upper part of the *Regressive Phase*. This vertebrate remains are found associated to tidal channel lag deposits (e.g., Scasso et al. 2012), and are exclusive to the upper part of the Puerto Pirámide Section and La Pastosa-Punta Delgada Section.

4 Neogene Paleobiodiversity and Paleoecology

4.1 Terrestrial and Marine Palynomorphs

Palynological analyses of the sedimentary successions exposed at the Península Valdés region provide a novel perspective in the interpretation of the paleovegetation–paleoclimate (spore and pollen analysis) and paleoceanographic conditions (fossil dinoflagellate cysts).

The palynological assemblages of the Gaiman Formation in the Istmo Carlos Ameghino (Palazzesi et al. 2006) suggest that they were deposited in a marine environment with abundant marine elements (dinocysts) joined with the large abundance of wind transported pollen grains (such as those of the Podocarpaceae) with shrubs and coastal taxa virtually absent. Of particular significance are the podocarpaceous gymnosperms *Dacrydium* and *Lagarostrobos* taxa, both extinct from America, but frequently found today in southwest Pacific. In Patagonia they probably occurred in the rainforest canopy, joined with other high humidity components such as *Gunnera* and tree fern genera (e.g., Cyatheaceae). Palms appear to have been other common element of the flora, probably spread on lowland regions close to the sea level. A comparable picture could be found today in the subtropical South America on lowland areas of the Atlantic Ocean.

Recently, conducted palynological studies on the Puerto Madryn Formation (Palazzesi and Barreda 2004; Palazzesi et al. 2014) identified a unique 10 m-thick layer corresponding to the *Maximum Flooding Phase*, that yielded abundant palynomorphs, with very well-preserved dinoflagellate cyst content. The botanical affinities of fossil pollen and spore species identified from the assemblage of the Puerto Madryn Formation include swamp and salt marsh communities that probably occupied tidally influenced grounds. The salt-tolerant Amaranthaceae (*Chenopodipollis chenopodiaceoides*), *Ephedra* (*Equisetosporites* spp.), and *Cressa* (*Tricolpites trioblatus*) are recognized as well as the fresh and brackish water Cyperaceae (*Cyperaceaepollis neogenicus*), Typhaceae (*Sparganiaceaepollenites* sp.), *Myriophyllum* (*Haloragacidites myriophylloides*), *Azolla* (water fern), and *Pediastrum* taxa. Semi-deciduous communities occupied well-drained and/or saline soils including the trees *Achatocarpus* (*Periporopollenites* sp.), *Prosopis* (*Rhoipites* sp.), *Celtis* (*Cicotriporites* sp.), and Anacardiaceae (*Striatricolporites* sp.), and probably some shrubs and herbs such as *Chuquiraga* (*Quilembaypollis stuessyi*), *Nassauvia* (*Huanilipollis*), and Calyceraceae (*Psilatricolporites protrudens*). The presence of some key neotropical lineages (e.g., *Schlechtendalia*, *Justicia*, Malpighiaceae, and Loranthaceae) indicate that temperatures rarely fell below freezing point; a recent analysis (Palazzesi et al. 2014) suggests that mean temperatures of the coldest season were between 11.4 and 16.9 °C (presently about 6.4 °C, see Chapter "The Climate of Península Valdés Within a Regional Frame"). Forest communities may have occupied upstream regions (e.g., Nothofagaceae, Podocarpaceae, Winteraceae) and may have grown as more or less discontinuous patches on mainland Patagonia. Also, the presence of *Saxegothaea* and *Araucaria* supports the presence of some wet-demanding trees.

Very well preserved marine palynomorphs include mostly dinoflagellate cysts and acritarchs, and prasinophyceae algae are subordinated. The dinoflagellate cyst assemblages show strong dominance of neritic taxa with high abundance of several species of the *Spiniferites/Achomosphaera* group, *Reticulatosphaera actinocoronata*, *Habibacysta tectata*, *Batiacasphaera minuta*, and *Melitasphaeridium choanophorum*, whereas neritic–oceanic species represented by *Nematosphaeropsis rigida* are scarce (Fuentes et al. 2016). The heterotrophic protoperidiniacean are well represented by round brown cysts attributed to *Brigantedinium* and by several species of *Selenopemphix* (*Selenopemphix brevispinosa* subsp. *brevispinosa*, *S. dionaeacysta*, *S. nephroides* y *S. quanta*). Cool water species such as *Habibacysta tectata*, *Bitectatodinium tepikiense* and *Filisphaera filifera* (e.g., Head et al. 1989; Versteegh 1995; De Schepper 2006) are abundant in the assemblage (Fuentes et al. 2016).

The dinoflagellate cyst stratigraphy of the Puerto Madryn Formation relies on the ranges of the key species *Labrynthodinium truncatum* subsp. *truncatum* and *Habibacysta tectata* (Fuentes et al. 2016) which constrain the age of the unit between the middle and late Miocene (see discussion in Sect. 5).

4.2 Marine Invertebrate Assemblages and Types of Accumulations

Longley referred in geological and paleontological literature as "Entrerriense or Paranense" fauna, some of its components are among the first marine fossils of Argentina in being discovered by Charles Darwin. Macroinvertebrates are dominated by a highly diverse and abundant molluscan assemblage (Figs. 6 and 7), followed by far, by bryozoans, decapods, brachipodos, and echinoids. Pioneer studies of this fauna included the description of bryozoans, corals, echinoids, and brachiopods. Modern paleontological knowledge was acquired in the 90s when the first global revision of bivalves and gastropods were carried out (e.g., del Río 1992; Brunet 1997; del Río and Martínez 1998). Apart from molluscs, only three modern papers deal with other invertebrate groups. One including the description of seven species of decapods (Casadío et al. 2005a), another one mentioning for the first time the presence of an ophiuroid in the Puerto Madryn Formation (Martínez and del Río 2008), and the third one, stating the presence of the warm water bryozoans (Casadío et al. 2005b). Although present, due to the poor quality of preservation of the macroinvertebrate fossils, there is a lack of published information for the Gaiman Formation in the Península Valdés region.

Microfaunistic content of the Puerto Madryn Formation is represented by 104 ostracods species and 67 foraminifera species, most of them benthic taxa that belong in the informal *Protelphidium tuberculatum* Zone, and few planktonic and agglutinated species (Marengo 2015 and reference therein).

Molluscan assemblage comprises 70 species distributed into 61 genera and represents the widest extensively distributed fauna that inhabited any marine littoral environment during Neogene times in Argentina, ranging from northeastern Patagonia northwards to the southern coast of Uruguay. One of its most important characteristic is the dominance of taxa with tropical affinities (i.e., *Chionopsis, Chama, Crassatella, Dinocardium, Lucinisca, Arca, Anadara, Amusium, Leopecten, Hexacorbula, Purpurocardia, Dosinia* and *Pteromyrtea,* among others), representing the first massive immigration of Caribbean components in Patagonia, and comprising the warmest water faunas that ever inhabited the Patagonian coast during

Fig. 6 Late Miocene tropical bivalves of the Puerto Madryn Formation. **a** *Glycymerita magna* del Río (1992) (CPBA 15522), **b–c** *Arca particularis* del Río (1992) (CPBA 12528), **d–e** *Hexacorbula caduca* del Río (1994) (CPBA 12933), **f–g** *Dosinia cuspidata* del Río (1992) (CPBA 15451), **h–i** *Purpurocardia leonensis* del Río (1986) (CPBA 11652), **j–k** *Crassatella kokeni* Ihering von Ihering (1899) (MACN-Pi 1502), **l** *Amusium paris* del Río (1992) (CPBA 15284), **m–n** *Chionopsis australis* del Río (1994) (CPBA 14556), **o** *Leopecten piramidesensis* von Ihering (1907) (CPBA 15195), **p–q** *Pteromyrtea danieli* del Río (1986) (CPBA 14041), **r–s** *Dinocardium novus* del Río (1994) (CPBA 14527). All specimens in natural size. CPBA = Cátedra de Paleontología de la Universidad de Buenos Aires (Repository); MACN-Pi = Museo Argentino de Ciencias Naturales B. Rivadavia (Buenos Aires), División Paleoinvertebrados (Repository)

Fig. 7 Late Miocene survivor genera (bivalves and gastropods) in the Puerto Madryn Formation. **a–b** *"Pitar" mutabile* del Río and Martínez (1998) (CPBA 15092), **c–d** *Felaniella villardeboaena* d'Orbigny (1842) (CPBA 9323), **e** *Aequipecten paranensis* d'Orbigny (1842) (CPBA 14932), **f** *Ameghinomya argentina* von Ihering (1897), **g** *Trophon leanzai* Brunet (1997), **h** *Retrotapes ninfasiensis* del Río (1997) (CPBA 13573), **i–j** *Caryocorbula pulchella* Philippi (1893) (CPBA 12932), **k** *Tegula patagonica* d'Orbigny (1842) (CPBA 12476), **l** *Cyrtopleura lanceolata* d'Orbigny (1842) (CPBA 14194). All specimens in natural size. CPBA = Cátedra de Paleontología de la Universidad de Buenos Aires (Repository)

Paleogene–Neogene times (del Río 1990) (Fig. 6). It constitutes part of the late Miocene Valdesian Molluscan Bioprovince identified by Martínez and del Río (2002a) and its development in this high latitudes corroborates that the cold oceanic Malvinas Current was not fully operating during that time along the Patagonian litoral (del Río 1988; Martínez and del Río 2002a).

When inferring the change in the composition of the late Miocene molluscan assemblage, it must be taken into account the global composition of the assemblages from Patagonia (Valdesian Bioprovince), from Buenos Aires and Entre Rios Provinces, and from Uruguay (Paranaian Bioprovince). On one hand, the peculiarity of this fauna is the abrupt loss of the Caribbean genera that withdrew their geographical range northward into tropical regions of the American continent, where survived until present days. On the other hand, this late Miocene assemblage has little in common with those Recent ones from the Argentinean and Magellanic

Bioprovinces, which suggests that it did not give rise to the Recent ones (Martínez and del Río 2002b). Of 1300 species inhabiting today the South West Atlantic, only a few are late Miocene survivors (*Leionucula puelcha, Lamellinucula semiornata, Adrana electa, Crassostrea rizhophorae, Felaniella villardeboaena, Tivela isabelleana, Mactra isabelleana, M. janeiroensis, Tellina gibber, Cyrtopleura lanceolata, Caryocorbula pulchella, C. caribaea, Tegula patagonica,* and *Halystilus columna*). A few others have closely related congeneric Miocene counterparts (*Glycymeris longior, Aequipecten tehuelchus, Trachycardium muricatum, Anomalocardia brasiliana, "Pitar" rostrata, Ameghinomya antiqua, Retrotapes exalbida, Tagelus plebeius, Trophon geversianus, Adelomelon beckii,* and *Pachycymbiola brasiliana*) (del Río et al. 2010). Cosmopolitan genera and endemic elements, most of which were already present in the area since late Oligocene times, are the only Miocene survivors into Recent times (i.e., *"Pitar", Retrotapes, Ameghinomya, Brachidontes, Cyrtopleura, Caryocorbula, Trophon, "Turritella", Adelomelon,* among others) (Fig. 7).

The striking and abrupt disappearance of the fauna at the end of the late Miocene involved the most severe extinction event of the Cenozoic interval, only comparable with that happened at the end of the Danian (del Río and Martínez 2015), or to the one that took place in the middle Miocene, when benthic faunas were completely renewed after the extinction of ten molluscan families (85% of the genera) (del Río 2004). Its short living lasting led to the redefinition of the Aonikense Molluscan Stage (Rovereto 1913) and to the recognition of the *Aequipecten paranensis* Coincident Biozone (del Río 1988) which is characterized, among other genera, by the novel occurrence of the pectinids *Amusium, Leopecten, Aequipecten,* and the endemic *"Chlamys"*. This association followed to the extinction of the typical pectinid genera that had shaped the early and middle Miocene Patagonian faunas (i.e., *Zygochlamys, Reticulochlamys, Jorgechlamys,* and *Nodipecten*; see del Río 2004).

4.3 Marine Vertebrate Assemblages

Marine vertebrate assemblages from the Miocene of Península Valdés and surroundings areas (e.g., west coast of Golfo Nuevo, Fig. 2) are exceptional not only for their species richness but also by the preservation of the fossil specimens. The great extension of Miocene marine deposits and the exceptional conditions for fossilization documented in the Península Valdés region are the proper scenario to record an important phase of diversification and radiation of many groups of marine vertebrates. The Gaiman Formation has yielded a poor marine vertebrate fauna, only represented by few records of elasmobranchs and osteichthyans fishes, birds, and cetaceans. In contrast the Miocene sediments that have yielded the most important number of marine vertebrates are part of the Puerto Madryn Formation, partly due to the broad exposures of this unit around the coast of Península Valdés. Of particular importance are the remains of fishes (elasmobranchs and osteichthyans), birds (spheniscids and anseriforms) and marine mammals (cetaceans and pinnipeds).

Marine birds of the Gaiman Formation in the Península Valdés region are represented by spheniscids, with a new species *Palaeospheniscus biloculata* known from an articulated skeleton (Acosta Hospitaleche 2007). The fossil cetaceans all belong to the clade of modern cetacean or Neoceti (Mysticeti and Odontoceti; Fordyce 2009). Odontocete remains belong to Physeteroidea and Delphinoidea. Physeteroids (or sperm whales) include two large-sized and well-preserved skulls which might belong to a new species. Delphinoids are represented by a new species of *Kentriodon* (Cione and Cozzuol 1990; Cione et al. 2011) known by an incomplete skull collected in Puerto Madryn city, to the west of Península Valdés.

The most significant fossils of marine vertebrates from the Puerto Madryn Formation in Península Valdés comes from the Istmo Carlos Ameghino Section and the lower and middle part of the Puerto Pirámide Section, and include fishes, birds, and marine mammals (Fig. 8). The paleoichthyofauna (elasmobranchs and osteichthyans) has an important fossil record. Elasmobranchs are less abundant than osteichthyans, and are represented by isolated teeth or tooth plates. The elasmobranchs remains include the otodontid *Carcharocles megalodon*, the lamnid "*Isurus*" *xiphodon* (Fig. 8a), the odontaspidid *Carcharias* sp., the squatinid *Squatina* sp., and the myliobatid *Myliobatis* sp. (Fig. 8b) (Cione et al. 2011;

Fig. 8 Reconstructions of marine vertebrates found in the Península Valdés region in the Miocene Puerto Madryn Formation. Elasmobranchs "*Isurus*" *xiphodon* A and *Myliobatis* sp. B, osteichthyans *Genypterus valdesensis* C, spheniscids *Madrynornis mirandus* D, cetaceans odontocetes *Notoziphius bruneti* E, cetaceans mysticetes, new species of Balaenidae F, pinniped phocid *Kawas benegasorum* G

Cabrera et al. 2012). The osteichthyans fossil record is most important and frequently preserved as articulated skeletons. They are represented by Silurifomes, Ophidiiformes, Gadiformes, Perciformes (Pinguipedidae, Pomadasidae), Pleuronectiformes, and Scorpaeniformes taxa (Cione et al. 2011). One of the most important remains of marine fishes of the Puerto Madryn Formation is *Genypterus valdesensis* (Fig. 8c) (new species, Riva Rossi et al. 2000), which is the first fossil ophidid reported from South America and represents the first well-preserved record of the genus worldwide. The specimen consists of a nearly complete articulated skull and some disarticulated postcranial elements.

Marine birds (spheniscids, anseriforms, and ciconiiforms) are very scarce in the Puerto Madryn Formation but some of them show extraordinary preservation and relevance in filling the gap in the evolutionary history of some groups (i.e., penguins). Spheniscids records include isolated bones assigned to *Paraptenodytes antarcticus* (Acosta Hospitaleche 2003), *Palaeospheniscus bergi* (Acosta Hospitaleche and Cione 2012), and indeterminate penguin species (Cione and Tonni 1981). The presence of *Paraptenodytes antarcticus* and *P. bergi* extends the temporal range of Miocene penguins to the late Miocene, suggesting that a high diversity still existed in this time (Acosta Hospitaleche and Cione 2012). Of particular importance are the remains of the new species *Madrynornis mirandus* (Fig. 8d), a close related species to the living penguins, known from a complete and articulated skeleton (Acosta Hospitaleche et al. 2007a). Other records of marine birds include Anseriformes, with the species *Dendrocygna* (Acosta Hospitaleche et al. 2007b).

The fossil cetaceans from the Puerto Madryn Formation in the Península Valdés area also belong to the group of modern cetaceans (Neoceti). Many specimens found in these sediments have been described as holotypes of new species or correspond to important fossil records for some groups of cetaceans. The mysticetes remains are represented mainly by balaenids and neobalaenines while balaenopteroids are rare. Balaenids are known by three specimens including well-preserved skulls belonging to a new species (Fig. 8f) (Buono 2014). The later correspond to small-sized balaenids which fills the temporal gap between the oldest right whale *Morenocetus parvus* (from the early Miocene Gaiman Formation in the Río Chubut valley; Cabrera 1926) and more modern lineages of balaenids. Another important mysticeti record corresponds to Neobalaeninae, with a complete mandible which represents the third report of a fossil neobalaenine worldwide, and the first fossil occurrence of this lineage in the southwestern Atlantic. It is also the oldest specimen reported so far, thus corroborating the idea of an early divergence time for neobalaenines and a possible austral origin of the living Pygmy right whale (*Caperea marginata*) (Buono et al. 2014). Balaenopteroids has not been previously reported for the Puerto Madryn Formation, however an incomplete skull and tympano-periotic complex recovered from the northwestern region of Península Valdés might represent the first record of this group from the Neogene of Patagonia. Other fossil records of Balaenopteroids also included isolated earbones recovered from different localities of Península Valdés. Odontocetes are represented by Ziphiids or beaked whales, with a new species recorded *Notoziphius bruneti*

(Fig. 8e), one of the most basal ziphiids and the first reported for the southwestern Atlantic Ocean (Buono and Cozzuol 2013).

Finally, pinnipeds are represented by the earliest record of a fossil seal in the southern hemisphere, *Kawas benegasorum* (Fig. 8g), known from an exceptionally preserved articulated partial skeleton from the Puerto Madryn Formation (Cozzuol 2001). These remains are the first record of Phocinae in these high latitudes (Cozzuol 2001); however the systematic affiliation of *Kawas* to this subfamily has been recently questioned based on the reinterpretation of some postcranial characters (Koretsky and Domning 2014).

The fossil marine vertebrate fauna from the Peninsula Valdés region document important evolutionary changes within this regional assemblage, with the diversification or declination of many lineages of marine vertebrates and the delineation of the extant diversity pattern. This process is part of a broader Miocene–Pliocene turnover among marine vertebrates also occurring in other parts of the Southern Hemisphere (Olson 1983; Valenzuela-Toro et al. 2013). The late Miocene marine vertebrates assemblage from Península Valdés and surrounding areas was characterized by a more modern-looking fauna than those presented in early Miocene sediments (Gaiman Formation) outcropping to the south, in the lower valley of Río Chubut, with the extinction or pseudoextinction of archaic lineages and the diversification of the modern ones. Among fishes, some pseudoextinction events occur in the South Atlantic Ocean in elasmobranchs (Heterodontidae, Pristiophoridae and Carcharhinidae) and osteichthyes families (*Oplegnathidae*), and toward the late Miocene the modern fish fauna began to establish (Cione et al. 2011). Miocene spheniscids fossil record documents an extraordinarily diverse and abundant fauna, especially during the early Miocene, in contrast to the present day one in the South Atlantic Ocean (e.g., Acosta Hospitaleche 2003; Acosta Hospitaleche et al. 2007a, b). South Atlantic Miocene penguins were represented by taxa not related to extant representatives, with some species extending their temporal range until the middle-late Miocene, and with the appearance of the modern penguin posterior to the middle Miocene (Acosta Hospitaleche and Cione 2012). Finally, the fossil record of cetaceans also documents change in the composition of the group, with the extinction of some archaic lineages (e.g., odontocetes Kentriodontidae and some Platanistoidea) and the diversification of the modern ones (e.g., odontocetes Physeteroidea, Ziphiidae, and Delphinoidea) between the middle and late Miocene (12–10 Ma; Fordyce 2009). This faunal turnover is also documented in other coeval areas of South America (e.g., Pacific coast of Chile and Peru; Gutstein et al. 2015) as well as in other localities worldwide (Fordyce 2009; Steeman et al. 2009).

4.4 Continental Vertebrate Assemblages

Although most of the continental vertebrate faunal assemblage from Península Valdés was discovered in the upper La Pastosa-Punta Delgada Section of the Puerto

Madryn Formation (Dozo et al. 2010; Scasso et al. 2012), additional individuals mentioned here correspond to isolated findings in other parts of Península Valdés, one in Punta Buenos Aires (Fig. 2a) (Noriega and Cladera 2008) and others in Punta Alt (Fig. 2a). The exhumed vertebrate fauna includes taxa typical of continental or freshwater environments, such as fishes, birds and mammals, of which the latter are most varied and abundant. The findings represent the first record of continental fossil vertebrates in the Puerto Madryn Formation, including Siluriformes and Percomorpha fishes, Gruiformes, Falconiformes, Anseriformes and Ciconiiformes birds, and Xenarthra, Litopterna and Rodentia mammals (Fig. 9).

Freshwater loricariid, pimelodid-like siluriforms (Fig. 9a) and undetermined percomorph fishes are the first record of these groups for Patagonia (Cione et al. 2005). The fossil materials are represented by vertebrae, fragmented spines and plates.

The Gruiformes birds include an undetermined Phorusrhacidae Psilopterinae, the smallest and more gracile phorusrhacids (Fig. 9b); some may even have retained their limited flying ability (Tambussi 2011). A fragment of a cranial roof belongs to a large-sized adult eagle (perhaps similar in size to the extant Black Chested

Fig. 9 Reconstructions of late Miocene continental vertebrates found in the Península Valdés region in the Miocene Puerto Madryn Formation. Freshwater siluriforms (*a*), gruiforms Phorusrhacidae Psilopterinae (*b*), falconiforms Accipitridae (*c*), anseriforms Anatidae Dendrocigninae (*d*), edentates pampatheriids *Scirrotherium carinatum* (*e*), edentates glyptodontids Palaehoplophorini (*f*), edentates Mylodontidae (*g*), litopterns Macraucheniidae (*h*), rodents Hydrochoeridae *Cardiatherium patagonicum* (*i*)

Buzzard Eagle *Geranoaetus melanoleucus*) and constitutes the first available fossil skull of Accipitridae in South America (Picasso et al. 2009) (Fig. 9c). This eagle and the Psilopterinae are the only predators known for the continental vertebrate association recovered in the Puerto Madryn Formation until now. The Anseriformes birds are represented by one humerus and one sinsacrum of Anatidae Dendrocygninae (Fig. 9d). These aquatic birds occur in lentic environments with dense surface vegetation, feeding mainly on the fruits of aquatic plants. These ducks would represent the oldest and southernmost record of a Dendrocygninae and some materials show considerable similarities with the living *Dendrocygna* (Dozo et al. 2010). The first South American record of Ciconiidae *Leptoptilos patagonicus* (Noriega and Cladera 2008) was recovered at the locality of Punta Buenos Aires (Noriega and Cladera 2008).

Among mammals, the Xenarthra include mainly osteoderms (some articulated) of Cingulata (Pampatheriidae and Glyptodontidae) and molariforms of Tardigrada (Mylodontidae). The pampatheriid *Scirrotherium carinatum* remains consist mainly of isolated osteoderms (Fig. 9e). This species is also mentioned for the "Conglomerado osífero" of the Ituzaingó Formation (late Miocene, Entre Ríos Province; Góis et al. 2013) for Argentina. The Glyptodontidae Palaehoplophorini are represented by articulated osteoderms from caudal ring, isolated osteoderms and femur (Fig. 9f). They are the predominant tribe in the Ituzaingó Formation and the specimens of the Puerto Madryn Formation would represent the first record for Patagonia. The Neuryurini glyptodontid is known by an isolated osteoderm; originally recorded in the Ituzaingó Formation, and is one of the least known groups of cingulates. This record, along with another one of early Miocene age from Santa Cruz province, are the first records of these glyptodonts for Patagonia (González Ruiz et al. 2011). A large mylodontidae is represented by a molariform; based on general comparisons, this specimen likely corresponds to *Ranculcus* or *Megabradys* (Fig. 9g). The former is present at the base of Ituzaingó Formation, at the base of Raigon Formation (late Miocene) (Perea et al. 1994) and in the Solimões Formation (late Miocene–Pliocene) of Acre State, Brazil (Latrubesse and Rancy 1994). The latter only occurs in the "Conglomerado osífero" (Ituzaingó Formation) (Cione et al. 2000).

The Litopterna are represented by postcranial elements (tibia–fibula, metapodial, etc.) of Macraucheniidae indet (Fig. 9h). Although a more precise determination is not possible, these specimens show the presence in the Puerto Madryn Formation of endemic South American ungulates.

Numerous teeth of different sizes corresponding to different ontogenetic stages of the new species of Rodentia Hydrochoeridae, *Cardiatherium patagonicum* (Fig. 9i) were collected from La Pastosa-Punta Delgada Section (Fig. 2) in southeastern Península Valdés (Vucetich et al. 2005). One remarkable finding at the same place are two nearly complete skulls of hydrochoerid rodents (Dozo et al. 2010). Based on morphological and morphometric characters of the molariforms, these are also assigned to *C. patagonicum*. This species is the most derived with regards to its dental morphology and represents the southernmost record of any capybara

(Vucetich et al. 2014). Finally, small isolated teeth of rodents in this area correspond to Chinchillidae and Caviidae (Dolichotinae).

The continental vertebrate assemblage of the Puerto Madryn Formation is the first recorded south of the Río Negro Province. It also includes the southernmost record of loricariid fishes, dendrocygnine birds, and hydrocherid rodents.

5 Chronological and Biostratigraphical Implications

The amount of information for the chronology of the major marine depositional events in Península Valdés is strongly biased toward that of the Puerto Madryn Formation.

According to Palazzesi et al. (2006) the maximum age limit for the Gaiman Formation is likely to be earliest early Miocene based on some of the pollen grains recovered from the Istmo Carlos Ameghino such as *Microcachryidites antarcticus*, *Arecipites subverrucatus* and *Haloragacidites trioratus*. These species are not recorded before early Miocene times and characterize the early Miocene assemblages of the San Jorge and Austral Basins (Barreda and Palamarczuk 2000). The Early Miocene C-T/L biozone erected by Barreda and Palamarczuk (2000) presents some typical age diagnostic species of the Istmo Ameghino section.

Since no isotopic or biostratigraphical data exists for the Gaiman Formation in Península Valdés, the age of this unit relay mainly in stratigraphic correlation to other better known regions of Patagonia. The *Patagoniense Transgression*, to which the Gaiman Formation is assigned, is dated as Aquitanian–Burdigalian (23–18 Ma) in the Austral Basin (Parras et al. 2012; Cuitiño et al. 2015a) and Burdigalian–Langhian (20–15 Ma) in the Golfo de San Jorge Basin (Cuitiño et al. 2015b). A similar early Miocene age is thought for the Gaiman Formation in Península Valdés.

The first isotopic $^{40}K/^{39}Ar$ age for the Puerto Madryn Formation comes from a volcanic ash layer from the uppermost beds of this unit at the southern margin of the Golfo Nuevo, yielding an age of 9.41 Ma (Tortonian) (Zinsmeister et al. 1981). This locality correlates with the uppermost beds of the Puerto Pirámide Section (Scasso and del Río 1987). This estimate is consistent with the $^{87}Sr/^{86}Sr$ isotopic age from pectinid and oyster shells collected from eight correlated shell beds at Península Valdés (Istmo Carlos Ameghino and Puerto Pirámide sections) which yielded a mean age of 10.1 Ma (Tortonian) (Scasso et al. 2001).

The Tortonian ages (late Miocene) attributed to the sediments of the Puerto Madryn Formation are further partially supported by its fossil content. The dinoflagellate cysts stratigraphy of the *Maximum Flooding Phase* of the Puerto Madryn Formation relies on the ranges of the key species *Labrynthodinium truncatum* subsp. *truncatum* and *Habibacysta tectata* which constrain the age of the unit between the Serravalian and the Tortonian (Fuentes et al. 2016). The palinologic study assigned a late Miocene age to the sediments of the Puerto Madryn Formation

(Palazzesi and Barreda 2004). The benthic foraminifera assemblage of this lithoestratigraphic unit belongs to the *Protelphidium tuberculatum* informal zone, of middle-late Miocene age (Caramés et al. 2004; Marengo 2015).

del Río (1988) assigned the molluscan assemblage to the middle Miocene, instead of a late Miocene age proposed by Martinez and del Río (2002b) based on the occurrence of peaks of rising temperature in the Southern Hemisphere during which the mollusc fauna could have established.

The marine vertebrate assemblage recovered from the Puerto Madryn Formation is represented by a more modern looking fauna than those observed in the Gaiman Formation. The comparison of this assemblage with that of the Paraná Formation in Entre Ríos Province (late Miocene, 9.47 Ma; Perez 2014) shows a difference in the taxonomic composition. In contrast to the Puerto Madryn Formation, the Paraná Formation assemblage is composed of modern taxa more related to living representatives (Cione et al. 2011). Based on this difference on fauna composition, a middle Miocene age has been proposed for the lower part of the Puerto Madryn Formation (where most of the marine vertebrates are collected), instead of a late Miocene age (Cozzuol 1996; Cione et al. 2000, 2011).

The continental mammals recovered from the upper levels of the Puerto Madryn Formation are correlated with the late Miocene assemblage in the "Conglomerado osífero" of the Ituzaingó Formation (Entre Ríos, Argentina). Particularly, the pampaterid *Scirrotherium carinatum* has an extensive distribution and has been found also in the late Miocene Solimôes Formation (Acre, Brazil) (Góis et al. 2013).

Based on the scarce paleontological and stratigraphical evidence from the Gaiman Formation, a preliminary early Miocene (Aquitanian-Burdigalian?) age is proposed. This age is supported by the marine vertebrate assemblage and in a lesser extent by the palynological data.

The proposed age for the Puerto Madryn Formation is based on isotopic dating, as well as palynological, invertebrate and vertebrate fauna. Most of the information (i.e., Sr isotopes, palynomorphs, and invertebrates) support a late Miocene (Tortonian) age for the unit. However, the information from marine vertebrates suggests a middle Miocene (Serravalian) age for the lower part of the unit.

6 Paleoenvironments and Landscape of Península Valdés During the Miocene

Global climate system has experienced important changes through the Neogene (e.g., Zachos et al. 2001, see Chapter "Climatic, Tectonic, Eustatic and Volcanic Controls on the Stratigraphic Record of Península Valdés"), driving the evolutionary patterns of terrestrial and marine biota, and hence the conformation of the extant diversity. Global Miocene climates were characterized by warmer temperatures than today, evidenced by the smaller volumes of Antarctic ice sheets. This warm climatic phase has a peak in the late middle Miocene (17–15 Ma) referred as the Mid Miocene Climatic Optimum and was followed by a gradual cooling and reestablishment of a major ice-sheet on Antartica by 10 Ma (Zachos et al. 2001).

Miocene paleoclimatic reconstructions in Patagonia are concordant with the global climatic conditions. The most relevant paleoclimatic evidence comes from continental vertebrates assemblages from Santa Cruz Formation in southern Patagonia and Sarmiento and Collón Cura formations in central Patagonia. According to Kay et al. (2012) the vertebrate assemblage of the early Miocene Santa Cruz Formation indicate that the climate was much warmer and wetter than today. Further north, the upper part of the Sarmiento Formation (early Miocene) bears a vertebrate assemblage that suggests a warm moist climate and the availability of lowland forest habitat (Madden et al. 2010). In addition, the paleontological and sedimentological evidence from the middle Miocene Collón Cura Formation in central Patagonia also indicate a warm and moist climate (Vucetich et al. 1993; Bilmes et al. 2014). During the late Miocene the paleobotanical information shows the diversification of the xerophytic groups and the forest were restricted to the Andean area (Palazzesi and Barreda 2007), delineating the late Neogene climatic trend.

Besides the global cooling trends during the late Neogene, an important regional process which determinate the Patagonian climate was the uplift of the Andes. This orographic element had not risen to a sufficient altitude during the early Miocene to block westerly winds and moisture from reaching the Atlantic coast. The evidences of the desertification process caused by the rain shadow effect are known from the late Miocene onwards (Blinsniuk et al. 2005; Palazzesi et al. 2014).

The paleoclimatic reconstruction of the Península Valdés during the Miocene is based on several lines of evidence, such as floristic and faunistic information, as well as the sedimentological context.

For the Gaiman Formation in Península Valdés, the scarce paleontological information precludes any precise inference about *paleoenvironmental* conditions. The only paleoclimatic evidence comes from the poor pollen assemblage which indicates high humidity components (Palazzesi et al. 2006). Sedimentological information from the Chubut River valley, suggest that widespread phosphogenesis was probably produced by cold and corrosive Antarctic waters (Scasso and Castro 1999). However, some marine vertebrates (i.e., fish) studied in the same region suggest warm temperature conditions (Cione et al. 2011).

On the contrary, the *paleoenvironmental* characterization for the Puerto Madryn Formation is better supported by an abundant paleontological and sedimentological dataset. *Facies Associations* (Sect. 3) determined for the lower part of the unit, in combination with taphonomic analysis of the molluscan assemblages (del Río et al. 2001) allowed determining a range of *paleoenvironmental* life habitats. These include mid-inner *shelf* suspension feeders, to the shallowest intertidal foreshore soft bottom infaunal suspension feeders. The *Transgressive Phase* of the Puerto Madryn Formation records an inner *shelf* environment under low sedimentation rates with strong storm waves and tidal currents that promote the accumulation of thick shell beds. The *Maximum Flooding Phase* represents the deepest marine environments for the Puerto Madryn Formation and has no records of macroinvertebrates. In turn, the *Regressive Phase* shows invertebrate accumulations related to shoreface sandbars and inner *shelf* environments (del Río et al. 2001) as well as impoverished accumulations related to brackish water and tidal channel erosion (channel lags) in the uppermost part of the unit.

The mollusc association present in the lower to middle part of the Puerto Madryn Formation indicate warmer marine temperatures than present day in Península Valdés. The late Miocene Valdesian Molluscan Bioprovince represents the first massive immigration of Caribbean components in Patagonia, and the warmest water faunas that inhabited the Patagonian coast during Paleogene–Neogene times (Martínez and del Río 2002a). However, other sources of evidence from the same levels, such as osteichthyans (Gosztonyi and Riva Rossi 2005) and cool water dinoflagellates (*Habibacysta tectata*, *Bitectatodinium tepikiense*, and *Filisphaera filifera*; Fuentes et al. 2016) might indicate relatively low temperatures during the deposition of at least part of the Puerto Madryn Formation. Cetaceans from this lithoestratigraphic unit are represented by lineages that today are distributed in temperate waters, which might also suggests lower temperatures during the late Miocene. In living cetaceans, the geographic ranges of many species are principally limited to specific water temperature ranges (Rice 1998; Kaschner et al. 2006), but other factors could be related to habitat preferences, as local bathymetry and a species' habitat preferences, limiting its ability to track its preferred temperature ranges (MacLeod 2009). In their evolutionary history, some groups of cetaceans have shown significant changes in temperature range over time, such as Monodontidae (now restricted to cold north polar waters but during Pliocene occurred in subtropical warm waters), and the extinct groups archaeocetes (with a change range from were subtropical–tropical to temperate regions) (Fordyce 2009), suggesting a certain plasticity to different range of temperatures over time.

The continental *paleoenvironmental* evidence is sourced from the upper beds of the Puerto Madryn Formation (La Pastosa-Punta Delgada Section) as well as from the palynological data recovered from the middle levels (Puerto Pirámide Section) of the unit. Seasonal swamps, ponds, and fluvial channels may have developed nearby the coast. There, some continental influence under warm and calm conditions is assumed according to the abundance of floating plants (e.g., *Azolla*) (Dozo et al. 2010) and other saltmarsh communities that occupied permanently wet soils (Palazzesi et al. 2014). Some fishes indicate warm continental temperatures, which are consistent with Cione's hypothesis that the range of Brazilian fish fauna (e.g., the presence of Loricarids) extended much more southwards during the Cenozoic (Cione et al. 2005, 2011). The depicted scenario is consistent also with inferred environmental preferences of some continental vertebrates reported here. Semiaquatic capybaras (Hydrochoeridae), ducks (Dendrocygninae) and freshwater fish (Siluriformes) records, in association with intertidal and tidal-fluvial channel deposits, suggest presence of mudflats, marshes, ponds, and tidal-fluvial meandering channels (Scasso et al. 2012). Fluvial influence in the coastal zone is evidenced in the form of seasonal *sedimentary structures* and impoverished trace fossils associations, which indicate connection between the coastal estuarine environments with an integrated fluvial drainage system. The record of large birds (psilopterins and accipitrids) and terrestrial mammals (pampatheriids, glyptodontids, mylodontids, chinchillids, caviids, and macrauchenids) indicate the proximity of the emerged areas, which according to the fossil pollen and spore species indicate the occurrence of open forest and shrubs.

The palynologically fertile 10 m-thick layer corresponds to the *Maximum Flooding Phase*, which is composed of laminated mudstones and siltstones. The taphonomic bias toward this part of the unit might be related to particular *paleoenvironmental* conditions that prevailed during its accumulation. Probable dysaerobic conditions in the sea bottom, supported by a monospecific suite of the ichnofossil *Chondrites* isp. and the lack of any other body or trace fossil, allowed organic matter to be accumulated and preserved, including the recovered pollen grains, spores, dinocyts, and algae. The record of terrestrial, aquatic, and marine palynomorphs strongly suggests that the marine accumulation of the Puerto Madryn Formation took place on the inner continental *shelf* and nearshore settings. This is also supported by the minor proportion of fine-grained *shelf* deposits and the dominance of open marine nearshore to estuarine deposits in the Puerto Madryn Formation. The botanical affinities of fossil pollen and spore species identified from the assemblage may have grown along an estuarine gradient from lowlands (Atlantic coastal region) to highlands (Patagonian inland). This fossil pollen reveals a significantly more diverse flora during the late Miocene than today, comparable with that of mid-latitudes on the Brazilian coast (Palazzesi et al. 2014). The presence of some key neotropical lineages (e.g., *Schlechtendalia*, *Justicia*, Malpighiaceae and Loranthaceae) indicate that temperatures rarely fell below freezing point. A recent analysis (Palazzesi et al. 2014) suggests that mean temperatures of the coldest season were between 11.4 and 16.9 °C (presently about 6.4 °C).

The sedimentological and paleontological evidence indicate that the Península Valdés was covered, during the Miocene, by a shallow sea with a gradation from open marine *shelf* to inner estuarine environments. Open marine settings were characterized by a diverse fauna which throve in shallow, normal marine salinity conditions under the action of wave processes. In contrast, the estuarine setting was dominated by tidal currents and influenced by freshwater discharge from the continent, bearing a poor diverse fauna.

The molluscan fauna show caribbean affinities suggesting warmer conditions than today, which is also supported by the associated marine vertebrates (specially fish) as well as the continental botanical and vertebrate information. However, some contradictory information derived from the dinoflagellates, some fish and cetaceans suggest the influence of colder water in this marine setting. A possible explanation for this phenomenon could be that this *shelf* area was intermittently influenced by Antarctic cold waters that interrupted the warm water conditions.

The paleoclimatic reconstruction resulted from the fossil evidence of the Península Valdés region parallels the global and regional (Patagonian) climatic conditions during the Miocene. The high marine faunal diversity registered during the late Miocene in Península Valdés might be a consequence of the influence of upwelling processes which, together with the wide *shelf* and strong tidal currents, promotes the primary productivity and the development of complex food webs. Similar environmental conditions prevail today along the coast of Península Valdés, promoting the unique faunal diversity, today protected as a Natural World Heritage Site by UNESCO.

7 Perspectives and Future Work

For the first time, in this contribution we integrate the vast amount of sedimentological and paleontological information, resulting in the reconstruction of the paleoenviromental conditions during the Miocene in the Península Valdés region. This integrative effort is the start point to future comparative studies with other coeval Miocene localities in Patagonia and other parts of South America in order to highlight regional environmental similarities or discrepancies. In addition, the integrative reconstruction of the Miocene environments allows comparisons to those of the present day. This includes the identification of geological processes as well as ecological factors controlling ancient and modern communities. This could finally contribute to the delineation of conservation strategies to preserve the exceptionally diverse biota and unique landscapes of Península Valdés.

This work highlighted some gaps in the geological and paleontological knowledge for the Miocene of the Península Valdés region. The most important points includes: (1) geochronology, (2) stratigraphic control of paleontological studies, and (3) the knowledge of poorly known key fossil groups.

At a first place, future work should be focused on the improvement of the geochronological information. An attempt to perform an integrated *paleoenvironmental* reconstruction of a sedimentary/paleontological record of a thick stratigraphic column requires the understanding of the velocity to which each layer or strata were deposited. The timing and rate of deposition determine the age differences between lower, middle, and upper strata of a stratigraphic column. The lack of a detailed chronological scheme for the Neogene units in Península Valdés hampers the possibility of understanding the time involved for each *paleoenvironmental* change recorded by sediments or fossils. Considering that the Miocene strata of Península Valdés holds many important fossil records, a confident chronological calibration is essential to better delineate the evolution of each fossil group.

Second, the lack of a precise stratigraphic control on the provenance of each paleontological record, precludes the integration of the *paleoenvironmental* information derived from each fossil finding. In addition, the patchy record of some fossils groups provides evidence for certain short time lapses. For example, the palynologic information is sourced from a 10 m-thick layer of the entire column of the Puerto Madryn Formation, which represents a geologically instantaneous picture of the landscape in eastern Patagonia.

Finally, this contribution reveals the disparity on the amount of information available for each fossil group. For example, there are many studies focused on the molluscan fauna, which has provided most of the evidence for the *paleoenvironmental* reconstructions. This contrasts to the little or lacking information available from other fossil groups, such as dinoflagellates and diatoms, which are key indicators of paleoceanographic conditions. We aim at exploring further these rising disciplines which will allow us to improve our knowledge about past environmental and climatic conditions that prevailed in Península Valdés during a crucial period of time.

Acknowledgements We thank at a first place to the editors (Pablo Bouza and Andrés Bilmes) for the kind invitation to contribute to this book; to the reviewers Sergio Vizcaíno and Sebastián Richiano, and the critical review by Sergio Martínez, for the useful suggestions to this chapter, which considerably improve its quality. Thanks to Jorge González for its excellent illustrations of the fossil fauna in their environment. Fieldworks were conducted under the authorization and permissions to collect fossil material from Secretaría de Cultura and Secretaría de Turismo y Áreas Protegidas, Chubut Province, Argentina. Finally, we express our gratitude to many people of Península Valdés, especially many generous estancia owners for their assistance and hospitality during the fieldworks.

Glossary

Assemblage (fossil)	Group of fossils found together in a given stratum or package of strata
Biostratigraphy	The discipline that study the spatial and temporal distribution of fossil organisms as a means of dating rock strata
Bioturbation	Term used to describe the disruption of sedimentary deposits by organism activity. Frequently used to describe bedding that has been highly disrupted but which doesn't show any specific traces
Bioturbation index	Rating system for the degree of *bioturbation* of sedimentary deposits. Higher ratings indicate greater amounts of disruption to primary *sedimentary structures*
Facies (associations)	Descriptive attributes that characterize a sedimentary deposit such as lithology, grain size or biota, which are used to interpret depositional processes. They can be grouped in *facies associations* to interpret depositional *paleoenvironments*
Maximum flooding phase	An interval of strata deposited at the time the shoreline is at its maximum landward position (i.e., the time of maximum transgression). It separates the transgressive and highstand phases

Paleoecology	The study of fossil organisms and their associated remains to interpret their life cycle, living interactions, natural environment, communities, and manner of death and burial. Such interpretations aid the reconstruction of the *paleoenvironment*
Paleoenvironment	Physicochemical conditions of an area in a certain period of the geological past. Mostly refers to topography/bathymetry, energy levels of waves, tides, winds, and fluvial currents, as well as temperature and salinity
Regressive phase	An interval of strata that form when sediment accumulation rates exceed the rate of relative sea level rise. The base of this phase is formed by the maximum flooding zone (or surface)
Sedimentary structures	Features of any sedimentary deposit, formed in response to the physical, biogenic or chemical processes that deposited the sediment (primary sedimentary structures) or modified them during or following deposition (secondary sedimentary structures)
Shelf (continental)	The flat portion of a continent that extends under a shallow sea (no deeper than 200 meters), from the coastline to the continental slope
Taxon	Any group or rank in a biological classification into which related organisms are classified
Transgressive phase	An interval of strata accumulated from the onset of coastal transgression until the time of maximum transgression of the coast, just prior to renewed regression

References

Acosta Hospitaleche C (2003) Paraptenodytes antarcticus (Aves: Sphenisciformes) en la Formación Puerto Madryn (Mioceno tardío temprano), provincia de Chubut, Argentina. Revista Española de Paleontología 18:179–183

Acosta Hospitaleche C (2007) Revisión sistemática de Palaeospheniscus biloculata (Simpson) nov. comb. (Aves, Spheniscidae) de la Formación Gaiman (Mioceno Temprano), Chubut, Argentina. Ameghiniana 44(2):417–426

Acosta Hospitaleche C, Cione AL (2012) The most recent record of † Palaeospheniscus bergi Moreno & Mercerat, 1891 (Aves, Spheniscidae) from the middle Miocene, northeastern Patagonia. Neues Jahrbuch Für Geologie Und Paläontologie - Abhandlungen 266:143–148

Acosta Hospitaleche C, Tambussi C, Donato M et al (2007a) A new Miocene penguin from Patagonia and its phylogenetic relationships. Acta Palaeontol Pol 52:299–314

Acosta Hospitaleche C, Tambussi C, Dozo MT (2007b) Dendrocygna Swinson (Anseriformes) en el Mioceno tardío de la Formación Puerto Madryn (Argentina): anatomía de la pelvis. XXIII Jornadas Argentinas de Paleontología de Vertebrados, Actas 4

Barreda VD, Palamarczuk S (2000) Estudio palinoestratigráfico integrado del entorno Oligoceno Tardío-Mioceno en secciones de la costa patagónica y plataforma continental argentina. In: Aceñolaza F, Herbst R (eds) El Neógeno de Argentina. INSUGEO, Serie Correlación Geológica, pp 103–138

Bilmes A, D'Elia L, Veiga GD et al (2014) Relleno intermontano en el antepaís fragmentado patagónico: Evolución neógena de la cuenca de gastre. Revista de La Asociacion Geologica Argentina 71(3):311–330

Blisniuk PM, Stern LA, Chamberlain CP et al (2005) Climatic and ecologic changes during Miocene surface uplift in the Southern Patagonian Andes. Earth and Planetary Science Letters 230:125–142

Brunet RFJ (1997) New species of Mollusca from the Entrerriense Formation (Upper Miocene) of Chubut Province, Argentina and species not previously reported from this formation. Part II Gastropoda. Tulane Studies in Geol 30(2):1–61

Buono MR (2014) Evolución de los Balaenidae (MAMMALIA, CETACEA, MYSTICETI) del Mioceno de Patagonia: Sistemática, Filogenia y Aspectos Paleobiológicos. Unpublished Ph.D. thesis, Universidad Nacional de La Plata, 324 pp. La Plata

Buono MR, Cozzuol MA (2013) A new beaked whale (Cetacea, Odontoceti) from the Late Miocene of Patagonia, Argentina. J Vertebr Paleontol 33:986–997

Buono M, Dozo MT, Marx FG et al (2014) A Late Miocene potential neobalaenine mandible from Argentina sheds light on the origins of the living pygmy right whale. Acta Palaeontol Pol 59:787–793

Cabrera A (1926) Cetáceos fósiles del Museo de La Plata. Revista del Museo de la Plata 29:363–411

Cabrera DA, Luis A, Cozzuol MA (2012) Tridimensional angel shark jaw elements (elasmobranchii, squatinidae) from the miocene of Southern Argentina. Ameghiniana 49(1):126–131

Caramés A, Malumián N, Náñez C (2004) Foraminíferos del Paleógeno del Pozo Península Valdés (PV.es-1), Patagonia septentrional Argentina. Ameghiniana 41:461–474

Carmona NB, Buatois LA, Mángano MG, Bromley RG (2008) Ichnology of the lower Miocene Chenque Formation, Patagonia, Argentina: animal—substrate interactions and the modern evolutionary Fauna. Ameghiniana 45(1):93–122

Casadío S, Feldmann RM, Parras A et al (2005a) Miocene Fossil Decapoda (Crustacea: Brachyura) from Patagonia, Argentina and their Paleoecological setting. Ann Carnegie Mus 74 (3):151–188

Casadío S, Hakansson E, Parras A et al (2005b) Free-living bryozoans in the Late Miocene Puerto Madryn Formation, Península Valdés: Paleoenvironmental and paleobiogeographical implications. Ameghiniana 42(4) (Suplemento: 20R. Buenos Aires)

Cione AL, Tonni EP (1981) Un pingüino (Aves, Sphenis- cidae) de la Formación Puerto Madryn (Mioceno tardío) de Chubut, Argentina. Comentarios acerca del origen, la paleo- ecología y zoogeografía de los Spheniscidae. Anais do Congreso Latinoamericano de Paleontología 2:591–604

Cione AL, Cozzuol MA (1990) Reidentification of Portheus patagonicus Ameghino, 1901, a supposed fish from the middle Tertiary of Patagonia, as a delphinoid cetacean. J Paleontol 64:451–453

Cione AL, Azpelicueta MM, Bond M et al (2000) Miocene vertebrates from Entre Ríos province, Argentina. Serie Correlación Geológica 14:191–238

Cione AL, Azpelicueta MM, Casciotta JR et al (2005) Tropical freshwater teleosts from Miocene beds of Eastern Patagonia. Southern Argentina. Geobios 38(1):29–42

Cione AL, Cozzuol MA, Dozo MT et al (2011) Marine vertebrate assemblages in the southwest Atlantic during the Miocene. Biol J Linn Soc 103(2):423–440

Cozzuol MA (1996) The record of the aquatic mammals in southern South America. Münchner Geowissenschaten Abhandlungen A30:321–342

Cozzuol MA (2001) A 'northern' seal from the Miocene of Argentina: implications for phocid phylogeny and biogeography. J Vertebr Paleontol 21:415–421

Cuitiño JI, Ventura Santos R, Alonso Muruaga PJ et al (2015a) Sr-stratigraphy and sedimentary evolution of early Miocene marine foreland deposits in the northern Austral (Magallanes) Basin, Argentina. And Geol 42(3):364–385

Cuitiño JI, Scasso RA, Ventura Santos R et al (2015b) Sr ages for the Chenque Formation in the Comodoro Rivadavia region (Golfo San Jorge Basin, Argentina): stratigraphic implications. Lat Am J Sed Basin Analysis 22:3–12

De Schepper S (2006) Plio-Pleistocene dinoflagellate cyst biostratigraphy and palaeoecology of the eastern North Atlantic and southern North Sea Basin. Unpublished Ph.D. thesis, Cambridge University, 253 pp

del Río CJ (1986) Bivalvos fósiles del Mioceno de península Valdés (Provincia el Chubut). 4th Congreso Argentino de Paleontologia y Bioestratigrafía. Actas 3:111–117

del Río CJ (1988) Bioestratigrafía y Cronoestratigrafía de Formación Puerto Madryn (Mioceno medio) - Provincia del Chubut-Argentina. Anales Academia Nacional de Ciencias Exactas Físicas y Naturales, Buenos Aires 40:231–254

del Río CJ (1990) Composición, Origen y Significado Paleoclimático de la Malacofauna "Entrerriense" (Miocene medio) de la Argentina. Anales Academia Nacional de Ciencias Exactas Físicas y Naturales, Buenos Aires 42:207–226

del Río CJ (1992) Middle Miocene Bivalves of the Puerto Madryn Formation (Valdés Península, Chubut Province, Argentina). Through Nuculidae to Pectinidae- Part I. Palaeontographica Abt A 225:1–58

del Río CJ (1994) Middle Miocene bivalves of the Puerto Madryn Formation (Valdés Península, Chubut Province, Argentina). Through Lucinidae to Pholadidae. Part II. Paleontographica Abt A 231:93–132

del Río C (1997) Cenozoic Biogeographic History of the Eurythermal Genus Retrotapes, New Genus (Subfamily Tapetinae) from Southern South America and Antarctica. The Nautilus, Florida 110:77–93

del Río CJ (2004) Tertiary marine Molluscan Assemblages of Eastern Patagonia (Argentina): a biostratigraphic analysis. J Paleontol 78(6):1097–1112

del Río CJ, Martínez SA (1998) Clase Bivalvia. In: del Río CJ (ed) Moluscos Marinos Miocenos de la Argentina y del Uruguay. Academia Nacional de Ciencias Exactas, Físicas y Naturales, Buenos Aires, Monografía 15(cap. 2):48–83

del Río CJ, Martínez SA (2015) Paleobiogeography of the Danian molluscan assemblages of Patagonia (Argentina). Palaeogeogr Palaeoc 417:274–292

del Río CJ, Martinez SA, Scasso RA (2001) Nature an origin of spectacular marine Miocene Shell beds of northeastern Patagonia (Argentina): paleoecological and bathymetric significance. Palaios 16:3–25

del Río CJ, Martinez S, Orensanz JM (2010) Tertiary roots in the Recent molluscan faunas of the Southwestern Atlantic Ocean. 3° International Paleontological Congress, London, p 140

d'Orbigny A (1842) Voyage dans l'Amerique meridionale (le Brásil, la Règle Orientale de l'Uruguay, la Règle Argentine, la Patagonie, la Règle du Chili, la Règle de Bolivie, la Règle du Pèrou), executè pendant less annè 1826–1833. P. Bertrand, Paris, 4 (Paleontologie), 188 p

Dozo MT, Bouza P, Monti A et al (2010) Late Miocene continental biota in Northeastern Patagonia (Península Valdés, Chubut, Argentina). Palaeogeogr Palaeoc 297:100–106

Fazio AM, Scasso RA, Castro LN et al (2007) Geochemistry of rare earth elements in early-diagenetic Miocene phosphatic concretions of Patagonia, Argentina: Phosphogenetic implications. Deep-Sea Res II 54:1414–1432

Feruglio E (1949) Descripción Geológica de la Patagonia. *Yac*imientos Petrolíferos Fiscales vol 2, 349 pp (Buenos Aires)

Fordyce RE (2009) Neoceti. In: Perrin WF, Thewissen JGM, Würsig B (eds) Encyclopedia of Marine Mammals. Elsevier, San Diego, pp 758–763

Fuentes SN, Guler MV, Cuitiño JI et al (2016) Bioestratigrafía basada en quistes de dinoflagelados del Neógeno en el Noreste de la Patagonia, Argentina. Revista Brasileira de Paleontologia 19(2):303–314

Góis F, Scillato-Yané GJ, Carlini AA et al (2013) A new species of Scirrotherium Edmund & Theodor, 1997 (Xenarthra, Cingulata, Pampatheriidae) from the late Miocene of South America. Alcheringa 37(2):177–188

González Ruiz LR, Zurita AE, Fleagle J et al (2011) The southernmost record of a Neuryurini Hoffstetter, 1958 (Mammalia, Xenarthra, Glyptodontidae). Paläontologische Zeitschrift. 85(3):155–161

Gosztonyi AE, Riva Rossi CM (2005) Los Peces Osteicteos del Mioceno Superior de Península Valdés. I° Simposio de Paleontología de Península Valdés, Puerto Madryn, Argentina, p 24

Gutstein CS, Horwitz FE, Valenzuela-Toro AM et al (2015) Cetáceos fósiles de chile: contexto evolutivo y paleobiogeográfico. In: Rubilar-Rogers D, Otero R, Vargas A, Sallaberry M (eds) Vertebrados fósiles de Chile. Museo Nacional de Historia Natural de Chile, vol 63, pp 339–383

Haller MJ (1979) Estratigrafía de la región al poniente de Puerto Madryn, provincia del Chubut, República Argentina. Actas 7° Congreso Geológico Argentino 1:285–297 (Buenos Aires)

Haller MJ, Mendía JE (1980) Las sedimentitas del ciclo Patagoniano en el litoral atlántico norpatagónico. In: Mendía JE, Bayarsky A (eds) Estratigrafía del Terciario en el valle inferior del río Chubut. Coloquio «R. Wichmann», Asociación Geológica Argentina. 8° Congreso Geológico Argentino, Actas vol 3, pp 93–606 (Buenos Aires)

Head MJ, Norris G, Mudie PJ (1989) New species of dinocysts and a new species of acritarch from the upper Miocene and lowermost Pliocene, ODP Leg 105, Site 646, Labrador Sea. In: Srivastava SP et al (eds) Ocean Drilling Program, Proceedings, Scientific Results, Leg 105, pp 453–466

Hernández RM, Jordan TE, Farjat AD et al (2005) Age, distribution, tectonics, and eustatic controls of the Paranense and Caribbean marine transgressions in southern Bolivia and Argentina. J South Am Earth Sci 19:495–512

Kaschner K, Watson R, Trites AW et al (2006) Mapping worldwide distributions of marine mammal species using a relative environmental suitability (RES) model. Mar Ecol Prog Ser 316:285–310

Kay RF, Vizcaino SF, Bargo MS (2012) A review of the paleoenvironment and paleoecology of the Miocene Santa Cruz Formation. In: Vizcaino SF, Kay RF, Bargo MS (eds) Early miocene paleobiology in Patagonia: high-latitude paleocommunities of the Santa Cruz Formation, pp 331–365

Koretsky IA, Domning DP (2014) One of the oldest seals (Carnivora, Phocidae) from the old world. J Vertebr Paleontol 34:224–229

Latrubesse E, Rancy A (1994) La Formación Solimões (Mioceno superior-Pleistoceno) de Amazonia sudoccidental; implicaciones paleoclimáticas y estratigráficas. Revista del Museo de Historia Natural de San Rafael 12:212

Lech R, Aceñolaza F, Grizinik M (2000) Icnofacies *Skolithos-Ophiomorpha* en el Neógeno del Valle inferior del Río Chubut, provincia de Chubut, Argentina. In: Aceñolaza FG, Herbst R (eds) El Neógeno de Argentina. Instituto Superior de Correlación Geológica, Serie Correlación Geológica vol 14, pp 147–161 (Tucumán)

MacLeod C (2009) Global climate change, range changes and potential implications for the conservation of marine cetaceans: a review and synthesis. Endangered Species Res 7:125–136

Madden RH, Kay RF, Vucetich MG et al (2010) Gran Barranca: a 23-million-year record of middle Cenozoic faunal evolution in Patagonia. In: Madden RH, Carlini AA, Vucetich MG, Kay RF (eds) The paleontology of Gran Barranca: evolution and environmental change through the Middle Cenozoic of Patagonia, pp 423–439

Malumián N, Náñez C (2011) The Late Cretaceous-Cenozoic transgressions in Patagonia and the Fuegian Andes: Foraminifera, palaeoecology, and palaeogeography. Biol J Linn Soc 103:269–288

Marengo H (2015) Neogene micropaleontology and stratigraphy of Argentina. The Chaco-Paranaense Basin and the Península de Valdés, Springer Briefs in Earth Systems Series 218 pp

Martínez SA, del Río CJ (2002a) Late Miocene Molluscs from the Southwestern Atlantic Ocean (Argentina and Uruguay): a paleobiogeographic analysis. Palaeogeogr Palaeoc 188:167–182

Martínez SA, del Río CJ (2002b) Las provincias malacológicas miocenas y recientes del Atlántico Sudoccidental. Anales Biol 24:121–130

Martínez SA, del Río CJ (2008) A new, first fossil species of *Ophioderma* Müller and Troschel, 1842 (Ophiuroidea: Echinodermata) (Late Miocene, Argentina). Zootaxa 1841:43–52

Noriega JI, Cladera G (2008) First record of an extinct marabou stork in the Neogene of South America. Acta Palaeontol Pol 53:593–600

Olson SL (1983) Fossil seabirds and changing marine environments in the Late Tertiary of South Africa. S Afr J Sci 79:399–402

Palazzesi L, Barreda V (2004) Primer registro palinológico de la Formación Puerto Madryn, Mioceno de la provincia del Chubut, Argentina. Ameghiniana 41:355–362

Palazzesi L, Barreda V (2007) Major vegetation trends in the Tertiary of Patagonia (Argentina): a qualitative paleoclimatic approach based on palynological evidence. Flora 202:328–337

Palazzesi L, Barreda VD, Scasso RA (2006) Early Miocene spore and pollen record of the Gaiman Formation (Northeastern Patagonia, Argentina): correlations and paleoenvironmental implications. 4th Latin American Congress on Sedimentology and 11th Argentinean Meeting of Sedimentology, Bariloche, Argentina, p 161

Palazzesi L, Barreda VD, Cuitiño JI et al (2014) Fossil pollen records indicate that Patagonian desertification was not solely a consequence of Andean uplift. Nat Com 5:3558

Parras A, Dix GR, Griffin M (2012) Sr-isotope chronostratigraphy of Paleogene-Neogene marine deposits: Austral Basin, Southern Patagonia (Argentina). J South Am Earth Sci 37:122–135

Perea D, Ubilla M, Martínez S et al (1994) Mamíferos neógenos del Uruguay: la edad Mamífero Huayqueriense en el "Mesopotamiense". Acta Geol Leopoldiense 17:375–389

Perez LM (2014) Nuevo aporte al conocimiento de la edad de la Formación Paraná, Mioceno de la Provincia de Entre Ríos, Argentina. Asoc Paleontol Arg Special Pub 14:7–12

Philippi RA (1893) Descripción de algunos fósiles Terciarios de la República Argentina. Anales del Museo Nacional de Chile (3rd edn), Mineralogía, Geología y Paleontología, pp 1–13

Picasso M, Tambussi C, Dozo MT (2009) Neurocranial and brain anatomy of a Late Miocene eagle (Aves, Accipitridae) from Patagonia. J Vertebr Paleontol 29(3):831–836

Rice DW (1998) Marine mammals of the world: systematics and distribution. Spec Publ No. 4, Soc Marine Mammalogy, Beaufort, NC

Riva Rossi CM, Gosztonyi AE, Cozzuol MA (2000) A Miocene cusk-eel (Ophidiiformes: Ophidiidae) from Península Valdés, Argentina. J Vertebr Paleontol 20:645–650

Rossetti DF, Bezerra FHR, Dominguez JML (2013) Late oligocene-miocene transgressions along the equatorial and eastern margins of brazil. Earth-Science Rev 123:87–112. doi:10.1016/j.earscirev.2013.04.005

Rovereto G (1913) La Penísola de Valdéz e la forme costiere della Patagonia settentrionale. Reale Academia dei Lincei. Estratto del vol 23, serie 5, 1 semestre (2), pp 103–105

Scasso RA, del Río CJ (1987) Ambientes de sedimentación, estratigrafía y proveniencia de la secuencia marina del Terciario superior de la región de Península Valdés, Chubut. Revista de la Asociación Geológica Argentina 42:291–321

Scasso RA, Castro LN (1999) Cenozoic phosphatic deposits in North Patagonia, Argentina: Phosphogenesis, sequence-stratigraphy and paleoceanography. J South Am Earth Sci 12:471–487

Scasso RA, Bellosi E (2004) Cenozoic Continental and Marine Trace Fossils at the Bryn Gwyn Paleontological Park (Chubut, Argentina). Guidebook for the First International Congress on Ichnology, Trelew 19 pp

Scasso RA, del Río CJ, Martínez S (1999) El contacto "Entrerriense"-"Patagoniense" en Península Valdés (Chubut). Examen de una discontinuidad. 14° Congreso Geológico Argentino, Actas vol 1, p 73

Scasso RA, McArthur JM, del Río CJ et al (2001) $^{87}Sr/^{86}Sr$ Late Miocene age of fossil molluscs in the "Entrerriense" of the Valdés Península (Chubut, Argentina). J South Am Earth Sci 14:319–329

Scasso RA, Cuitiño JI, Escapa I (2010) Mesozoic-Cenozoic basins of Central Patagonia with emphasis in their tidal systems. In: del Papa C, Astini R (eds) Field excursion guidebook, 18th International sedimentological congress. Mendoza, Argentina, pp 1–43

Scasso RA, Dozo MT, Cuitiño JI et al (2012) Meandering tidal-fluvial channels and lag concentration of terrestrial vertebrates in the fluvial-tidal transition of an ancient estuary in Patagonia. Lat Am J Sediment Basin Analysis 19:27–45

Scasso RA, Cuitiño JI, Bouza PJ (2015) Miocene and modern tidal deposits of the Valdés Península. Field guide of the field trip 3, 9th international conference on tidal sedimentology, Puerto Madryn, Argentina. 55p

Steeman ME, Hebsgaard MB, Fordyce RE et al (2009) Radiation of extant cetaceans driven by restructuring of the oceans. Syst Biol 58:573–585

Tambussi C (2011) Palaeoenvironmental and faunal inferences based on the avian fossil record of Patagonia and Pampa: what works and what does not. Biol J Linn Soc 103:458–474

Valenzuela-Toro AM, Gutstein CS, Varas-Malca RM et al (2013) Pinniped turnover in the South Pacific ocean: new evidence from the Plio-Pleistocene of the Atacama desert, Chile. J Vertebr Paleontol 33:216–223

Versteegh GJM (1995) Palaeoenvironmental changes in the Mediterranean and North Atlantic in relation to the onset of Northern Hemisphere glaciations (2.5 Ma B.P.). A palynological approach. Unpublished Ph.D. thesis, University of Utrecht, 134 p, 6 pl. The Netherlands

von Ihering H (1897) Os molluscos dos terrenos terciarios de Patagonia. Revista del Museo Paulista, San Pablo 2:217–382

von Ihering H (1899) Die Conchylien der Patagonischen Formation. Neues Jahrbuch für Mineralogie, Geologie und Palaeontologie 2:1–41

von Ihering H (1907) Les Mollusques fossiles du Tertiaire et du Cretace superieur del'Argentine. Anales del Museo Nacional Buenos Aires (series 3) 7:1–611

Vucetich MG, Mazzoni MM, Pardiñas UF (1993) Los roedores de la Formación Collón Cura (Mioceno Medio) y la Ignimbrita Pilcaniyeu. Cañadón del Tordillo, Neuquén. Ameghiniana 30:361–381

Vucetich MG, Deschamps CM, Olivares AI et al (2005) Capybaras, size, shape and time: a model kit. Acta Palaeontol Pol 50(2):259–272

Vucetich MG, Deschamps CM, Vieytes EC et al (2014) Late Miocene capybaras (Rodentia, Cavioidea, Hydrochoeridae): skull anatomy, taxonomy, evolution and biochronology. Acta Paleontol Pol 59:517–535

Zachos J, Pagani M, Sloan L et al (2001) Trends, rhythms, and aberrations in global climate 65 Ma to present. Science 292:686–693

Zinsmeister WJ, Marshall LG, Drake RE et al (1981) First radioisotope (Potassium-Argon) age of marine Neogene Rionegro beds in northeastern Patagonia, Argentina. Science 212:440

The Climate of Península Valdés Within a Regional Frame

Fernando Coronato, Natalia Pessacg and María del Pilar Alvarez

Abstract Peninsula Valdés shares with the whole of Eastern Patagonia the main features of the regional climate, i.e. scarce rainfall, strong winds and cool-temperate temperatures. Not with standing it has an ill-defined climate because of its geographical location not far from the transitional area, where Pacific and Atlantic air masses merge. Also, because of its latitude (42°–43° S), the southward migration of the subtropical anticyclones is still noticeable over the area in summer. This chapter aims to explain the interplay between large scaled factors as the above-mentioned, and local ones as the almost insularity of the study area. A concise description of the climate is presented through the usual basic elements, temperature, precipitation, and wind. The maritime influence upon these variables is evaluated. It is shown that although mostly commanded by the rain-shadowed westerlies as the entire Patagonia, the climate of Peninsula Valdés has singularities that make it a less arid, more even, and milder climate which presents some Mediterranean features. Historic trends of rainfall and temperature are discussed and appear to be in agreement with global warming projections, according to which future scenarios would be drier and warmer in the Península Valdés region.

Keywords Climatic gradient · Temperature · Precipitation · Wind · Oceanity

F. Coronato (✉) · N. Pessacg · M. del Pilar Alvarez
Instituto Patagónico para el Estudio de Ecosistemas Continentales
(IPEEC) - Consejo Nacional de Investigaciones Científicas y Técnicas
(CONICET) - CCT Centro Nacional Patagónico (CENPAT),
Boulevard Almirante Brown 2915, U9120ACD Puerto Madryn,
Chubut, Argentina
e-mail: coronato@cenpat-conicet.gob.ar

N. Pessacg
e-mail: pessacg@cenpat-conicet.gob.ar

M. del Pilar Alvarez
e-mail: alvarez@cenpat-conicet.gob.ar

1 The Climate of Patagonia

Patagonia, the southern tip of South America, extends from 40° S—where the continent is 1000 km width—and gradually narrows until disappear in Cape Horn at 56° S. This region is the only continental mass in the Southern Hemisphere that intersect the mid-latitude *westerlies* (Fig. 1a, b), which strongly influences the atmospheric circulation at lower and upper levels and consequently the climate of the region (Zhu et al. 2014).

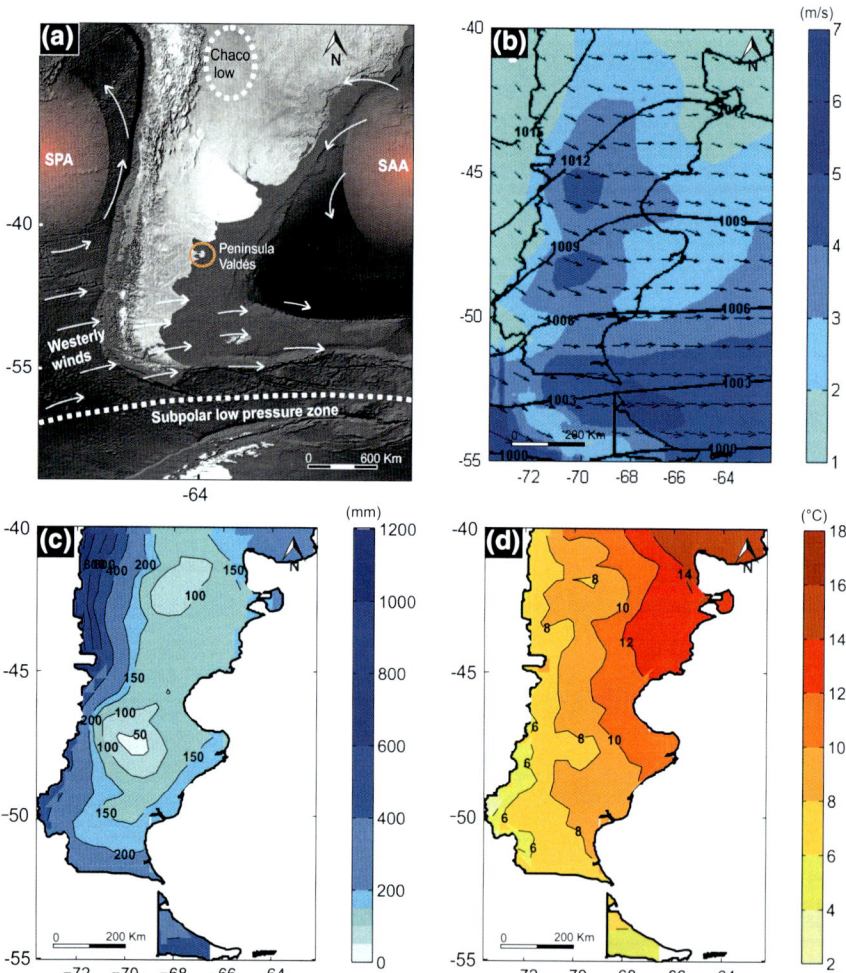

Fig. 1 **a** Major climate features of southern South America; **b** wind direction (*arrows*), intensity (m/s, shaded), and mean sea level pressure (hPa, *black line*) from ERA-Interim reanalysis (European Centre for Medium-Range Weather Forecast—ECMWF-), average 1980–2009; **c** precipitation (mm/a) **d** temperature (°C) at 2 m from CRU dataset (Climate Research Unit, University of East Anglia), average 1980–2009

Patagonia is also located north of the subpolar low pressure trough and between two semi-permanent anticyclones systems at 30° S approximately. One is located over the South West Pacific and the other over the South Atlantic Ocean (Prohaska 1976; Paruelo et al. 1998; Fig. 1a). The strong pressure gradient between both belts generates the strong upper-air westerly jet (Hobbs et al. 1998; Lenaerts et al. 2014).

The climate of regions located at mid-high latitudes is primarily influenced by the physical features of the surface such as the topography, distribution of land and sea, and the extent and concentration of sea ice. In Patagonia, the Andes Mountains strongly affect the regional climate by blocking the disturbances embedded in the westerly flow (see Chapter "Climatic, Tectonic, Eustatic and Volcanic Controls on the Stratigraphic Record of Península Valdés"), producing the precipitation over this area and influencing wind patterns and precipitation in the whole region (Insel et al. 2010). South of 40° S the Andes rarely exceed 3 km height and consequently the Pacific air masses are dominant in the region (Labraga and Villalba 2009). Uplift on the west side of the Andes leads to hyper humid conditions, while downslope subsidence dries the eastern plains leading to arid and highly evaporative conditions (Garreaud et al. 2013). At the same latitude of the Peninsula Valdes but in western Patagonia, there is a very strong precipitation gradient, ranging from 6000 mm/a on the Chilean coast to less than 300 mm/a in Argentina, just a few tens of kilometres leeward the Andes (Smith and Evans 2007). Thus, in Argentine Patagonia the maximum precipitation is located close to the Andes, between 40 and 45° S, and the west-east gradient in this region is from 1500 mm/a close to the mountain range to 150 mm/a the Atlantic coast (Fig. 1c). As in most arid and semiarid mid-latitudes regions of the world, precipitation is related to *frontal activity* (Jobbágy et al. 1995; Bell 1981). Yet, since the forcing of frontal rainfall is dynamical, the annual cycle of precipitation in Patagonia is not as pronounced as at lower latitudes, with thermodynamic forcing (Garreaud and Aceituno 2007).

In Patagonia as a whole, precipitation is concentrated from May to October and, on average, 46% of it falls during the three winter months, i.e. June–August (Labraga and Villalba 2009; Jobbágy et al. 1995). Such concentration is related to the intensification and equatorward shift of the *stormtrack* (preferred path of *synoptic-scale disturbances*) and the subtropical anticyclones during this season, which increment the frequency of Pacific synoptic perturbations (Garreaud and Aceituno 2007). There is a clear relationship between winds and precipitation, since stronger westerlies increase the local precipitation in western Patagonia whereas decrease local precipitation east of the Andes (Garreaud et al. 2013).

In northeastern Patagonia, where some influence of Atlantic air masses is noticeable, precipitations are more evenly distributed throughout the year (Jobbágy et al. 1995). The South Atlantic Ocean may also be an additional source of moisture during intense precipitation events over this region (Agosta et al. 2015).

In regard to surface temperature, the spatial pattern of this variable is characterized by a gradient northeast-southwest (Fig. 1d), related to both latitude and altitude effects. Mean annual values range from 15 °C in the northern sector to 4.5 °C in Tierra del Fuego and southern Andes. East to the Andes, the north-south temperature gradient is stronger in summer due to the stronger continental warming in the north, leading to a north-south pattern of temperature variability (Berman 2012). On the contrary, in winter, the high-spressure systems which expand to the east of the Andes lead to intense southerly advections of cold air over most of east Patagonia and thus the temperature variability east of the Andes is less marked during this season (Müller et al. 2005; Berman 2012).

In much of Patagonia temperature pattern is correlated to wind. Persistent and strong westerlies throughout the year will result in a decrease of the amplitude of the local air temperature annual cycle. Besides, the strong winds reduce the perceived temperature (i.e. wind chill), and because of higher wind speed in summer the wind-chill effect is more pronounced during this season driving to cool felt summers (Coronato 1993).

The cloud coverage in Patagonia is almost permanently high, particularly over the southern part of the region during summer due to the intensification of the westerlies. The southward shift of the subtropical anticyclone leads to a decrease of cloud coverage over northern Patagonia during summer (Castañeda and González 2008).

1.1 Influence of Large-Scale Variability Modes

Understanding the climate variability in Patagonia is relevant on one hand due to effect in the mechanisms that govern the circulation of the extratropical Southern Hemisphere, and on the other hand because of the impact of such variability in the *cryosphere*, biosphere and society (Garreaud et al. 2013). In particular, the variability of rainfall in dry areas of Patagonia induces changes in different processes as desertification, water erosion and soil compaction, which influence the human activities and increase the climate vulnerability of the region (Berman et al. 2013).

The *Southern Annular Mode* (SAM) is the leading pattern of year-to-year circulation variability at high latitudes in the Southern Hemisphere. The SAM describes the north–south movement of the westerly wind belt surrounding Antarctica (negative and positive phase of the SAM, respectively) (Kidson 1988; Karoly 1990). The changes in the position of the westerly wind belt influences the strength and position of cold fronts which in turn influences the seasonal and annual mean temperatures in the east of Patagonia (Gillett et al. 2006; Garreaud et al. 2008; Silvestri and Vera 2009). The positive phase of the SAM is related to anomalously warm and dry conditions in southern Argentina south of 40° S, related with the southward shift of the stormtrack (Gillett et al. 2006). Garreaud et al. (2013) found a correlation between the SAM and the circumpolar anomalies of zonal flow at mid-and high latitudes which in turn lead to precipitation anomalies in Patagonia during summer. In contrast, during winter there is a low correlation between SAM and precipitation anomalies.

The other two leading patterns of circulation variability in the Southern Hemisphere are the Pacific South American modes PSA1 and PSA2, respectively

(Kidson 1988). Both models are wave trains extending from tropical Pacific to the extratropical regions of the Southern Hemisphere. These modes of variability show a large correlation with *El Niño-Southern Oscillation* (*ENSO*). In fact, some authors described that the dynamics of the tropical signal ENSO is transported to the higher latitudes via the PSAs modes (Turner 2004).

Different studies have found that the interannual variability of temperature in Patagonia is related to the ENSO phases. The El Niño events are negatively correlated with the temperature in Patagonia during spring (Garreaud 2009). This is in good agreement with the results of Rusticucci and Vargas (2002), which showed a persistence of cold air masses over Patagonia during these events.

On the other hand, González and Vera (2010) and González et al. (2010) found an influence of the Pacific sea surface temperature in the interannual variability of winter precipitation over northwest of Patagonia. Besides, Garreaud (2009) found a positive correlation between precipitation in Patagonia and El Niño events for the period March–November. This is in line with a simultaneous and positive correlation between the streamflow and El Niño events for the rivers of north of Patagonia (Campagnucci and Araneo 2007). The variability of precipitation over the Andes of Patagonia during ENSO events is in part explained by changes in westerlies during these periods (Schneider and Gies 2004).

Berman et al. (2013) found that the precipitation in central-north Patagonia is positively correlated with precipitation over the surrounding Atlantic and southeast of South America. The increase of rainfall over the central-north Patagonia is related to the weakened westerly flow over the region and the consequent increasing of Atlantic air flow.

Temperature in Patagonia also exhibits variability on intraseasonal scales (1–3 months). Jacques-Coper et al. (2015a) found that the intraseasonal variability modulates the strong and persistent warm conditions in the east of Patagonia. Besides that, Jacques-Coper et al. (2015b) analyzed the relationship between the intraseasonal air temperature variability in eastern Patagonia and the Madden–Julian Oscillation (Madden and Julian 2971)—main mode of intraseasonal variability of the coupling between atmospheric circulation and tropical convection. The authors showed evidence that the amplitude of the intraseasonal component of temperature in eastern Patagonia is modulated in 1.5 °C approximately for the Madden–Julian Oscillation. Besides that, the authors found that one of the eight phases of this oscillation induces warm conditions and favour heat waves in eastern Patagonia.

1.2 Climate Change in Patagonia

Many studies in the last years have addressed the analysis of the climate global change observed during the twentieth century and projected for the twenty-first century, mainly as a consequence of the anthropogenic activity. These activities result in a constant increase in the greenhouse gases concentration in the atmosphere, which in turn drive to rising global annual mean surface temperature (IPCC 2013).

In most of the Patagonia there is a clear warming since 1950 (Rosenblüth et al. 1997; Boninsegna et al. 2009; Villalba et al. 2003; Vincent et al. 2005). This warming would have been higher than in the rest of Argentina (Barros et al. 2014). In average, the mean annual temperature increased 0.4 °C in the last 50 years, being the rising up to 1 °C in some areas, whereas minima and maxima temperatures increased, respectively, 0.4–0.8 °C and 0.5 and 1 °C depending on the area. However, in the northeast, where the Peninsula Valdés is located, Rusticucci and Barrucand (2004) and Barros et al. (2014), observed a negative linear trend for the minimum temperature and a positive one for the maximum temperature. Coronato and Bisigato (1998) showed that the weather station of Trelew, at 150 km approximately of the study area (Fig. 2) is the one that better depicts the year-to-year trend of temperature in the entire Patagonia. Also, clear regional trends in extreme temperatures were found, evidenced by a decrease in the number of cold days and nights and an increase of warm days and nights since 1960 (Rusticucci and Barrucand 2004).

With reference to precipitations, not significant changes where observed in the last 50 years in most of Patagonia and throughout the region interannual variability overrides changes in long term. However two exceptions appear: decreasing of precipitation over the northern Andes and increasing rainfall in the northeast of the

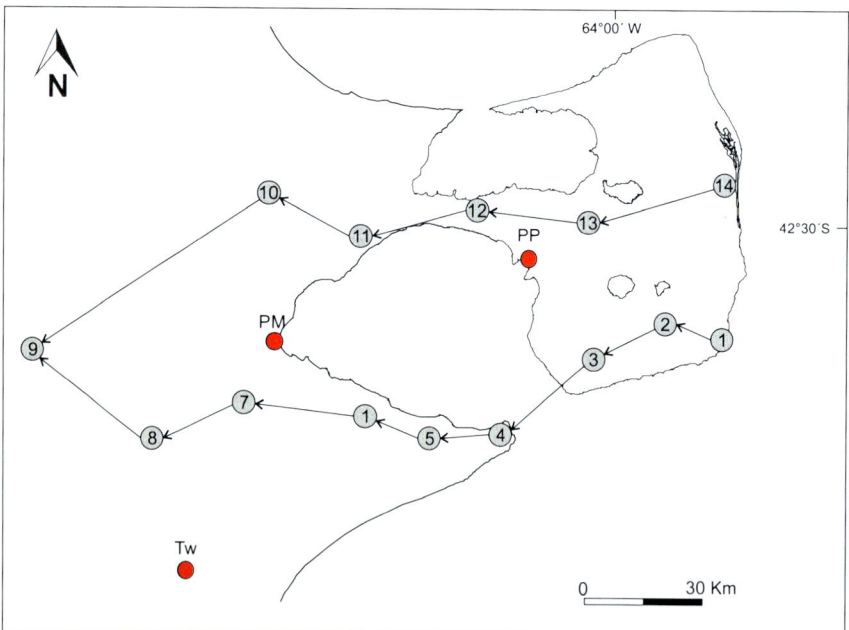

Fig. 2 Location map of two east-west transects of the interpolated points (*grey dots*) listed in Table 2. *Red dots* correspond to: *PP* Puerto Pirámides; *PM* Puerto Madryn; *Tw* Trelew

region including the Península Valdés (Castañeda and González 2008; Masiokas et al. 2008).

The projections performed with different models (General Circulation Models and Regional Climate Models) and for different scenarios, estimate for Patagonia a warming between 0.5 and 1 °C for the near future (2015–2039) and a warming between 2.5 and 3 °C for the far future (2075–2099) (Barros et al. 2014). The greatest warming is projected for February while the smallest is for April (Nuñez et al. 2009).

Regarding the precipitation, the different scenarios and models project less precipitation over Patagonia for both near and far future in the order of 0–10 and 10–20%, respectively. Diminishing is important in the Andes over the climatological rainfall maximum, and it is related to the shift of the Pacific Ocean stormtrack (Blázquez et al. 2012). Moreover, less rainfall in the Andes is important in terms of hydrological ecosystem services since the major river basins of Patagonia have their headwaters in that mountain range.

In the last decades, a few studies analyze the climate change in Patagonia through fossil records. Among them, Villalba et al. (2012) performed reconstructions of SAM indices from tree rings in the Southern Hemisphere extratropics. These authors found that the SAM pattern has a high frequency of positive phases since the 1950s that is exceptional in the past 600 years and is connected with a south shift of the westerlies. This change in the winds pattern impacts in the temperature and precipitations pattern. Besides, the tree growth is related not only to the increase of greenhouse gases but also to the ozone depletion in the stratosphere at high latitudes and the internal variability of the atmosphere (Ablaster and Meehl 2006; Deser et al. 2012; Gonzalez et al. 2014).

2 Climate of Peninsula Valdés

2.1 Climatic Singularities of Peninsula Valdés

Not many geographical features in the world meet so accurately the etymology of the word "peninsula" such as Valdés does. Indeed, since Peninsula Valdés is "almost an island" it is not surprising that its climate displays insular characteristics that differ from the adjacent mainland. According to its latitudinal location, the Peninsula Valdés (42–43° S) falls within the southern westerlies belt that spread Pacific influence all over Patagonia. However, latitude is not high enough so as to avoid a quite evident subtropical anticyclonic influence during the summer, noticeable up to 45° S (Fig. 1a). Because of sunnier and dryer summers some authors consider that northeastern Patagonia has a marginal Mediterranean climate (Le Houerou 2005). Yet, owing to the eastern location within the Patagonian context, stronger anticyclonic (Atlantic) influence means increasing flux of NE winds on the area, which may carry episodic heavy rains, penetrating little inland. So, on one hand there is the Mediterranean-like dry summer affecting northern

Table 1 Maritime influence as reflected by temperatures (°C)

Location	Mean temperature	Annual temperature range	Days of frost	Absolute minimum
Punta Delgada	12.5	10.4	6.4	−4.9
Puerto Madryn	14.0	14.0	43	−11.6

Patagonia from coast to coast, while on the other hand, the most eastern locations (i.e. the Península Valdés) may benefit of sporadic rainfalls of Atlantic origin.

As result of the twofold origin of rainfall in the area, the intra-annual rainfall is more evenly distributed than in westwards locations, even if (as elsewhere in central Patagonia) the autumn peak is still noticeable.

Rather than the rainfall distribution along the year, it is the thermal regime that better reflects the influence of the Atlantic Ocean on the climate of the Península Valdés. Every temperature-based index shows clearly a strong gradient between the more exposed eastern locations and the inner ones. A cursory comparison between basic temperature data (series 1959–1968) from Punta Delgada and Puerto Madryn (Fig. 2) is enough to grasp the climatic differences at both ends of a 110 km-long transect (Table 1).

More detailed transects can be drawn from De Fina et al. (1968; Table 2, Fig. 2). These authors worked with carefully interpolated data checked with local and official records (SMN, Servicio Meteorologico Nacional) from the period 1951–1960. Both east-west transects show increasing continentality towards the west as well as decreasing yearly rainfall amount. It can be noted that no point within the Peninsula Valdés is located above 100 ma.s.l. (no climatic data from central depression [−40 mb.s.l.] are available; Table 2). Owing to the flat or gently undulating relief, no marked topoclimatic influence exists and observed differences stem on larger scale factors, mainly distance to the sea. Only further inland, westwards of the Peninsula Valdés, the gradually increasing altitude may become a noticeable topoclimatic factor, yet partially counteracted by the plateau-like relief (Coronato 1994). This could be seen in the change in the trend to higher summer temperature observable in the west end of the transects presented on Table 2.

Trend towards less maritime conditions westwards are observed not only in temperature patterns but also in less evenly distributed and less abundant rainfall (Table 2).

The maritime influence all over the Peninsula Valdés and its swift decreasing further inland are clearly evidenced by the diverse isolines mapped in Fig. 3. This is noticeable not only in the plain annual temperature range (Fig. 3a) but also in the more subtle asymmetry and phase lag in the temperature curve, usually assessed by simple difference between April and October mean temperatures (Fig. 3b), as well as in more elaborated calculations as Daget Continentality index (Fig. 3c) and even in no temperature-related parameters, as absolute humidity (Fig. 3d). The Daget Continentality Index (CI) (Daget 1968) is defined as follows:

Table 2 Calculated figures based on Punta Delgada and Puerto Madryn data. Ea: Estancia (farm, inhabited spot)

		Location	Mean January temperature (°C)	Mean July temperature (°C)	Elevation (m.a.s.l.)	Annual rainfall (mm)	Winter months rainfall (mm)
South transect	1	Punta Delgada	17.4	6.8	54	236	64
	2	Ea. La Pastosa	17.8	6.9	40	230	62
	3	Ea. La Cantábrica	18.3	6.7	30	220	60
	4	Ea. El Pedral	19.0	6.7	18	206	57
	5	Ea. Muzio	19.1	6.3	78	196	52
	6	Ea. Urtazun	19.5	6.3	83	182	46
	7	Km 11	20.0	6.0	113	169	40
	8	Ea. El Confort	19.9	5.3	160	160	33
	9	Sierra Chata	20.3	5.2	187	149	33
North transect	10	Ea. Valdés Creek	18.3	7.0	30	251	64
	11	Ea. La Ernestina	18.6	6.8	53	203	53
	12	Ea. Iriarte	19.6	7.0	10	168	44
	13	El Desempeño	19.7	6.5	87	163	41
	14	Ea. Dos Naciones	20.5	6.3	95	158	38
	9	Sierra Chata	20.3	5.2	187	149	33

Data from De Fina et al. (1968)

$$CI = [1.7\,A/\sin(\varphi + 10 + 9\,h)] - 14$$

where A = annual temperature range; φ = latitude; h = absolute altitude in km.

The values of every one of the parameters above over the Península Valdés are among the highest figures along the Patagonian Atlantic coast north of 48° S according to the maps presented by Coronato (1994). Sensible maritime influence coupled to mild temperatures make climatic risk much lower than in the rest of Patagonia to the greatest advantage of sheep farming in the area.

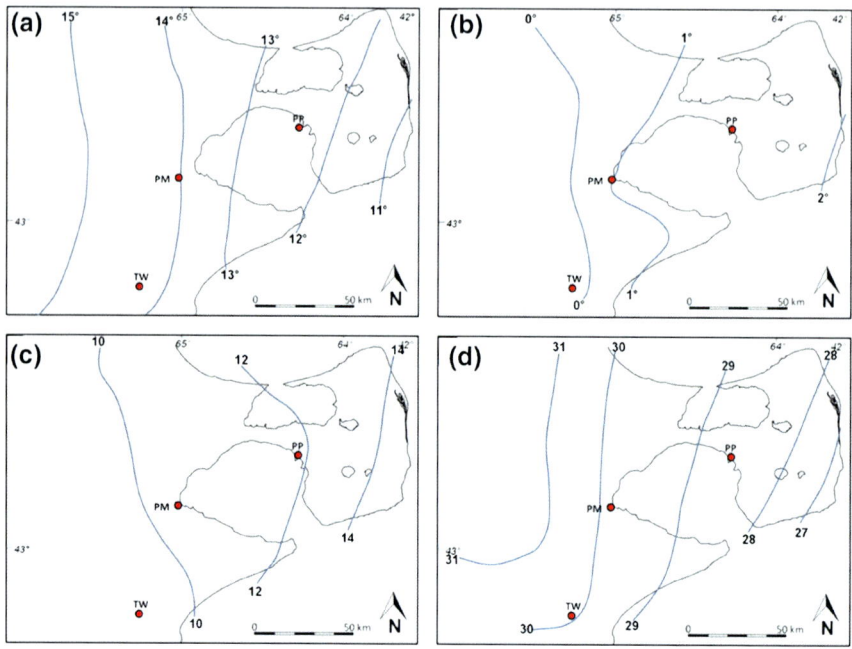

Fig. 3 Maritime influence on Península Valdés climate. **a** Annual temperature range; **b** Autumn-Spring thermal difference; **c** Mean vapor pressure (Jan, hPa); **d** Daget's Continentality Index (from Coronato 1994); *PP* Puerto Pirámides; *PM* Puerto Madryn, *TW* Trelew

2.2 Characterization of Península Valdés Climate

2.2.1 Precipitation

Spatial distribution (1979–1992) of rainfall over the Península Valdés presents a west-east gradient since precipitation increases towards the Atlantic shore; annual average is 218 mm at the isthmus (Ea. Iriarte) and from there gradually increases eastwards beyond 260 mm on the external shore (Fig. 4). It should be noted that the meteorological station located in Puerto Madryn, has always recorded accumulated rainfall averages lower than those of the stations in the Península Valdés regardless of the selected period. This increase towards the Atlantic has been interpreted as due to the influence that adjacent water bodies (Atlantic Ocean, Nuevo Gulf and San José Gulf), generated on the rainfall amount (Barros et al. 1979). The prevailing S and SW winds running through the gulfs and the ocean are moisture loaded and thus generate some increase of rainfall amount, respect to the values registered in the stations westwards in the study area.

The longest rainfall record of the region (Ea. La Adela, 1912–2014) shows a mean annual precipitation of 230.8 mm and a pattern that highlights the high interannual variability (Fig. 4). Although most of the mean annual precipitation

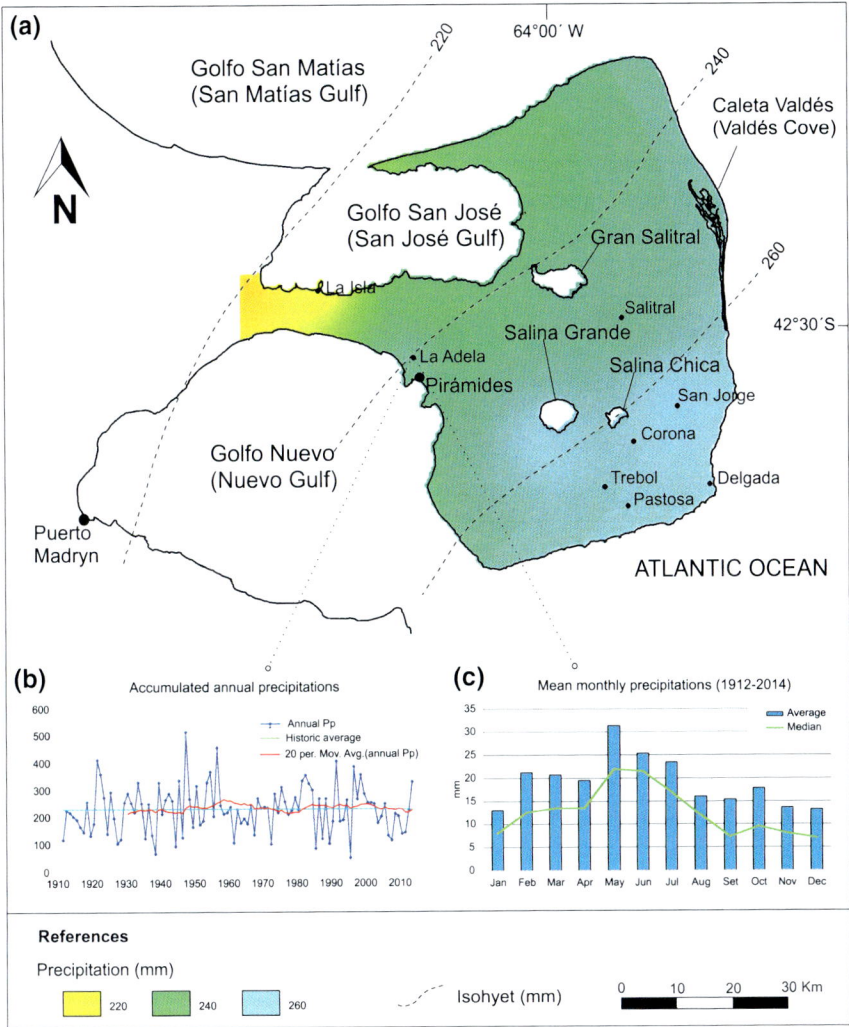

Fig. 4 a Isohyetal map of the Península Valdés region, **b** precipitation trend in Ea. La Adela, **c** Mean monthly precipitations in Ea. La Adela. The climatological records gathered by Instituto Nacional de Tecnología Agropecuaria (INTA), Servicio Meteorológico Nacional (SMN), Centro Nacional Patagónico (CENPAT) and from the mapped farms in the area

values are within the 100–300 mm/a range, there are some years of very low precipitation (about 50 mm) and other with exceptionally high values (>500 mm; Fig. 4). As an example, an extreme case of variability occurred in 1947/1948, when the variation from 1 year to the next was 388 mm (126 and 514 mm, respectively). This versatility in the rainfall amount is typical of the arid climates, in which storm effect can markedly alter the expected annual value.

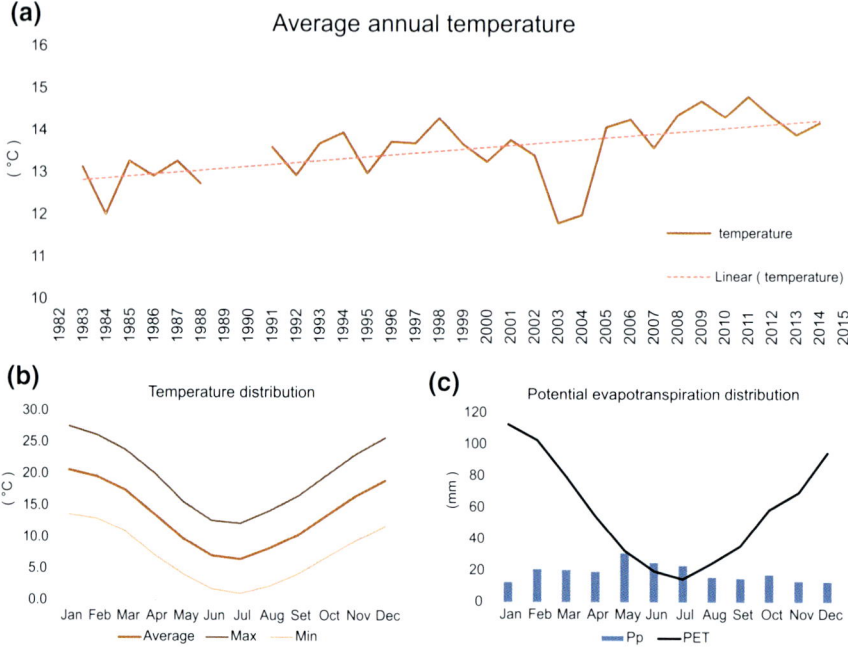

Fig. 5 a Temperature historic trend of Península Valdés region, b Mean monthly temperatures, c Mean monthly Potential Evapotranspiration (PET). Data from the Centro Nacional Patagónico (CENPAT) station

A 20-year moving average line was included in the graph (Fig. 5) to smooth out short-term fluctuations and highlight longer term trends. It can be observed that there is not a uniform tendency for the whole period, but the alternation of humid and dry periods can be recognized all along the record.

The annual distribution of the mean monthly precipitations shows a wettest period in May–July and a drier one in November–January, even if from September the values are already low. The existence of a high variability in the records for the same month within the analyzed period, makes the distribution of the mean monthly precipitations being affected by extraordinary storms. A strong single precipitation event could generate great space and time variations of the mean values distributions. To avoid this, the analysis of median values allows to smooth the variations and confirms the existence of a wetter period between the autumn–winter seasons. Similar monthly precipitation patterns are recorded in other farms of Península Valdés, showing in some cases that the rainy period is slightly displaced to March and April.

2.2.2 Temperature

Temperature records are much scarcer than the rainfall ones and only the Puerto Madryn station has a record long enough to be representative (more than 20 years); the official meteorological station located in within the Península Valdés itself (i.e. Ea. Punta Delgada) has a 10-year long record but stopped in 1968. According to Puerto Madryn records, the average annual value (1983–2014) is 13.6 °C, and the mean annual temperatures show a clear upward trend, since values were between 12 and 13 °C in the beginning of the period but reached 14 and 15 °C by the end of the period (Fig. 6a), which is in agreement with the Patagonian warming mentioned above. As for the annual distribution of mean monthly temperatures, the coldest month correspond to July, with an historical average of 6.4 °C, while the warmest correspond to January, with an historical average of 20.4 °C. As can be seen in the thermogram (Fig. 5b), the average values of the maximum and minimum monthly temperatures accompany the annual average temperature cycle. July has the low minimum temperature: 1 °C, and January the high maximum temperature: 27.5 °C. Absolute minimum temperatures at both ends of the study area were shown in Table 1. With respect to the average temperature range, it is lower in the winter months, with values of 10.8 and 11.1 °C for June and July, respectively, and reaches the maximum values in summer, with values of 14.0 and 13.9 °C for December and January.

Regarding the temperature records of the stations located within the Península Valdés (Punta Delgada excepted), although they are very restricted in terms of

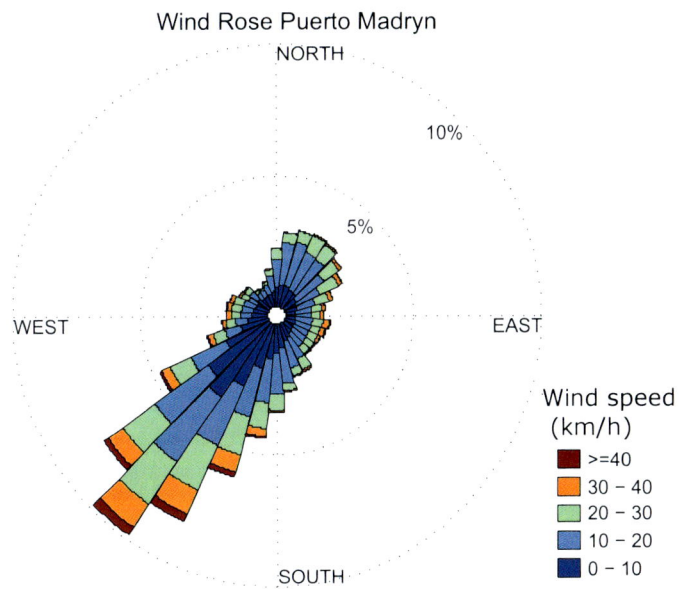

Fig. 6 Puerto Madryn wind rose (Data from CENPAT station: 1982–2010)

length, the annual distribution of the figures is the same as for Puerto Madryn (Alvarez 2010). This is due to the fact that the latter is located at the same latitude and that the temperature variable is much more homogeneous than the precipitation. The only discernible differences stem on the influence of the sea, as seen in Sect. 2.1.

2.2.3 Wind

As mentioned above, the prevalent atmosphere circulation in the region is strongly influenced by the presence of two large high-pressure systems or quasi-stationary anticyclones located on both sides of the continent, and a belt of low pressure, located at about 60° S (see Fig. 1a). The persistence of these pressure systems throughout the year determines an overwhelming proportion of westerly winds (NW-W-SW), although the frequent change of position of the migratory cyclones produce fluctuations in wind direction and intensity (Labraga and Davies 2014).

Although there are no records long enough in the Península Valdés, the wind trend of Puerto Madryn (CENPAT station records) shows the effect of these pressure systems (Fig. 6). The southwestern winds dominate throughout the year, even though it is possible to observe the incidence of the Atlantic anticyclone by the presence of NE winds (Fig. 6). With respect to the velocities, the annual average is 16.2 km/h with a maximum in December of 19.4 km/h and a minimum in May 11.4 km/h (Labraga and Davies 2014).

A study of the migration of the dunes of the southern sector of the peninsula (Del Valle et al. 2008), based on the analysis of its geomorphology, together with the interpretation of wave trains from ERS 1–2 imagery, allowed to define the existence of a bi-directional pattern in which the high prevalence of N wind is not found elsewhere in Patagonia. The orientation of the dune fields in the Valdes Peninsula (WNW–ESE) responds to intermediate conditions in wind direction, between the west wind prevailing towards the west and the north wind prevailing towards the east. In agreement with its easternmost location, Punta Delgada records (1959–1968) show 37% of N and NE winds.

2.2.4 Evapotranspiration

In the Península Valdésthe annual *potential evapotranspiration* (PET) estimated by Thornthwaite-Mather method (1957) is 704 mm/a. Its intra-annual distribution is in direct proportion to the march of the monthly average temperatures, thus the highest values correspond to the summer months, and the lowest to the winter months (Fig. 5b, c).

When the goal is the estimation of the aquifer recharge, instead of the climate characterization, the real evapotranspiration should be calculated differently (see Chapter "Groundwater Resources of Península Valdés"). In that case, as the storm events have a high influence in the recharge phenomena, a soil-water balance in daily step should be carried on to estimate the daily real evapotranspiration (Carrica 2009). Otherwise from the monthly balance could erroneously be conclude that there is no water recharge given that there are no excesses in any month of the year because the PET is always higher than the precipitations (Alvarez et al. 2013).

2.3 Climate Classification

Patagonia fits especially bad in general climatic classifications mainly due to the uniqueness of being a definitely arid region located on the eastern (lee) coast of a continent at mid-latitudes (Coronato et al. 2008). This arrangement of factors, not repeated anywhere else in the world, can certainly be attributed to the rain-shadow effect (see Chapter "Climatic, Tectonic, Eustatic and Volcanic Controls on the Stratigraphic Record of Península Valdés") created by the Andean Range and extending to the Atlantic shores.

The other major geographic factor affecting Patagonian climate is the Drake Passage that significantly warms the Antarctic air masses that reach South America, which, for this reason, has much milder winters than homologous east-located areas in North America at the same latitudes.

As already seen, the Península Valdés exhibits an original set of climatic parameters within the already original Patagonian climate. The persistent westerly flux, the scarcity of rainfall and the quite marked range of temperature that are common throughout the Patagonian foreland, are noticeably attenuated because of the maritime (Atlantic) influence resulting from the indented coastline.

In most of Eastern Patagonia, there is no doubt that—according to Köppen-Geiger-Pohl (Geiger and Pohl 1953) Köppen's classification—climate is B (arid) in first term and k (cool) in the third term. Referring to the second term, it is certainly W (desert) in drier areas but becomes S (steppe) in the sub-Andean fringe, and towards the Magellan Straits (when eventually first term becomes C since climate turns to Cfc (i.e. temperate well-watered climate with cool summers). In the Península Valdés, all the climatic parameters related to moisture (Fig. 3) reflect less arid conditions compared to the adjacent mainland. Because of this decrease in aridity some maps of the Köppen-Geiger climatic classification consider the Península Valdés as BSk while the adjacent mainland falls within the BWk class. Beyond the transitional climatic position of the Península Valdés, Viedma (<150 km towards northeast) is definitely considered BSk in every map and/or classification.

Climate differences between the Península Valdés region and the adjacent mainland are also observed by analyzing the values of aridity index (AI = p/ETP; where p = precipitation and ETP = potential evapotranspiration). Comparison of the aridity index in Punta Delgada and Puerto Madryn (0.26 and 0.19 respectively) reflects that both points fall in different ranks, considering that the upper limit of arid conditions is AI = 0.20 (MAB 7 1977). However, other authors place the arid/semi-arid limit at AI = 0.3 (Le Houerou 2005), and therefore range differences disappear. All in all, the lack of climatic definition showed by the two examples above, allows confirming the transitional characteristic of the Península Valdés climate, which, in fact, may vary from year-to-year from arid to semi-arid and vice versa. These transitional features, not only in rainfall amount but also in seasonal distribution, may explain the Península Valdés ill-defined belonging to one or another phytogeographical district, namely Monte or Patagonia (León et al. 1998).

As stated before, because of the prevalent trend of winter rains manifested throughout Patagonia north of 47° S, the regional climate may be considered like a (cooler, drier and more windy) Mediterranean bioclimate. According to the "Mediterraneity Index" (MI) defined as the ratio between winter trimester rainfall and summer trimester rainfall[1] (Le Houérou 2004), Punta Delgada has a sub-Mediterranean climate (MI = 1.9) while Puerto Madryn and Trelew have an attenuated Mediterranean climate (MI 1,1.0 and 1.4, respectively). Needless to say, higher figures (MI > 2) are found towards west and south, where rains from the Atlantic Ocean become negligible.

3 Perspectives and Future Work

The transitional climate of Península Valdés, not far from the boundary between Atlantic and Pacific circulation realms, makes the area an interesting spot to monitoring global climate change and regional vegetation shift. Thus, the major concern should be to overcome the current lack of long-term systematic observations in the area. Doubtless, continuity and consistence in meteorological records are paramount in any future work on climatic or ecological research in Península Valdés.

Acknowledgements We acknowledge Dr. Federico Robledo (FCEN-UBA) and Dr. Juan Rivera (IANIGLA-CCT Mendoza-CONICET) for the revision of our manuscript and their valuable suggestions.

Glossary

Cryosphere	Are those portions of Earth's surface, where water is in solid form, i.e. water bodies ice, snow cover, glaciers, ice sheets and frozen ground. It plays a significant role in the global climate
ENSO	(El Niño–Southern Oscillation) Is an irregularly periodical variation in winds and sea surface temperatures over the tropical eastern Pacific Ocean, affecting much of the tropics and subtropics

---

[1]Ranges of "Mediterraneity" are: MI > 2 Mediterranean climate; 2 > MI > 1.5 sub-Mediterranean; 1.5 > MI > 1 attenuated Mediterranean; MI < 1 non-Mediterranean.

Frontal activity	Sharp discontinuities of temperature, moisture and wind properties caused by the displacement of boundaries between air mass (fronts), set in motion by low pressure cells and often entailing storms and rainfall
PET	(Potential Evapotranspiration) Maximum quantity of water capable of being evaporated in a given climate from a continuous stretch of vegetation covering the whole ground and well supplied with water
Storm Tracks	Relatively narrow zones in oceans where cyclone-generated storms travel driven by the prevailing winds. Cyclones from the circumpolar storm track in the Antarctic Ocean may derive northward and affect Patagonia
Synoptic-scale disturbances	Are those fit in a horizontal length scale of the order of 1000 km or more. This corresponds to a horizontal scale typical of mid-latitude depressions
Westerlies	Prevailing flux from the west in the mid-latitudes emanating from the polarward sides of the subtropical high-pressure cells. Those of the southern hemisphere are stronger and more constant

References

Ablaster J, Meehl G (2006) Contributions of external forcings to southern annular mode trends. J Clim 19:2896–2905

Agosta E, Compagnucci R, Ariztegui D (2015) Precipitation linked to atlantic moisture transport: clues to interpret patagonian palaeoclimate. Clim Res 62(3):219–240

Alvarez MP et al (2013) Estimación de recarga en zonas áridas según distintos métodos. Área medanosa del sur de Península Valdés (provincia de Chubut). In: González N et al (eds) Agua subterránea recurso estratégico, vol 1, pp 46–51

Alvarez MP (2010) Investigación geohidrológica en un sector de Península Valdés, provincia de Chubut. Ph.D. Thesis. Universidad Nacional de La Plata

Barros V, Scian B, Mattio H (1979) Mapas de precipitación de la Provincia de Chubut. CENPAT-Recursos Hídricos de Chubut, Rawson

Barros V, Vera C (coordinators) and collaborators, Secretaría de Ambiente y Desarrollo Sustentable de la Nación (2014) Tercera Comunicación Nacional sobre Cambio Climático. Cambio Climático en Argentina; Tendencias y Proyecciones (CIMA), Buenos Aires

Bell N (1981) Precipitation. In: Goodall D, Perry R (eds) Arid land ecosystems. Cambridge University Press, Cambridge, pp 373–393

Berman A (2012) Análisis de la variabilidad climática en la Patagonia Argentina. Facultad de Ciencias Exactas y Naturales. Universidad de Buenos Aires. http://digital.bl.fcen.uba.ar/Download/Tesis/Tesis_5123_Berman.pdf

Berman A, Silvestri G, Compagnucci R (2013) On the variability of seasonal temperature in southern South America. Clim Dyn 40(7–8):1863–1878

Blázquez J, Nuñez M, Kusunoki S (2012) Climate projections and uncertainties over South America from MRI/JMA global model experiments. Atmos Clim Sci 2:381–400

Boninsegna J et al (2009) Dendroclimatological reconstructions in South America: a review. Palaeogeogr Palaeoclimatol Palaeoecol 281:210–228

Campagnucci R, Araneo D (2007) Alcances de El Niño como predictor del caudal de los ríos andinos argentinos. Ingeniería Hidráulica en México 22(3):23–35

Carrica J (2009) Cálculo de la recarga en zonas áridas y semiáridas. Recarga de acuíferos. Aspectos generales y particulares en regiones áridas. VI Congreso Argentino de Hidrogeología, Santa Rosa, La Pampa, pp 71–80

Castañeda M, González M (2008) Statistical analysis of the precipitation trends in the Patagonia region in southern South America. Atmósfera 21(3):303–317

Coronato F (1993) Wind chill factor applied to Patagonian climatology. Int J Biometeorol 37:1–6

Coronato F (1994) Influence of the eastern central Patagonia plateaus on the oceanic characteristics of the climate. Anales del Instituto de la Patagonia: Serie Ciencias Naturales Punta Arenas (Chile) 21:131–146

Coronato F, Bisigato A (1998) A temperature pattern classification in Patagonia. Int J Climatol 18:765–773

Coronato A et al (2008) Physical geography of Patagonia. In: Rabassa J (ed) The Late Cenozoic of Patagonia and Tierra del Fuego. Development in Quaternary Sciences, vol 11. Elsevier, pp 13–55

Daget Ph (1968) Quelques remarques sur le degré de continentalité des climats de la région holarctique CNRS-CEPE

De Fina A et al (1968) Difusión de cultivos índices en la provincia del Chubut. Publicación N 110. INTA, Buenos Aires

Del Valle H et al (2008) Sand dune activity in north-eastern Patagonia. J Arid Environ 72:411–422

Deser C et al (2012) Uncertainty in climate change projections: the role of internal variability. Clim Dyn 38:527–546

Garreaud R, Aceituno P (2007) Atmospheric circulation and climate variability. In: Veblen T, Young K, Orme A (eds) The physical geography of South America. Oxford University Press, pp 45–59

Garreaud R et al (2008) Present-day South American climate. PALAEO3 Special Issue (LOTRED South America) 281:180–195

Garreaud R (2009) The Andes climate and weather. Adv Geosci 7:1–9

Garreaud R et al (2013) Large-scale control on the Patagonian climate. J Clim 26:215–230

Geiger R, Pohl W (1953) Revision of the Köppen-Geiger Klimakarte der Erde. Erdkunde 8:58–61

Gillett N, Kell N, Jones P (2006) Regional climate impacts of the southern annular mode. Geophys Res Lett 33:L23704. Doi:10.1029/2006GL027721

González M, Vera C (2010) On the interannual wintertime rainfall variability in the southern Andes. Int J Climatol 30:643–657

González M, Skansi M, Losano F (2010) Statistical study of seasonal winter rainfall prediction in the Comahue region (Argentina). Atmósfera 23:277–294

González P et al (2014) Stratospheric ozone depletion: a key driver of recent precipitation trends in South Eastern South America. Clim Dyn 42:1775–1792

Hobbs J, Lindesay J, Bridgman H (1998) Climates of the southern continents: present, past, and future. Wiley, New Jersey, USA

Insel N, Poulsen C, Ehlers T (2010) Influence of the Andes Mountains on South American moisture transport, convection, and precipitation. Clim Dyn 35(7):1477–1492

IPCC (2013) Climate change 2013: the physical science basis. In: Stocker T et al (eds) Contribution of working group I to the fifth assessment report of the intergovernmental panel on climate change. Cambridge University Press, Cambridge, UK

Jacques-Coper M et al (2015) Summer heat waves in southeastern Patagonia: an analysis of the intraseasonal timescale. Int J Climatol. Doi:10.1002/joc.4430

Jacques-Coper M et al (2015b) Evidence for a modulation of the intraseasonal summer temperature in Eastern Patagonia by the Madden-Julian oscillation. J Geophys Res Atmos. Doi:10.1002/2014jd022924

Jobbágy E, Paruelo J, León R (1995) Estimación del régimen de precipitación a partir de la distancia a la cordillera en el noroeste de la Patagonia. Ecol Austral 5:47–53

Karoly D (1990) The role of transient eddies in low-frequency zonal variations of the southern hemisphere circulation. Tellus Ser A 42:41–50

Kidson J (1988) Interannual variations in the southern hemisphere circulation. J Clim 1: 1177–1198

Labraga J, Davies E (2014) CENPAT-Unidad de Investigación de Oceanografía y Meteorología. http://www.cenpat.edu.ar/fisicambien/climaPM.htm. Accessed 18 Oct 2014

Labraga J, Villalba R (2009) Climate in the Monte Desert: past trends, present conditions, and future projections. J Arid Environ 73(2):154–163

Le Houerou H (2005) The isoclimatic Mediterranean biomes: bioclimatology, diversity and phytogeography. Montpellier, France, 766 p

Le Houérou H (2004) An agro-bioclimatic classification of arid and semiarid lands in the isoclimatic Mediterranean Zones. Arid Land Res Manag 18:301–346

Lenaerts J et al (2014) Extreme precipitation and climate gradients in Patagonia revealed by high-resolution regional atmospheric climate modeling. J Clim 27:4607–4621

León R et al (1998) Grandes unidades de vegetación de la Patagonia Extra Andina. Ecol Aust 8:125–144

MAB 7 (1977) Map of the world distribution of arid regions. UNESCO, Paris, p 55

Masiokas M et al (2008) 20th-century glaciar recession and regional hydroclimatic changes in the northwestern Patagonia. Global Planet Change 60:85–100

Müller G, Ambrizzi T, Nuñez M (2005) Mean atmospheric circulation leading to generalized frosts in central southern South America. Theor Appl Climatol 82:95–112

Nuñez M, Solman S, Cabré M (2009) Regional climate change experiments over southern South America. II: climate change scenarios in the late twenty-first century. Clim Dyn 32:1081–1095

Paruelo J et al (1998) The climate of Patagonia: general patterns and controls on biotic processes. Ecol Aust 8:85–101

Prohaska F (1976) The climate of Argentina, Paraguay, and Uruguay. In: Schwerdtgefer W (ed) Climates of Central and South America, vol 12. World Survey of Climatology, Elsevier, pp 57–69

Rosenblüth B, Fuenzalida H, Aceituno P (1997) Recent temperature variations in Southern South America. Int J Climatol 17:67–85

Rusticucci M, Vargas W (2002) Cold and warm events over Argentina and their relationship with the ENSO phases: risk evaluation analysis. Int J Climatol 22:467–483

Rusticucci M, Barrucand M (2004) Observed trends and changes in temperature extremes in Argentina. J Climate 17:4099–4107

Schneider C, Gies D (2004) Effects of El Niño-Southern oscillation on southernmost South America precipitation at 53° S revealed from NCEP–NCAR reanalysis and weather station data. Int J Climatol 24:1057–1076

Silvestri G, Vera C (2009) Nonstationary impacts of the southern annular mode on southern hemisphere climate. J Clim 22:6142–6148

Smith R, Evans J (2007) Orographic precipitation and water vapor fractionation over the Southern Andes. J Hydrometeorol 8:3–19

Thornthwaite C, Mather J (1957) Instructions and tables for computing potential evapotranspiration and water balance. Publ Cimatol 10:185–311

Turner J (2004) Review: the El Niño-Southern oscillation and Antarctica. Int J Climatol 24:1–31

Villalba R et al (2003) Large-scale temperature changes across the southern Andes: 20th-century variations in the context of the past 400 years. Clim Change 59:177–232

Villalba R et al (2012) Unusual Southern Hemisphere tree growth patterns induced by changes in the southern annular mode. Nat Geosci 5(11):793–798

Vincent L, Peterson T, Barros V (2005) Observed trends in indices of daily temperature extremes in South America 1960–2000. J Climate 18:5011–5023

Zhu J et al (2014) Climate history of the Southern Hemisphere Westerlies belt during the last glacial–interglacial transition revealed from lake water oxygen isotope reconstruction of Laguna Potrok Aike (52° S, Argentina). Clim Past 10:2153–2169

Late Cenozoic Landforms and Landscape Evolution of Península Valdés

Pablo Bouza, Andrés Bilmes, Héctor del Valle
and César Mario Rostagno

Abstract The present landscape of the Península Valdés is the result of a complex interrelation between climatic (aeolian deposition, windblown processes, glacial and interglacial cycles, pluvial and fluvial processes), tectonic, and *eustatic controls* that had work in the Andean foreland during the late Cenozoic. Based on a geomorphological approach, which includes new descriptions, interpretations, and hierarchically classification of the main landforms of this region, together with previous geomorphological surveys, the Península Valdés area was grouped in three major geomorphologic systems: Uplands and Plains, Great Endorheic Basins, and Coastal Zone. Based on the interrelationship among these three geomorphological systems the landscape evolution of the late Cenozoic of Península Valdés could be summarized in five main stages: (1) development of fluvial and alluvial systems during the Pliocene early Pleistocene; (2) closed basin formation associated to tectonic processes during the early middle Plesitocene; (3) first marine transgressions during the late Pleistocene; (4) flooding of the gulfs and construction of the peninsula in the late Plesitocene–Holocene; (5) final flooding in the region during the middle Holocene.

P. Bouza (✉) · H. del Valle · C.M. Rostagno
Instituto Patagónico para el Estudio de los Ecosistemas Continentales (IPEEC),
Consejo Nacional de Investigaciones Científicas y Técnicas (CONICET)—CCT
Centro Nacional Patagónico (CENPAT), Boulevard Brown 2915,
ZC: U9120ACD Puerto Madryn, Chubut, Argentina
e-mail: bouza@cenpat-conicet.gob.ar

H. del Valle
e-mail: rostagno@cenpat-conicet.gob.ar

C.M. Rostagno
e-mail: delvalle@cenpat-conicet.gob.ar

A. Bilmes
Instituto Patagónico de Geología y Paleontología (IPGP), Consejo Nacional de
Investigaciones Científicas y Técnicas (CONICET)—CCT Centro Nacional Patagónico
(CENPAT), Boulevard Brown 2915, ZC: U9120ACD Puerto Madryn, Chubut, Argentina
e-mail: abilmes@cenpat-conicet.gob.ar

Keywords Miocene · Quaternary · Landscape evolution model · Patagonia · Endorheic basins

1 Introduction

The landscape of Península Valdés, as well as the Extra-Andean Patagonia region (Fig. 1), is characterized by arid-semiarid conditions. Low rainfall and sparse vegetation cover are typical features of this region and are of considerable importance for the operation and development of landforms (Thomas 1997). Many of these landforms have large patches of bare soils and so are exposed to wind erosion, raindrop impact, and surface runoff.

Although wind is an important geomorphological agent that has deeply modified the Península Valdés landscape, water erosion is the most severe geomorphic process, either as *raindrop-splash* and *laminar runoff* (interril) or as concentrated flow erosion, in the form of rills or gullies. In addition, as the Península Valdés is bounded by the Atlantic Ocean and the Nuevo and San José Gulfs (Golfo Nuevo and Golfo San José), distinctive characteristics related to coastal processes also imprint the landscape of this region. Previous studies in the Península Valdés region based on different geomorphological approaches (aeolian, physiographical and coastal) were performed by Rostagno (1981), Beltramone (1983), Codignotto and Kokot (1988) and del Valle et al. (2008).

The objective of this chapter is to perform an updated characterization of the main geomorphological units of Península Valdés by describing, interpreting, and hierarchically classifying the main landforms of the study area.

The result of this work expects to develop a useful tool in Península Valdés to better understand the Neogene–Quaternary landscape evolution, soil genesis and soil distribution, hydrogeological characteristics, distribution patterns of vegetation, geoecological functions and processes, and distribution of archaeological material.

The present chapter follows a hierarchically classification adapted from Peterson (1981), Iriondo and Ramonell (1993) and Súnico (1996) that uses the geographic scales, genetic relationships and shapes of the topographic forms. Following this scheme, the Península Valdés region (defined in the hierarchy classification of this work as a *super system*) was grouped in three major *systems* (Fig. 2). These systems are then successively divided into smaller and genetically more homogeneous classes defined as *units* that could be subdivided, respectively, into *landforms elements*. In a descending order these categories are as follows:

Super system: the geological dimension (space-time insight). It takes into account the regional geology from a geodynamic evolution viewpoint (see Chapter "Climatic, Tectonic, Eustatic, and Volcanic Controls on the Stratigraphic Record of Península Valdés"). Plains and basins in a continental-marine setting are named to as distinctive landscapes affected by a relative rise of the land followed by the formation of closed depressions, surrounded by a marine environment.

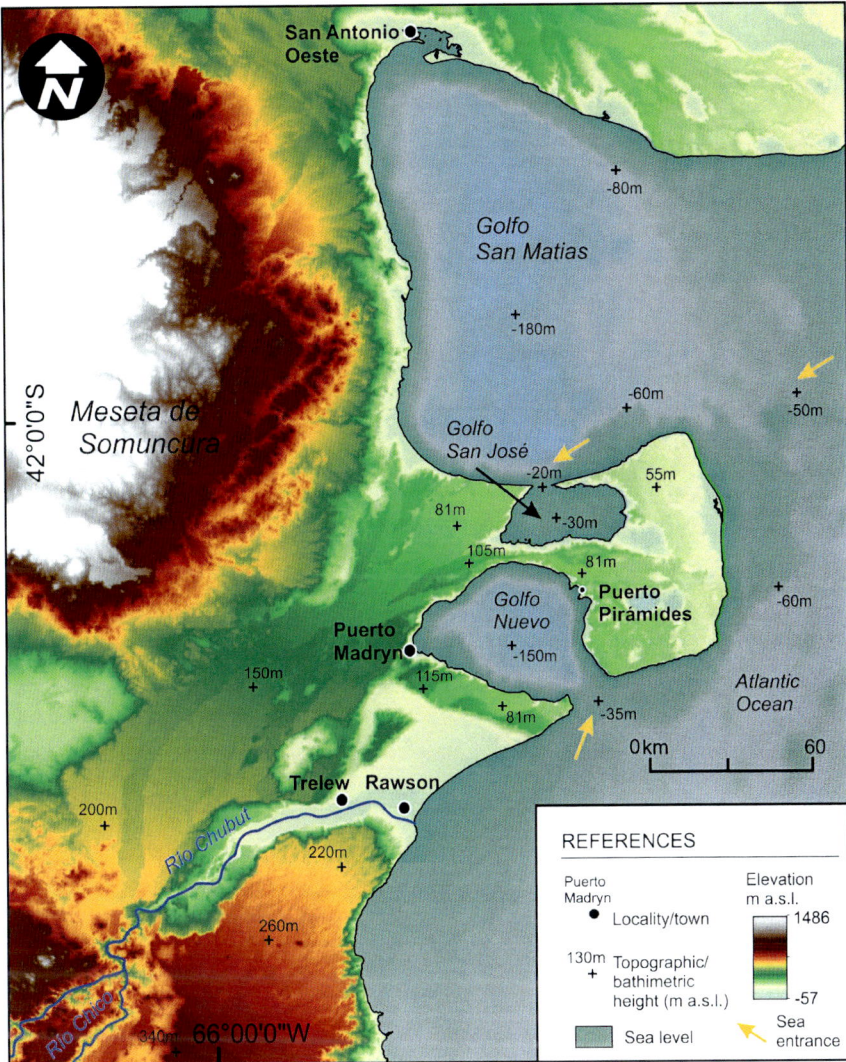

Fig. 1 Digital elevation model (DEM) of northeastern Chubut province

System: Uplands and Plains, Great Endorheic Basins, and Coastal Zone landforms are the main geomorphological settings based on the dominant landform processes within the system.

Unit: is the minor geomorphological landform defined by morphogenetic criteria and space-time relationships. In this category, relict landforms are included (e.g., fluvial terrace levels and beach ridges sequences), especially to define remaining parts of a same geomorphic surface which has been otherwise degraded by erosion. An erosional relict must be older than the destructive erosion cycle. Recognition of relict

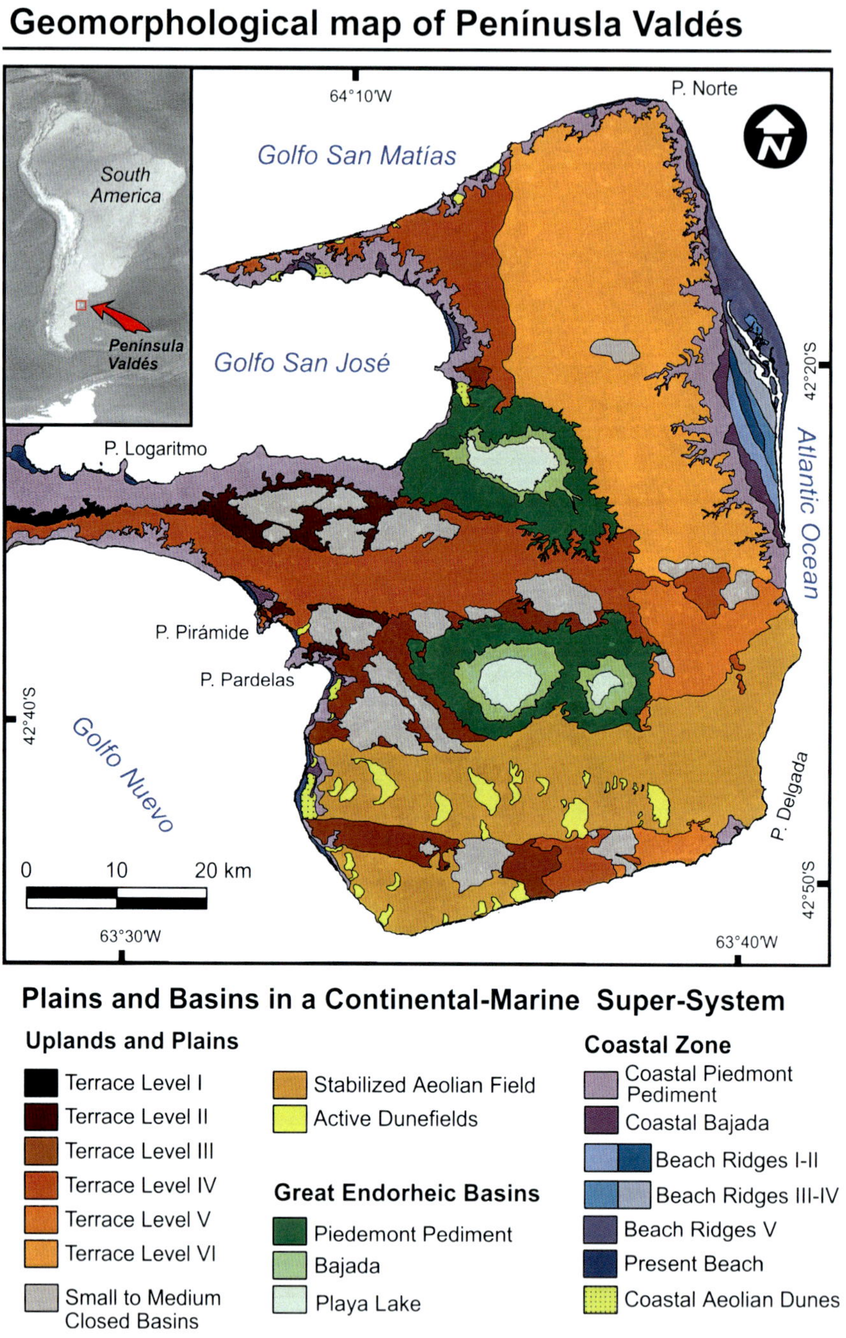

Fig. 2 Geomorphological sketch of the Península Valdés

land surfaces is the basic tool for establishing relative ages of the different surfaces (Peterson 1981). Some examples of this category are relict fluvial Terrace Levels and Stabilized Aeolian Field, Small to Medium Closed Basins and Beach Ridges I–IV. *Landform elements*: This description level is used on some geomorphological units to separate different geomorphic processes observed within a Unit (e.g., shallow *pans* by deflation, discontinuous patches of sandy aeolian mantles, *nebkas*, gullies, *desert pavement* patches). Landform elements are normally not represented in regional maps mapped only at high scale surveys.

2 System A: Uplands and Plains

This system corresponds to a landscape consisting of wide plateau-like plains make up of Pliocene–Pleistocene fluvial terraces of the *Rodados Patagónicos lithostratigraphic unit* (see Chapter "Geology of Península Valdés"). Other geomorphological units (e.g., small to medium closed basins, stabilized aeolian fields and active dunefields) are superimposed the regional landscape.

2.1 Terrace Levels I–VI

The six Terrace Level units are included in the Plio-Pleistocene *Rodados Patagónicos* lithoestratigraphic unit of Plio-Pleistocene age (Fidalgo and Riggi 1970), also known in other Patagonian areas as the Patagonian Shingle Formation (Darwin 1846). The term "Rodados Patagónicos" is used in different parts of the Patagonian foreland and includes terraces of gravel deposits of varied genesis and age (Martínez et al. 2009). In all the cases, the terrace levels were affected by different geomorphic processes and are highly strongly interacting with the other geomorphological units of this system. The terrace levels are dissected by centripetal drainage networks and the retreat and coalescence of erosion scarps of small/medium closed basins; many of them (i.e., Terrace Levels III, IV, and V) are discontinuously covered by Stabilized Aeolian Fields and the Active Dunefields Units (del Valle et al. 2008) (Fig. 3d).

In the study area, terrace level deposits are 1–6 m thick, composed of sandy gravel sediments with a maximum clast size of 6 cm (pebble-size clasts). Rhyolites are the main component of the gravel size fraction, and in smaller proportion andesites and basalts. The accumulation of these deposits, related to fluvial plains generated during Neogene and Pleistocene glaciations, occurred in an arid periglacial environment (Mercer 1976). While all the terrace levels could be seen as a whole unit (i.e., *Rodados Patagónicos*; see Chapter "Geology of Península Valdés") some diagnostic properties of each terrace level allow to differentiate six mayor units (i.e., Terrace Levels I–VI). Each terrace level was identified by a Roman numeral according to the descending order of relative age. The terrace levels were separated

Fig. 3 a Outcrop of the Rodados Patagónicos (Terrace Level VI), Calcrete zone (CZ); cryogenic morphologies: columns (C) and windows (W), bar scale = 2 m. **b** and **c** Epigenetic fossil ice-wedge casts in RP deposit (Terrace Level I), the marine sedimentites of Puerto Madryn Formation (Miocene) affected by cryogenic processes is observed. **d** Active dunefield, complex dome-shaped dunes and transverse ridges and

on the basis of two relative age parameters: morphostratigraphic position and soil profile development (e.g., carbonate accumulation rate; see Chapter "Soil–Geomorphology Relationships and Pedogenic Processes in Península Valdés").

The morphostratigraphic position parameter includes topographic heights and surface area of the terraces. Generally, the older are the terrace levels, the higher are their heights, and smaller are their surface areas. In the study region maximum heights of the terrace levels range between 104 and 50 m a.s.l. having topographic heights differences of more than 5 m among them (Table 1). The altitude of the terrace levels progressively increases to the south of the study area, reaching the 700 m a.s.l., 400 km SW from Península Valdés. The surface area of the different units are directly related to the degree of dissection and consequently with their age. Older terraces, as Terrace Level I and II, are highly dissected and exhibits several gullies and ravines developed by fluvial and hillslope processes. With the exception of Terrace Level V, the smaller is the terrace level surface, the higher is the degree of dissection and oldest is the terrace (Table 1, Fig. 2).

Table 1 Topographic heights measured using SRTM data

Supersystem	Area (km^2)	Max. height (m)	Min. height (m)	Avg. height (m)
System A: uplands and plains				
Terrace Level I	67.07	104	91	99
Terrace Level II	82.24	99	75	91
Terrace Level III	232.08	89	69	81
Terrace Level IV	595.08	84	57	71
Terrace Level V	151.71	60	45	55
Terrace Level VI	838.05	59	38	51
Small to middle closed basins	313.95	96	22	67
Stabilized Aeolian field	644.77	86	38	66
Active Dunefields	66.83	90	53	73
System B: great endhoreic basins				
Piedemont Pediment	344.37	83	−23	36
Bajada	90.74	18	−40	−8
Playa lake	70.07	Salina grande (−43 m)		
		Salina Chica (−19 m)		
		Gran Salitral (0 m)		
System C: coastal zone				
Coastal Piedmont Pediment	491.17	92	3	38
Coastal Bajada	48.91	22	0	13
Beach Ridges I–IV	54.32	20	2	12
Beach Ridges V	67.45	12	0	7
Present Beach	25.71	2	0	1
Coastal Aeolian Dunes	14.65	84	3	28

Soil profiles on terraces allows terrace levels differentiation. As soil evolution involves time-depending properties (Gile et al. 1966; Bachman and Machette 1977; Machette 1985; Bouza et al. 2007; Bouza 2012), the rate and depth of pedogenic carbonate accumulation, the occurrence of *argillic horizons*, and clay-minerals transformation and neoformation could be used as tool for correlation of soils and unconsolidated deposits (Table 1, see Chapter "Soil–Geomorphology Relationships and Pedogenic Processes in Península Valdés"). The soils developed on Terrace Levels I and II correspond to a Haplocalcids–Petrocalcids complex, while on the younger terrace levels, a Haplocalcids–Natrargids–Natrigypsids complex occurs.

Finally, *ice-wedge casts* and cryoturbations structures allow to differentiate terrace levels affected by persistent arid periglacial conditions in the Península Valdés region. During Greatest Patagonian Glaciation (Rabassa and Clapperton 1990), that occurred ca 1.0–1.16 Ma (Ton-That et al. 1999), the older deposits were affected by cryogenic processes, which are registered by fossil ice-wedge casts and a three-dimensional reticulate structure (network system) resembling columns (carbonate cements) and windows (sediments without cements) (Trombotto 1998, 2008). Examples of Late Cenozoic deposits affected by cryogenic processes are the Miocene marine sediments of the Puerto Madryn Formation (see Chaps. "Geology of Península Valdés" and "Miocene Marine Transgressions: Paleoenvironments and Paleobiodiversity") and Terrace Level I, where fossil ice-wedge casts are prominent (Fig. 3b, c), while on younger terrace levels only columns and windows structure are observed (Fig. 3a). The columns and windows structure is apparently associated with cryogenic conditions when it is observed transversally to the net-like structure revealing a classical polygonal pattern (Trombotto 1996).

The tread of these geomorphic surfaces are gently sloping, mostly descending to the east. The surface treads are linked through vegetated surface risers, which generally at the junction with younger terrace treads, a series of small to medium closed basins are commonly observed (see Sect. 2.2).

Small pans at the bottom of the closed basins generated by deflation are randomly distributed on these geomorphic surfaces. In other cases, the pans have a distribution following a paleo-channels pattern of Rodados Patagónicos surfaces.

2.2 Small to Medium Closed Basins

This geomorphological unit corresponds to closed basins with a surface area below 100 km^2 with complex genesis, consisting of a combination of water and wind erosion and hillslope (raindrop impact and surface water runoff) and *mass wasting processes*. As it is explained for the genesis of the Great Endorheic Basins System (see Sect. 3) a geodynamic factor might have contributed.

Both small and medium closed basins develop both, on and between levels of terraces (Fig. 2). In Península Valdés region they occupy 315 km^2, reaching with a maximum diameter of 10 km and depths up to 25 m. The evolutions of these closed basins produce a dissection of the terrace levels by the retreat and coalescence of erosion scarps.

2.3 Stabilized Aeolian Field

This geomorphological unit is represented by relict landforms that are no longer active. Two areas with stabilized aeolian landforms are distinguishable, both located in the southern part of the Península Valdés. The largest one is located in the central area forming a corridor that stretches from the west to the east coast (i.e., from the Golfo Nuevo to the open Atlantic coast), and the smaller one is a fringe-like dunefield in the southwest corner of the Peninsula (Fig. 2). The two stabilized aeolian fields cover an area of approximately 645 km^2, where a well-developed vegetation cover of grasses (mainly *Sporobolus rigens*) and shrubs (principally *Hyalis argentea*) occurs on the inactive dunes. The stabilized aeolian landforms are represented by a 0.4–2 m thick sandy layer. Local relief shows an undulating appearance with some depressed areas that may reach the substratum (Miocene marine sedimentites or terrace level deposits).

Many stabilized linear (longitudinal) forms are distinguishable. The linear forms develop as the marginal sand dunes of the advancing the active dunefield unit. They are observed in the northern and eastern sectors (Fig. 3d), and are bounded by an homogeneous sand mantle toward both sides as northern and southern winds redistribute part of the eastward transported sand. These recently deposited sand layers change the original Xeric Haplargids soils into Arenic Haplargids (see Chapter "Soil–Geomorphology Relationships and Pedogenic Processes in Península Valdés"). *Blowouts* are common in the linear dunes and new and smaller sand dune fields in the stabilized sand mantles. Longitudinal dunes exhibits parallel to subparallel crests, west–east oriented with a variable thick, in function of the degree of degradation. Dune crest heights range between 0.6–1.5 m thick and 10–15 km in length. Separations between longitudinal dunes are irregular and go from 60 to 150 m in the west part of the Península Valdés to more than 1000 m in the east part (Súnico 1996). While longitudinal dunes are strongly modified by present aeolian erosion (blow-out formation) and superimposed by the younger active dunefield unit, it is possible to distinguish that they converge to "Y" shaped junctions (Fig. 3d). The west–east orientation of the stabilized longitudinal dunes suggests that they were formed by predominant winds from the west quadrant. However, to establish the relationship between landform, vegetation, aeolian materials and the frequency and intensity of the winds, a detailed study is needed (Súnico 1996).

2.4 Active Dunefields

These landforms are developed in the west coast of Península Valdés and moves westward mostly above the stabilized aeolian lanform units.

The general orientation of the dunefields in Península Valdés is in agreement with the prevailing regional wind flow from the WNW, and active dunes show a coincident trend. Notwithstanding active dunes present a variety of forms which reflect local variations in the wind flow. In the area of this study, mesoscale wind circulation is strongly influenced by the shape of the coastline of Golfo Nuevo. The sources of windblow sediment are the extensive sandy beach located on southwestern coast of Península Valdés, where a continued supply of loose, sand-sized sediment is available transported inland by the prevailing westerly winds (Haller et al. 2000). Sand is also derived from aeolian erosion of friable marine sediments of Puerto Madryn Formation (see Chapter "Geology of Península Valdés") exposed at the cliffs along the shoreline (Súnico 1996; del Valle et al. 2008).

Trend of the displacement of the dunes is W–NW with a migration rate of 8–10 m/a (Annual rate 1969–2002; del Valle et al. 2008). Active dunefields is the smallest unit of the upland and plains system and occupy a total area of 67 km^2, covering 7% of the southern portion of the Península Valdés. Three areas with active dunefield are observed. A smaller one in the central west of the Península Valdés, a larger one in the south-central area and a medium one in the southwest corner of the Peninsula (Fig. 2). Active dunefields observed in the coastal zone system (i.e., coastal aeolian dunes) are considered in another unit and will be described later in the Sect. 4.6.

The dune types of the Península Valdés are far from showing a simple pattern, regardless of the sector in which they are located. Two main types of active sand dune are distinguished: (1) compound dome-shaped dunes, and (2) complex dome-shaped dunes (del Valle et al. 2008). Besides these dominant forms, a few scattered parabolic dunes, barchans, sand sheets, and shrub dunes (*nebkas*) are also present in the area. Compound dome-shaped dunes (2–20 m high) occur in a wide area mainly in the south-central and southwest areas (Fig. 3d). Transverse sand ridges are locally superimposed on the domes and may modify them greatly; the resulting dune form is transverse domal-ridge (row of connected domes).

Complex dome-shaped dunes (4–20 m high) are located exclusively in the south-central area. Transverse ridges and star dunes coalesce or grow together. Network dunes consist of NW–SE trending main ridges and nearly vertical secondary ridges (NE–SW orientation).

There have been no final conclusions for the period when the dunes and their morphology developed as the age of the Península Valdés has not been studied systematically.

3 System B: Great Endorheic Basins

This system corresponds to landforms developed over the uplands and plain system. Although many landforms and processes are similar to those described in small–medium closed basins units (Fig. 2) the intensity of the different geomorphic processes is more accentuated in this system, allowing to clearly distinguishing three geomorphologic units: *Piedmont Pediment, Bajada, and Playa lake*.

The Great Endorheic Basin System is characterized by a typical centripetal drainage network. This drainage dissects the piedmont *pediments units*, give rise to alluvial–fluvial fans that form the *bajada unit*, and ends at the bottom of the basin in the *playa lake unit*. In the Península Valdés this system represent more than 20% of the entire area, and includes the Gran Salitral, the Salina Grande, and the Salina Chica basins (Fig. 2). Salina Grande and Salina Chica are located in two different basins, though both share the same 25 km W–E elongated depression occupying 250 km^2. They are separated by an interfluve (5 m a.s.l.) of tertiary and quaternary sediments 50 m below the surrounding terrace level units. The Salina Grande lies at 43 m below sea level, one of the lowest areas of Patagonia. The playa lake of Salina Chica is at 19 m below sea level. In the Salina Grande and Salina Chica, differences of elevation between the playa lakes and the upper surrounding terrace levels are around 120 and 100 m, respectively. The Gran Salitral, a closed basin adjacent to the Golfo San José, extends in a W–E direction 15 km N of the Salina Grande and Salina Chica depressions. It is shallower (0 m a.s.l.) and presents a rectangular shape with almost straight W, S, and N borders; to the E, an extended drainage network has enlarged the basin in what seems to be the capture of a pre-existent closed basin.

In order to analyze the Great Endorheic Basin System a description of the main geomorphological units will be presented.

3.1 Piedmont Pediment

The Piedmont Pediment unit (also defined in the literature as flank pediments, Fidalgo and Riggi 1970) is defined by a gently and short slope transport surfaces of bedrock, covered by a thin alluvium, developed between an upland area where erosion dominates (i.e., the erosion scarps of Terrace levels) and a lower plain where active aggradation dominates (i.e., *Bajadas*) (Fig. 4a, b). Dohrenwend and Parsons (2009) defined this sequence of landforms and processes on hillslope as pediment association.

In the Piedmont Pediment unit pediments are the dominant landforms that develop in the large closed basins. Changes in base level throughout the genesis of the large depressions that contains the closed basins of the Salina Grande, Salina Chica, and Gran Salitral gave rise to different pediment surfaces developed in ancient *bajadas* (Súnico 1996; Alvarez et al. 2008).

Piedmont pediment-surfaces are carved on marine sandstone of Puerto Madryn Formation and are densely incised by the drainage network that locally is reactivated by the lowering of the base level that accelerates the erosion process (Súnico 1996). In the three Great Endorheic Basins of the study area, there is more than one pediment level, which reflects variations at the base level (Súnico 1996; Alvarez et al. 2012).

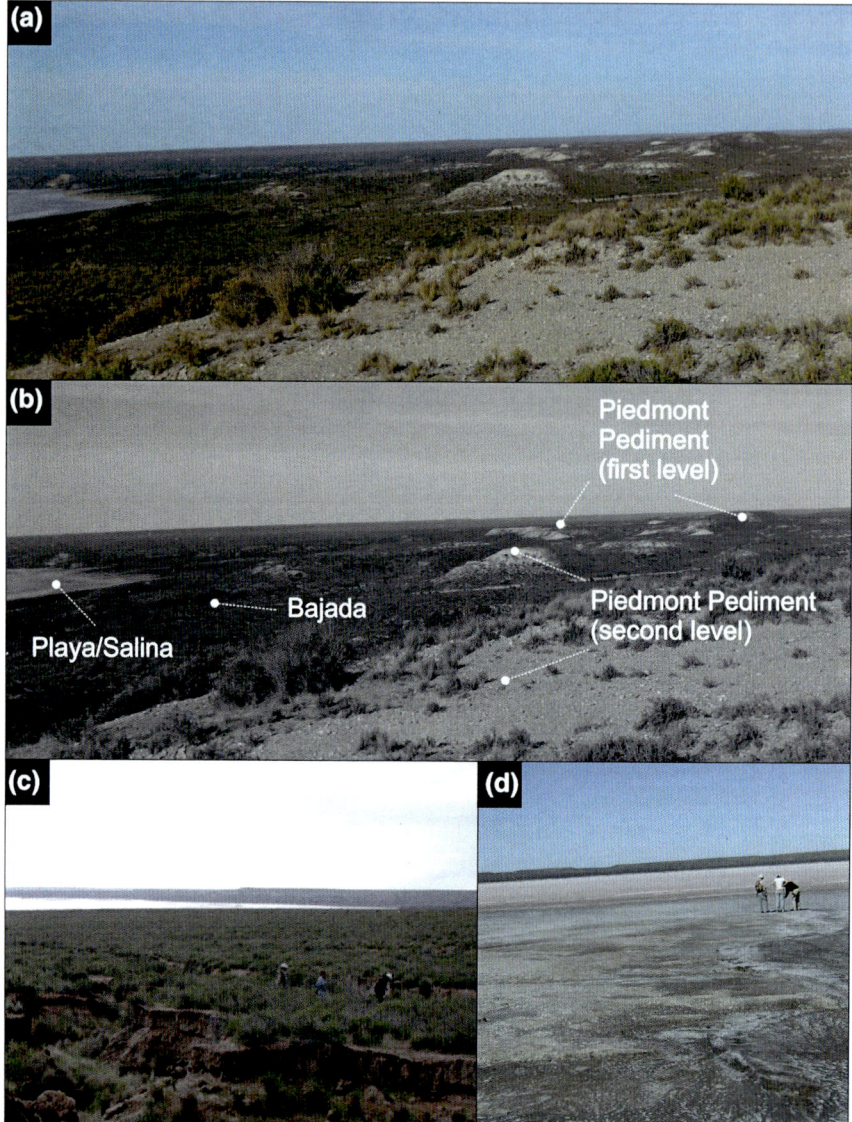

Fig. 4 Great Endorheic Basin System; **a** and **b** geomorphological sequence on hillslope setting: erosion escarps, piedmont pediment levels, *bajada* (pediment association) and playa Lake; **c** and **d** details of *bajada* and playa lake units, respectively

The springs of the Salina Grande and Salina Chica basins give rise to wetlands in otherwise dry landscapes. Spring waters flows from the southern scarps of the Salina Grande and Salina Chica basins. These springs originate in the sand dune areas that cover part of the southern Península Valdés (Alvarez et al. 2008) where water recharge occurs. The soils developed on Piedmon Pediment Unit correspond to

Natrargids, Calciargids, complex (Bouza et al. 2008; Blanco et al. 2010, see Chapter "Soil–Geomorphology Relationships and Pedogenic Processes in Península Valdés").

3.2 Bajadas

In the different pediment levels, sheet erosion dominates, although a network of gullies has developed and, in the lower portion of the piedmont pediments units, they deposit most of the sediment charge and form a gentle sloping depositional surface.

The coarse sediments are first deposited and build the alluvial fan that connects the pediment to the playa lake; the finer sediments are deposited at the playa. This aggradational landform is named *bajada* and consists of a series of coalescing alluvial fans. This unit is linked to drainage systems that incise the piedmont pediment and is represented by sandy gravel deposits. In the Great Endorheic Basins system, the *bajadas* occupy a small portion adjacent to the playa (90.74 km^2; Table 1). The soils developed on *bajadas* correspond to Torriorthents (Blanco et al. 2010, see Chapter "Soil–Geomorphology Relationships and Pedogenic Processes in Península Valdés").

3.3 Playa Lakes

The playa lake unit represents the base level of the Great Endorheic Basin Systems (Fig. 4d). They are uniform flat areas (slope <0.01°) were the water table is generally near the surface. Due to the influence of the water table near the surface, the soils correspond to suborder Aquents. The vegetation is developed on the perimeter the playa lake and consists predominantly of the genus *Distichlis* and *Sarcocornia.*

In the Península Valdés, the playa lake units constitute shallow ephemeral lakes or salinas of a few km^2 in extent up to 70 km^2 (i.e., Salina Chica, Salina Grande and Gran Salitral). These units are composed of laminated clays and silts, interbedded with fine- to medium-grained sands that form a network of cracks during dry conditions. In the Salina Grande and Salina Chica, where water is continuously flowing from springs located in the upper pediments, important salt accumulation occurs at the bottom of the basin (see Chapter "Geology of Península Valdés").

4 System C: Coastal Zone

The Coastal Zone of Península Valdés includes the littoral region of Golfo Nuevo and Golfo San José and the eastern coastline of the Atlantic Ocean. The coastal landform in the mentioned gulfs is characterized by an alternation of headlands and bays, where due to the process of water wave diffraction erosion predominates on cliffs and wave-cut platforms and accretion on the beaches. Whereas that on the Atlantic coast, except in the Caleta Valdés area, the marine erosion is more prominent, reaching rectification of the coastline.

As in the case of the Great Endorheic Basins System, on the coastal area of the Golfo Nuevo and Golfo San José, there are two levels of pediments caused by local changes in base level developed when these gulfs were also formed as Great Endorheic Basins, presumably during early middle Pleistocene. During the late Glacial transgression, approximately between 15 and 10 ka BP (Mouzo et al. 1978; Ponce et al. 2011) Great Endorheic Basins were flooded by the sea. Changes of base level caused by sea level rise are recorded by oldest shorelines mainly represented by beach ridges of the middle Holocene (San Miguel Formation) and some relics of wave-cut platforms.

On the Atlantic coast, there is also a sequence of beach ridges corresponding to the Caleta Valdés Formation, representing the sea level during late Pleistocene (see Chapter "Geology of Península Valdés" and Sect. 5).

4.1 Coastal Piedmont Pediment

The coastal slopes of the Golfo Nuevo and Golfo San José has the same geomorphological units and genesis than Great Endorheic Basins System: the erosion scarp, two levels of ancient pediments and *bajadas*. The integration of these geomorphological units with the coastline gives an irregular shape, where levels of ancient pediments culminate abruptly in cliffs and associated wave-cut platforms (coastal erosion), whereas, accretion prevails in the distal areas of *bajadas* (beaches and salt marshes; Fig. 5b). Dominant soils are Natrargids, Calciargids, Torriorthens complex with a shrub steppe vegetation community. Holocene stream valleys dissected coastal pediments on the eastern slopes of Península Valdés region. These valley were filled with aeolian sandy sediments where the soils correspond to Haplocalcids with a grassland vegetation community (see Chapter "Vegetation of Península Valdés: Priority Sites for Conservation").

4.2 Coastal Bajada

This unit connect the coastal piedmont pediment with the Pleistocene and Holocene beach ridges. Geomorphic properties, vegetation patterns, and soil development have the same characteristics that those observed in Great Endorheic Basin System (see Sect. 3.2).

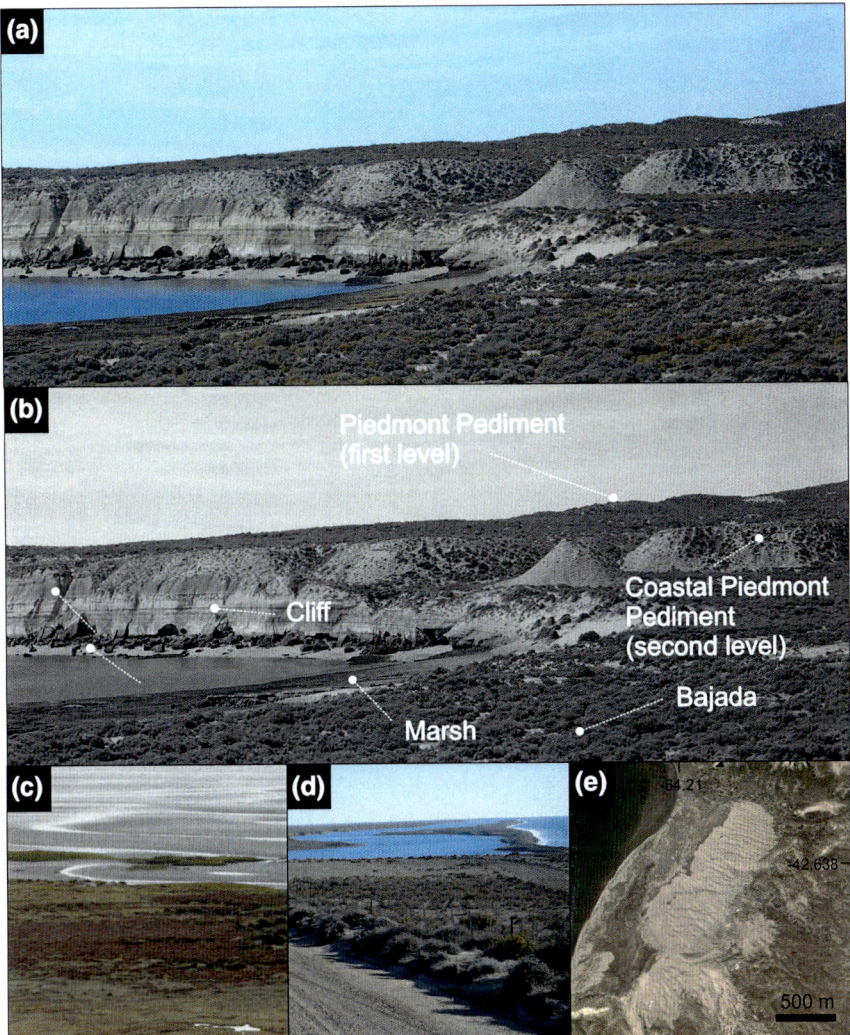

Fig. 5 Coastal zone system; **a** and **b** geomorphological sequence on littoral hillslope setting: erosion escarps, cliff (and wave-cut platforms associated), coastal piedmont pediment levels, *bajada* (pediment association), and present beaches (with salt marshes associated); **c** details of present beach and salt marsh; **d** coastal lagoon of Caleta Valdés site, BR I–IV Pleistocene Beach ridges (Caleta Valdés Formation), CS cuspate spit, BR V Holocene beach ridge (spit, San Miguel Formation)

4.3 Beach Ridges I–IV

On the easternmost side of the Península, in Caleta Valdés, a Pleistocene beach ridge sequence has long been recognized and described (Feruglio 1949–1950;

Fasano et al. 1983; Codignotto and Kokot 1988; Rutter et al. 1989; Súnico 1996; Rostami et al. 2000; Brückner et al. 2007; Pedoja et al. 2011; Bouza 2014) (Fig. 2). These deposits correspond to Caleta Valdés Formation (Haller et al. 2000; see Chapter "Geology of Península Valdés") and are composed of pebbly conglomerates with sandy matrix and few banks of calcareous skeletal remains of fossil mollusk. Nonevidences of Pleistocene shorelines are observed in other parts of Península Valdés (Fig. 2).

The beach ridge sequence has four paleoshorelines informally named with Roman number (i.e., I–IV) in descending order of elevation (Fig. 2). The highest marine terrace, beach ridge I, reach over 20 m a.s.l. dropping to about 14 m a.s.l. seaward, which is about the same reaches as beach ridges II, III, and IV. Although identification is tenuous because the four shorelines have roughly the same level elevation, they are separated on the basis of elongated paleo-coastal lagoons, vegetation patterns, soil development, and minor elevation differences.

Numerical ages obtained using ^{14}C, U/Th, ESR, and amino acid racemization give a middle to later Plesitocene age for this deposits (Rutter et al. 1989; Rostami et al. 2000; Schellmann and Radtke 2000; Brückner et al. 2007). These ages in addition to stratigraphic correlations with other beach ridges of the Atlantic Patagonian coast allow correlating them with specific Marine Isotope Stages (MIS). Formation of shorelines II, III, and IV are correlated to the interglacial period of the MIS 5 (80–130 ka), whereas shoreline I could be correlated to both the MIS 5 or even to the MIS 7 interglacial (230–190 ka).

As sea levels during MIS 5 and MIS 7 were around to ±4 m a.s.l. (e.g., Hearty and Kindler 1995; Schellmann and Radtke 2000; Pedoja et al. 2011) the higher elevations of beach ridges I–IV were used to demonstrate that the Atlantic Patagonian coast had a large-scale uplift during the Quaternary (Rostami et al. 2000; Pedoja et al. 2011). The hypothesis concerning this uplift is that it was and is still being generated by the subduction of the Chile ridge and the associated dynamic uplift (Guillaume et al. 2009; Pedoja et al. 2011; see Chapter "Climatic, Tectonic, Eustatic, and Volcanic Controls on the Stratigraphic Record of Península Valdés").

4.4 Beach Ridges V

This geomorphological unit is the youngest raised beach located between 5 and 10 m a.s.l. (Fig. 2). It is a barrier spit formed by Holocene beach ridges composed of coarse and very coarse gravel with a small sandy matrix and shells. The unit is included in the San Miguel Formation (Haller 1981) and has long been studied in previous works (Feruglio 1949–1950; Codignotto and Kokot 1988; Rutter et al. 1989; Codignotto et al. 1992; Súnico 1996; Monti 1997; Rostami et al. 2000; Kokot et al. 2005; Brückner et al. 2007, Schellmann and Radtke 2010; Pedoja et al. 2011). This geomorphological unit is developed in the Caleta Valdés area, to the east of the Pleistocene shorelines and also as pocket beaches in the Golfo San José and in the

Golfo Nuevo (Fig. 2). Due to the soil parent materials of beach ridge deposit is relatively young, the soils have a weak development, presumable corresponding to Torriorthens.

Absolute ages are based on ^{14}C ages and give a Holocene age, between 6500 (at the uppermost beach ridges) to 2.2 ka ^{14}C yr BP (at the lowest beach ridges) (Codignotto and Kokot 1988; Rutter et al. 1989; Codignotto et al. 1992; Monti 1997; Rostami et al. 2000; Brückner et al. 2007). Thus, beach ridge IV formation is correlated to the MIS 1 and more specifically to the end of the Holocene Climatic Optimum or the mid-Holocene Thermal Maximum (5–6 ka BP, *sensu* Briner et al. 2006).Therefore, it can be assumed that the uppermost beach ridges, which are 6.5–5.2 ka BP, were deposited during the maximum Holocene transgression when relative sea level was higher than present. Sea level later declined, reaching the present beach level as early as 2.2 ka BP. As south Patagonian Atlantic coast did not undergo significant uplift during the Holocene (Schellmann and Radtke 2010; Pedoja et al. 2011), surface elevations of these beach ridges have been strongly dominated by eustatic sea level variations (see Chapter "Climatic, Tectonic, Eustatic, and Volcanic Controls on the Stratigraphic Record of Península Valdés").

4.5 Present Beach

In the same way than the San Miguel Formation the present beaches are sandy and gravelly pocket beaches, formed between headlands of rocky (Miocene sedimentites) shorelines.

In Caleta Valdés site, the San Miguel Formation (beach ridge V) is constituted by a set of gravel beach ridges of middle Holocene age. Longshore drift has been predominantly north to south for the last 4–5 ka BP (Codignotto et al. 1992; Rostami et al. 2000). Evolution of the area has been monitored by Kokot et al. (2005). In this period, the northern spit has been growing southward, and its rate has increased 25 m/a (1971–87), 89 m/a (1987–96), and 167 m/a (1996–1999). These beach ridges form the V system proposed by Fasano et al. (1983, Fig. 5d). In the coastal lagoon of Caleta Valdés, barrier islands, tidal plains, salt marshes, and cuspate spit are observed (Codignotto and Kokot 1988).

The barrier island genesis is probably due to the segmentation from a previous spit by channels generated by the action of the sea. The origin of cuspate spit is due to longshore drift operating on a coastline from two different directions as occurs in tidal currents.

On some intertidal areas, small coastal salt marshes occur. These landforms develop in the intertidal zone where a generally muddy substrate supports varied and normally dense stands of halophytic plants (Allen and Pye 1992). The main salt marshes in Península Valdés area develop in Golfo San José. Riacho San José marsh (see Chapter "Vegetation of Península Valdés: Priority Sites for Conservation") is located to the west of the Istmo Carlos Ameghino. This wetland is classified as salt-restricted-entrance embayment salt marsh, characterized by a sandy-loam sediment gain-size, and protected from the wave action by sandy and/or gravel spits (Bouza et al. 2008). Dominant plant species in Riacho marsh is the genus *Spartina*, where *S. alterniflora* installs at the lowest marsh level and *S. densiflora*, accompanied by *Limonium brasiliense*, *Sarcocornia perennis*, and *Atriplex sp.*, extends at the highest marsh level. Fracasso marsh is located to the northeast of Istmo Carlos Ameghino and was classified as an open coast salt marsh with a predominantly silty grain size at the highest marsh level and sandy at the lowest marsh level due to the marine influence. Dominant plant species in Fracasso marsh is *Sarcocornia perennis*, accompanied by patches of *Limonium brasiliense* (installed on the marsh levees of the tidal creeks) and isolated plants of *Spartina densiflora*. *Spartina alterniflora* extends over a thin patch parallel to the coastline at lowest marsh level and on the point bars of the tidal creeks (Idaszkin et al. 2014; see Chapter "Vegetation of Península Valdés: Priority Sites for Conservation"). The soils are Aquents, where in the lowest marsh levels, sulfidic material occurred, being these soils classified as Sulfaquents (see Chapter "Soil–Geomorphology Relationships and Pedogenic Processes in Península Valdés").

In some erosional coasts in both Golfo Nuevo, as in the Gofo San José, rocky marshes are developed on top of wave-cut platforms, being specially dominated by *Spartina alterniflora* (Bortolus et al. 2009).

4.6 Coastal Aeolian Dunes

Sand dunes on the coast of Península Valdés are the typical small ridges of sand found at the top of the sand beaches, and above the usual maximum reach of the waves, in the transition area with coastal *bajada* (i.e., foreshore zone). In preferential deposition sites as small low areas, coastal dunes can be installed on coastal pediments. In these cases, the great coastal aeolian landforms of Península Valdés are located in Golfo San José and Golfo Nuevo on windward coast, where both compound linear and transverse dunes were identified (Fig. 5e).

Compound linear dunes (7–13 m high) are located at the NW coastal dunefield. Most linear dunes along the escarpment adopt a compound form owing to the NNE winds, attached to the cliff headland. A linear dune characteristic is that adjoining ridges often branch or merge at a "Y" junction. In the study area, junctions are most common where the ridge deflection occurs.

5 Landscape Evolution of the Late Cenozoic of Península Valdés

The present landscape of the Península Valdés is the result of a complex interrelation between climatic, tectonic, volcanic, and eustatic controls that had work in the Andean foreland during the late Cenozoic. (See Chapter "Climatic, Tectonic,

Eustatic, and Volcanic Controls on the Stratigraphic Record of Península Valdés"). These interrelations had induced base level changes that favored the erosion of many geomorphological units (e.g., Terraces Levels I–VI) but that also the development of many others (e.g., *bajada*). Based on the interrelationship among the geomorphological units of the Upland and Plains System, the Great Endorheic Basin System and the Coastal Zone System, the landscape evolution of the late Cenozoic of Península Valdés could be reconstructed.

During the Pliocene to early Pleistocene ($\sim$5 to $\sim$1 Ma) the sea coast was 100 km far east from the present coast line (Fig. 6a). At that moment, the landscape of the Península Valdés region was quite different, characterized by fluvial systems (i.e., *Rodados Patagónicos*) developed from the SW to the NE (Cortés 1981; González Díaz and Di Tommaso 2011) and alluvial and fluvial system connected to the Meseta de Somuncura (Fig. 1). Terrace Levels I–VI were deposited in that period, indicating that at least six important base level variations has caused fluvial system modifications.

Later, in a period not clearly defined, closed basins start to developed (Fig. 6b). These depressions were not flooded by the sea and include both, the major depressions and the gulfs of Península Valdés (e.g., Golfo Nuevo, Salina Grande depressions). The period of closed basin formation could be roughly estimated based on the stratigraphic position of the Great Endorheic Basins System units. Geomorphological units of this system are developed over terraces I–VI, thus an age younger than 1 Ma is proposed for this period (youngest age of the *Rodados Patagónicos*). On the other hand as pediments of the Great Endorheic Basins are correlated with coastal pediments which are older than the Beach ridges I–VI, the closed basin formation period would have been not older than 230 ka (maximum age of the MIS 7 obtained in beach ridge I). This indicates that during the lower (i.e., Calabrian stage) to middle Pleistocene the Golfo Nuevo, Golfo San José, Gran Salitral, Salina Chica, and Salina Grande depressions started to develop. Regarding the causes and trigger mechanisms involved in the closed basin formation, there are many controversies. Whereas, wind erosion was proposed to explain the close basins formation (Mouzo et al. 1978; González Díaz and Di Tommaso 2011); tectonics was also proposed (Roveretto 1921; Kostadinoff 1992; Isla 2013) and a combination of both processes was also suggested (Kostadinoff 1992; Haller et al. 2000). In this chapter, we arrive to the conclusion that while wind erosion could have been important, tectonic activity related to fault blocks was probably the trigger mechanism for the formation of the great closed basins. This idea is supported by (1) the occurrence of closed basins formed on pebble gravel deposits that cannot be removed by deflation, (2) borders of the major depressions are straight and in many cases match with subsurface faults (see Kostadinof 1992), (3) post Miocene faults were observed in the region (Haller et al. 2000, Chapter "Geology of Península Valdés"). Further work is needed to bring some light to this issue.

During the late Pleistocene in the Penísnula Valdés region a landscape similar than today's start to configure. At that time, sea level raised during many interglacial periods, some of them remarkably registered in Península Valdés (MIS5/7). However, nonevidence of Pleistocene marine ingressions (Beach ridges I–IV) is

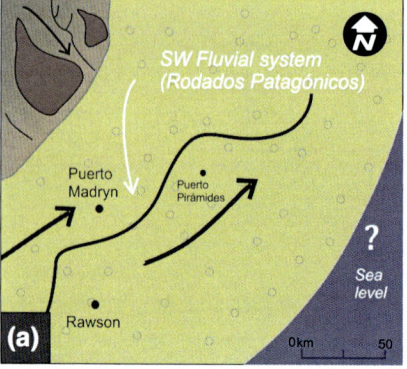
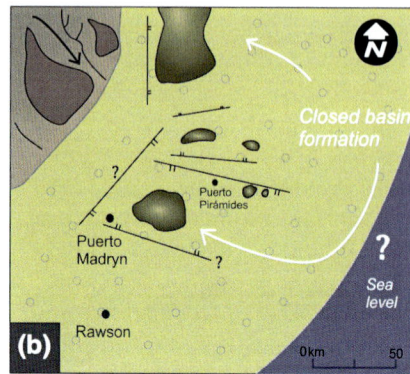
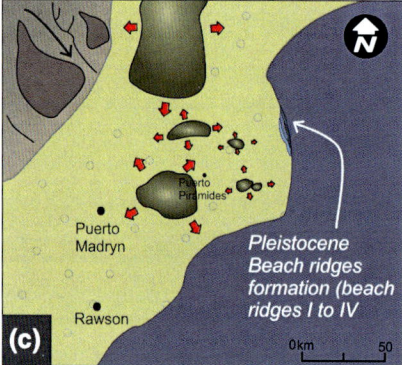
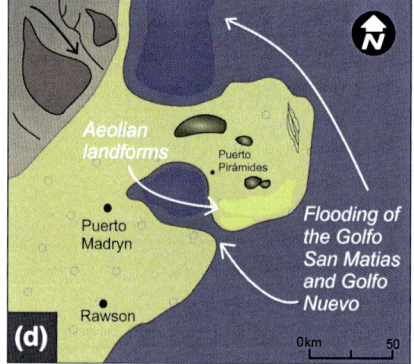
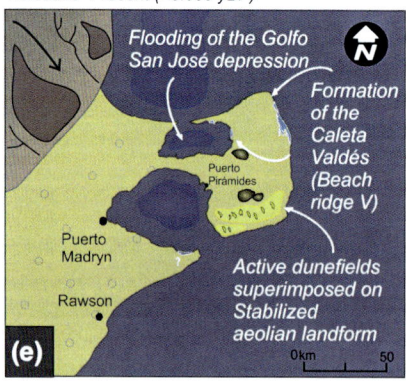
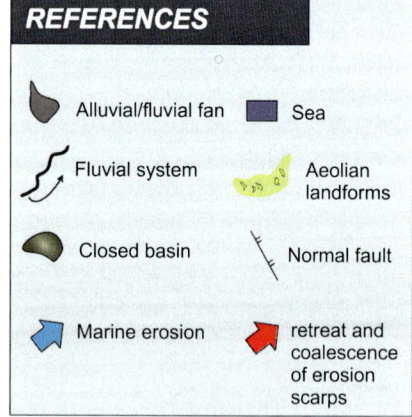

Fig. 6 Landscape evolution of the Península Valdés area

registered inside the Golfo Nuevo and the Golfo San José, indicating that gulf flooding did not occur in the Pleistocene (at least the final part of the MIS5 ~80 ka), but probably during the Mid-Holocene Thermal Maximum (5–6 ka BP). Topographic/bathymetric data of the Golfo San Matías and Golfo Nuevo indicate depression borders much deeper than the topographic borders of the Golfo San José (−55, −35 and −20 m a.s.l., respectively). Thus, Golfo San Matías and Golfo Nuevo would have been flooded before the Golfo San José (Fig. 6c, d), as previous studies suggested (i.e., 10–19 ka BP; Codignotto 2008; Ponce et al. 2011; Isla 2013). Finally, during the last millennium, sand dunes formation in the upland and plain system and present beach/marsh formation in the coastal zone system took place, resulting in the current landscape of The Península Valdés (Fig. 6e).

As a final remark, it is possible to suggest that, despite many authors had contributed to the knowledge of this region, important uncertainties related to the landscape evolution of the Península Valdés still exist. Understanding the formation of the closed depressions, to determine the timing of sea ingressions in the gulfs or to comprehend the trigger mechanisms of relative sea level changes are far from being clear and offer significant potential for further studies.

Acknowledgements The authors would like to thank the helpful reviews of Professors Marcelo Zárate and Alejandro Monti which improved the final version of this manuscript. This research has been funded by the CONICET (PIP 0190 and PIP 0632) and Agencia Nacional de Promoción Científica y Tecnológica (PICT 1876 and 2167).

Glossary

Bajada	Alluvial plain formed at by the coalescing of several alluvial fans
Barchan dunes	Barchan or barkhan dune is a crescent-shaped dune. Barchans face the wind, appearing convex and are produced by wind action predominately from one direction
Blowouts	Depressions or hollows formed by wind erosion on a preexisting sand deposit, generally in vegetation-stabilized dune fields
Closed basin	A basin draining to some depression or pond within its area, from which water is lost only by evaporation or percolation. A basin without a surface outlet
Complex dunes	Complex dunes consist of two or more different types of simple dunes which have coalesced or are superimposed, usually of barchanoid or linear shapes
Compound dunes	Compound dunes consist of two or more dunes of the same type which have coalesced or are superimposed

Desert pavement	A stony surface generally composed of a layer of angular or subrounded gravels one or two stones thick developed on a mantle of finer stone free material
Dome-shaped dunes	Dome dunes are the rarest type of dune. They are circular and do not have a slipface. The wind can blow material onto the dune from any side
Ice-wedge casts	The filling of sand or other materials that replaces former ice wedges
Linear dunes	Linear dunes form straight or nearly straight lines. Some linear dunes are shaped like a wiggling snake, with regular curves. Linear dunes develop where wind pressures are nearly equal on both sides of a dune
Mass wasting processes	The failure and movement by gravity of a volume of soil or rock to a downslope site
Nebkas	A small landform that forms mainly around shrubs, built, and shaped by the action of wind
Parabolic dunes	If strong winds erode a section of the vegetated sand (commonly referred to as a blowout), a parabolic dune may form
Pediments	A gently sloping erosional surface developed on bedrock or older uncosolidated deposits subaerially exposed or covered by a discontinuous to continuous veneer of alluvial deposits
Playa lake	A near level area at the bottom of a closed basin, sometimes temporarily covered with water
Salt marshes	A flat or gently sloping vegetated wetland in the upper intertidal zone on sheltered parts of the coast (estuaries, inlets, lagoon shores)
Simple sand dunes	Simple dunes consist of individual dune forms which are spatially separate from nearby dunes. These sand dunes are small in most cases, with wavelengths (shortest distance from one dune crest to the other) of 10–500 m
Star dunes	Star dunes have pointed ridges and slip faces on at least three sides. Star dunes develop where winds come from many different directions
Transverse dunes	A large, strongly asymmetrical, elongated dune lying at right angles to the prevailing wind direction
Wave-cut platforms	A flat to gently sloping surface at the base of a cliff, formed by erosion by waves. It develops as a result of wave abrasion and is also called abrasion platform

References

Mouzo F et al (1978) Rasgos de la geología submarina del Golfo Nuevo (Chubut). Acta Oceanographica Argentina, 2(1):69–70.

Bouza PJ et al (2008) Estudio de caso Chubut: Centro-Norte Península Valdés. In: Cantú MP, Becker AR, Bedano JC (eds) Evaluación de la sustentabilidad ambiental en sistemas agropecuarios. Fundación Universidad Nacional de Río Cuarto, Río Cuarto, pp 165–181

Bouza PJ et al (2008) Geomorfología y Características morfológicas y fisicoquímicas de suelos hidromórficos de marismas patagónicas. 21° Congreso Argentino de la Ciencia del Suelo. Actas 450, Potrero de los Funes, San Luis

Guillaume B et al (2009) Neogene uplift of the central eastern Patagonia: dynamic response to active spreading ridge subduction? Tectonic, 28 (TC2009). Doi:10.1029/2008TC002324

Blanco P et al (2010) Synergistic use of Landsat and Hyperion imageries for ecological site classification in rangelands. WHISPERS, June 2010, Actas: 51. Reykjavik, Iceland, pp 1–4. Doi:10.1109/WHISPERS.2010.5594878

Allen JRL, Pye K (1992) Coastal saltmarshes: their nature and importance. In: Allen JRL, Pye K (eds) Saltmarshes: morphodynamics, conservation and engineering significance. Cambridge University Press, Cambridge, pp 1–18

Alvarez MP, Weiler NE, Hernández MA (2008) Geohidrología de humedales cercanos a la costa con cota bajo nivel del mar, Península Valdés, Argentina. Rev Latinoam Hidrogeologia 6 (1):35–42

Alvarez MP et al (2012) Groundwater flow model, recharge estimation and sustainability in an arid region of Patagonia, Argentina. Environ Earth Sci 66(7):2097–2108

Bachman GO, Machette MN (1977) Calcic soils and calcretes in the southwestern United States. U.S. Geological Survey, Open-File Report 77-794, p 163

Beltramone C (1983) Rasgos fisiográficos de Península Valdés (Chubut, Argentina) Terra Aridae 2 (1):168–188

Bortolus A, Schwindt E, Bouza PJ, Idaszkin YL (2009) A characterization of Patagonian salt marshes. Wetlands 29:772–780

Bouza PJ (2012) Génesis de las acumulaciones de carbonatos en Aridisoles Nordpatagónicos: su significado paleopedológico. Revista de la Asociación Geológica Argentina 69:298–313

Bouza PJ (2014) Paleosuelos en cordones litorales de la Formación Caleta Valdés, Pleistoceno superior, NE del Chubut. Revista de la Asociación Geológica Argentina 71(1):1–10

Bouza PJ et al (2007) Fibrous-clay mineral formation and soil evolution in Aridisols of northeastern Patagonia, Argentina. Geoderma 139:38–50

Briner JP et al (2006) A multi-proxy lacustrine record of Holocene climate change on northeastern Baffin Island, Arctic Canada. Quaternary Res 65:431–442

Brückner H et al (2007) Erste Befunde zu Veränderungen des holozänen Meeresspiegels und zur Größenordnung holozäner 14C-Reservoireffekte im Bereich des Golfo San José (Península Valdés, Argentinien). Bamberger Geographische Schriften 22:93–111

Codignotto JO, Kokot RR (1988) Evolución geomorfológica holocena en Caleta Valdés, Chubut. Revista de la Asociación Geológica Argentina 43(4):474–481

Codignotto JO, Kokot RR, Marcomini SC (1992) Neotectonism and sea level changes in the coastal zone of Argentina. J Coastal Res 8:125–133

Codignotto J (2008) Península valdes: entre el mar y la tierra. In: Codignotto J (ed) Sitios de interés geológico de la República Argentina. SEGEMAR, Buenos Aires, pp 683–696

Cortés LM (1981) Estratigrafía Cenozoica y estructura al oeste de la Península Valdés, Chubut. Consideraciones tectónicas y paleogegráficas. Revista de la Asociación Geológica Argentina 36(4):424–445

Darwin CR (1846) Geological observations on South America. Being the third part of the geology of the voyage of the Beagle, under the command of Capt. Fitzroy, R.N. during the years 1832 to 1836. Smith Elder and Co., London

del Valle HF et al (2008) Sand dune activity in north-eastern Patagonia. J Arid Environ 72: 411–422

Dohrenwend JC, Parsons AJ (2009) Pediments in arid environments. In: Parsons A, Abrahams A (eds) Geomorphology of desert environments. Springer, Netherlands, pp 377–411

Fasano JL, Isla FI, Schnack EJ (1983) Un análisis comparativo sobre la evolución de ambientes litorales durante el Pleistoceno tardío-Holoceno: Laguna Mar Chiquita (Buenos Aires)-Caleta Valdés (Chubut). Simposio Oscilaciones del nivel del mar durante el último hemiciclo deglacial en la Argentina. IUGS-UNESCO N° 61, Actas: 27–47. Mar del Plata

Feruglio E (1949-1950) Descripción geológica de la Patagonia. Dirección General de Yacimientos Petrolíferos Fiscales, Buenos Aires, p 1114

Fidalgo F, Riggi JC (1970) Consideraciones geomórficas y sedimentológicas sobre los Rodados Patagónicos. Revista de la Asociación Geológica Argentina 25:430–443

Gile L, Peterson F, Grossman RB (1966) Morphological and genetic sequences of carbonate accumulation in desert soils. Soil Sci 101:347–360

González Díaz EF, Di Tommaso I (2011) Evolución geomorfológica y cronología relativa de los niveles aterrazados del área adyacente a la desembocadura del río Chubut al Atlántico (Provincia del Chubut). Revista de la Asociación Geológica Argentina 68(4):507–525

Haller M, Monti A, Meister C (2000) Hoja Geológica 4363-1, Península Valdés, Provincia del Chubut. Secretaría de Energía y Minería, Servicio Geológico Minero Argentino, Boletín 266, Buenos Aires, p 34

Hearty P, Kindler P (1995) Sea-level highstand chronology from stable carbonate platforms (Bermuda and The Bahamas). J Coastal Res 11:675–689

Iriondo MH, Ramonell C (1993) San Luis. In: Iriondo M (ed) El Holoceno en la Argentina. CADINQUA 2, pp 131–162

Isla FI (2013) The flooding of the San Matías Gulf: the Northern Patagonia sea-level curve. Geomorphology 203:60–65

Kokot RR, Monti AJ, Codignotto JO (2005) Morphology and short-term changes of the Caleta Valdés Barrier Spit, Argentina. J Coastal Res 215:1021–1030

Kostadinoff J (1992) Estudio geofísico de la Península de Valdés y los golfos nordpatagónicos. Revista de la Asociación Geológica Argentina 47:229–236

Machette MN (1985) Calcic soils of the Southwestern United States. Geol Soc Am Spec Pap 203:1–21

Martínez O, Rabassa J, Coronato A (2009) Charles Darwin and the firstscientific observations on the Patagonian Shingle Formation (Rodados Patagónicos). Revista de la Asociación Geológica Argentina 64(1):90–100

Mercer JH (1976) Glacial history of southernmost South America. Quaternary Res 6:125–166

Monti AJA (1997) Morfodinámica y ciclicidad de la acreción en depósitos costeros del Holoceno: Chubut, Argentina. Doctoral Thesis, Universidad de Buenos Aires, p 160

Pedoja K et al (2011) Uplift of quaternary shorelines in eastern Patagonia: Darwin revisited. Geomorphology 127(3–4):121–142

Peterson F (1981) Landforms of the Basin and Range Province defined for soil survey. Max C. Fleischmann College of Agriculture, Agricultural Experiment Station, Technical bulletin 28, p 52

Ponce JF et al (2011) Palaeogeographical evolution of the Atlantic coast of Pampa and Patagonia from the last glacial maximum to the Middle Holocene. Biol J Linn Soc 103:363–379

Rabassa J, Clapperton CM (1990) Quaternary glaciations of the southern Andes. Quaternary Sci Rev 9:153–174

Rostagno CM (1981) Reconocimiento de los suelos de la Península Valdés.. Centro Nacional Patagónico, Chubut Argentina (inédito), Publicación 44, Puerto Madryn, p 24

Rostami K, Peltier WR, Mangini A (2000) Quaternary marine terraces, sea-level changes and uplift history of Patagonia, Argentina: comparisons with predictions of the ICE-4G (VM2) model of the global process of glacial isostatic adjustment. Quaternary Sci Rev 19:1495–1525

Roveretto G (1921). Studi di geomorfología argentina. V: La Penisola Valdés. Vol Soc Geol Italiana 30:1–47

Rutter N, Schnack EJ, Rio J, Fasano JL, Isla FI, Radtke U (1989) Correlation and dating of Quaternary littoral zones along the Patagonian coast, Argentina. Quaternary Sci Rev 8: 213–234

Schellmann G, Radtke U (2000) ESR dating stratigraphically well-constrained marine terraces along the Patagonian Atlantic coast (Argentina). Quatern Int 68–71:261–273

Schellmann G, Radtke U (2010) Timing and magnitude of Holocene sea-level changes along the middle and south Patagonian Atlantic coast derived from beach ridge systems, littoral terraces and valley-mouth terraces. Earth Sci Rev 103:1–30

Súnico A (1996) Geología del Cuaternario y Ciencia del Suelo: relaciones geomórficas-estratigráficas con suelos y paleosuelos. Doctoral Thesis, Facultad de Ciencias Exactas y Naturales, Departamento de Graduados, Universidad Nacional de Buenos Aires, p 227

Thomas D (1997) Arid zone geomorphology. process, form and change in drylands, second edition. Wiley, Chichester, p 713

Ton-That T et al (1999) Datación por el método ^{40}Ar/^{39}Ar de lavas basálticas y geología del Cenozoico Superior en la región del Lago Buenos Aires, provincia de Santa Cruz, Argentina. Revista de la Asociación Geológica Argentina 54:333–352

Trombotto D (1996) The old cryogenic structures of Northern Patagonia: The cold episode Penfordd. Zeitschrift für Geomorphologie 40(3):385–399

Trombotto D (1998) Paleo-permafrost in Patagonia. Bamberger Geographische Schriften 15: 133–148

Trombotto D (2008) Geocryology of Southern South America. In: Rabassa J (ed) The Late Cenozoic of Patagonia and Tierra del Fuego. Elsevier, Amsterdam, p 513

Vegetation of Península Valdés: Priority Sites for Conservation

Mónica B. Bertiller, Ana M. Beeskow, Paula D. Blanco,
Yanina L. Idaszkin, Gustavo E. Pazos and Leonardo Hardtke

Abstract This chapter describes the main vegetation units of Península Valdés at scale 1:250,000 with emphasis on relevant physiognomic and floristic characteristics. Based on photogrammetry (aerial photograph pairs 1:60,000) and ground check, 18 dominant singular plant species arrangements (vegetation units) were identified reflecting the variety of environmental conditions at a mesoscale (1:250,000) within Península Valdés. At sites selected for ground check, floristic–physiognomic census including a complete floristic plant species list with the relative abundance of each species were performed. After that, censuses of species abundance were ordered by principal component analysis. The layer structure, the main life forms and the dominant species for each identified and mapped vegetation unit were described. Among them, we identified shrubby vegetation units at northern and central Península Valdés and, grassy vegetation units at southern Península Valdés. A map of vegetation units and some pictures of the most representative vegetation units complete the vegetation description. Moreover, this chapter includes a detailed description of the plant communities (resolution scale 1:1) characterizing four sites identified as priorities for ecosystem conservation.

M.B. Bertiller (✉) · A.M. Beeskow · P.D. Blanco · Y.L. Idaszkin · G.E. Pazos · L. Hardtke
Instituto Patagónico para el Estudio de los Ecosistemas Continentales (IPEEC)-Consejo Nacional de Investigaciones Científicas y Técnicas (CONICET) - CCT Centro Nacional Patagónico (CENPAT), Boulevard Brown 2915, 9120 Puerto Madryn, Chubut, Argentina
e-mail: monika@cenpat-conicet.gob.ar

A.M. Beeskow
e-mail: beeskow@cenpat-conicet.gob.ar

P.D. Blanco
e-mail: blanco@cenpat-conicet.gob.ar

Y.L. Idaszkin
e-mail: idaszkin@cenpat-conicet.gob.ar

G.E. Pazos
e-mail: gpazos@cenpat-conicet.gob.ar

L. Hardtke
e-mail: hardtke@cenpat-conicet.gob.ar

© Springer International Publishing AG 2017
P. Bouza and A. Bilmes (eds.), *Late Cenozoic of Península Valdés, Patagonia, Argentina*, Springer Earth System Sciences,
DOI 10.1007/978-3-319-48508-9_6

Priority sites for conservation are located in Salt marshes, Uplands and Plain Systems and Endorheic Basins. Some contrasts between conserved and degraded community states are also exemplified.

Keywords Desert steppes · Grass steppes · Patagonia · Salt marshes · Sandy grasslands · Sheep grazing · Shrub steppes

1 Introduction

The phytogeographical identity of the vegetation of Península Valdés has long been discussed by various authors due to the presence of characteristic *floristic elements* of two different *Phytogeographical Provinces*. Some specialists included the vegetation of Península Valdés in the Patagonian Phytogeographical Province (Lorentz 1876; Holmberg 1898; Kühn 1922, 1930; Cabrera 1976; Soriano 1956; Morello 1958; León et al. 1998). Other researchers referred the vegetation of this portion of Patagonia to the Monte Phytogeographical Province (Hauman 1920, 1926, 1931, 1947; Parodi 1934, 1945, 1951; Castellanos and Pérez Moreau 1944). Moreover, Frenguelli (1940) and Soriano (1950) defined the vegetation of Península Valdés as an ecotone between the vegetation of both Provinces. This debate continued since more recently, Roig and Martínez Carretero (1998) included vegetation of Península Valdés in the Monte while León et al. (1998) in the Patagonian Phytogeographical Province. In this contribution, we adopt the criteria of León et al. (1998).

Probably, all these discrepancies are due to the fact that despite sharing *floristic elements* characteristic of both Patagonian and Monte, most of the vegetation of Península Valdés lacks of *Larrea* species (main floristic components of the Monte) and has an abundant and extended presence of *Chuquiraga* species (main floristic components in the Central District of the Patagonian Province).

2 Main Plant Adaptive Strategies

The Península Valdés is characterized by vast terraces (Uplands and Plains System, see Chapter "Late Cenozoic Landforms and Landscape Evolution of Península Valdés") covered by discreet plant life forms arranged in a patchy structure. Native species present today are the result of plant adaptive responses to the prevailing arid conditions after the Andean uplift (see Chapter "Climatic, Tectonic, Eustatic, and Volcanic Controls on the Stratigraphic Record of Península Valdés"). These adaptive responses to aridity along with high intensity and frequency of dry winds, and relative low temperatures (Ares et al. 1990; see Chapter "The Climate of Península Valdés Within a Regional Frame") resulted in particular structural and functional adaptations. The main life forms in Península Valdés are shrubs (evergreen and deciduous), bunch perennial grasses, and forbs (Sala et al. 1989;

Golluscio and Sala 1993; Bertiller et al. 2006). Dominant evergreen shrubs show adaptations such as cushion form, small leaves, and green leafless stems. Epidermal pubescence is another adaptive feature giving a typical grey or green opaque colour to leaves. Drought deciduous shrubs may also be present in the shrubby canopy. Vegetation canopy in the northern-central Península Valdés consists of a patchily arranged shrubby matrix. Less conspicuous grass species and forbs are scattered in almost all inter-patch areas or associated to shrub patches. In contrast, sandy soils in southern Península Valdés are mainly covered by rhizomatous perennial grasses forming a continuous grass stratum with interspersed shrubs.

Most of the plant activity in the Península Valdés depends on soil water accumulated during the winter-spring precipitation period although precipitation pulses could also occur in autumn and summer (Ares et al. 1990; Barros and Rivero 1982, see Chapter "The Climate of Península Valdés Within a Regional Frame"). Increasing temperature and day length coupled with high soil moisture conditions in early spring leads to the reactivation of the vital functions in most species. This activity period is extended up to early–late summer depending on the rooting depth of the species (Bertiller et al. 1991; Campanella and Bertiller 2008). In this sense, the activity of deep rooted shrubs, both deciduous and evergreen, is more extended into the dry summer compared to that of perennial grasses with shallow rooting depth. The timing of flowering, with some exceptions, has a marked seasonality occurring only once in the year.

3 Description of the Main Vegetation Units (Scale 1:250,000)

Eighteen dominant singular plant species arrangements (vegetation units) reflect the variety of environmental conditions at a mesoscale (1:250,000) within Península Valdés (Bertiller et al. 1981). Based on photogrammetry (aerial photograph pairs 1:60,000) and ground check, 18 main vegetation units were identified in Península Valdés. At sites selected for ground check, floristic–physiognomic census (Mueller-Dombios and Ellemberg 1974) including a complete floristic plant species list with the relative abundance of each species were performed. After that, censuses of species abundance were ordered by principal component analysis. The layer structure, the main life forms and the dominant species for each identified and mapped vegetation unit were described (Bertiller et al. 1981). Nomenclature followed the Flora Patagónica (Correa 1969, 1971, 1978, 1984a, b, 1988, 1998, 1999), and was updated by the Data Base Flora Argentina (http://www.floraargentina.edu.ar/).

These dominant vegetation units range from shrubby steppes in the north and centre of Península Valdés to perennial grass steppes dominating in the south of Península Valdés (Fig. 1). The north of Península Valdés is mostly characterized by terraces and piedmont slopes (Uplands and Plains System, see Chapter "Late Cenozoic Landforms and Landscape Evolution of Península Valdés") covered by shrub and grass-shrub steppes, dominated by shrubs of the genus *Chuquiraga* and perennial grasses of the genus *Nassella* and *Jarava*. The Great Endorheic Basins System characterizing the central portion of the Península Valdés are covered by shrub and the shrub-perennial grass steppes with predominance of shrubs (*Chuquiraga* spp., and *Cyclolepis genistoides*), and the perennial grass *Nassella tenuis*. The south portion of the Península Valdés is characterized by terraces covered in some places by West–East stabilized aeolian field. Terraces are colonized by patchy steppes consisting of a mosaic of perennial grass- and shrubby steppes. Stabilized aeolian field deposits are covered by perennial grass steppes with isolated shrubs and shrub steppes of *Hyalis argentea*. Moreover, active dunefield and playa lakes deposits without vegetation alternate with these vegetation units. Finally, coastal landforms around Península Valdés are covered by shrubby or grass-shrubby steppes.

3.1 Vegetation of System A: Uplands and Plains

3.1.1 Vegetation of the Terrace Level I

Shrub steppe of Chuquiraga avellanedae and Schinus johnstonii (VU17, Fig. 1)

This shrubby steppe is the dominant vegetation unit in the Terrace level 1 occupying a narrow band in the Istmo Carlos Ameghino. The total vegetation cover is 40–50% consisting of a dominant shrubby stratum (ca. 40% cover, 50–200 cm tall) mostly represented by *C. avellanedae* and *S. johnstonii* along with *Condalia microphylla* and *Prosopidastrum globosum*. The other plant layers are a dwarf shrub stratum (1–5% cover, 5–10 cm tall) dominated by *Acantholippia seriphioides* and an herbaceous layer dominated by herbaceous perennial forbs: *Boopis anthemoides*, *Hoffmannseggia trifoliata*, and *Perezia recurvata* and by perennial and annual grasses *N. tenuis*, *Pappostipa humilis*, *Jarava neaei*, *Pappostipa speciosa*, *Poa ligularis* and *Schismus barbatus*.

3.1.2 Vegetation of the Terrace Level II

Shrub steppe of Chuquiraga erinacea ssp. hystrix and C. avellanedae with perennial grasses (VU13, Figs. 1 and 2a)

This vegetation unit is mostly characteristic of central Península Valdés dominating the Terrace level 2 as well as piedemont pediments and bajadas (System B, see Sect. 3.2). The total plant cover is 60–80% and consists of a mosaic of shrubby and perennial grass patches. Shrubby patches are dominated by a tall layer (50% cover, 50–180 cm tall) represented by *C. erinacea* ssp. *hystrix*, *C. avellanedae* along with *Lycium chilense*, *S. johnstonii*, *C. microphylla*, and *Brachyclados megalanthus* with dwarf shrub and herbaceous perennial species (5–10 cm tall) covering 5% of the

Vegetation map of Península Valdés

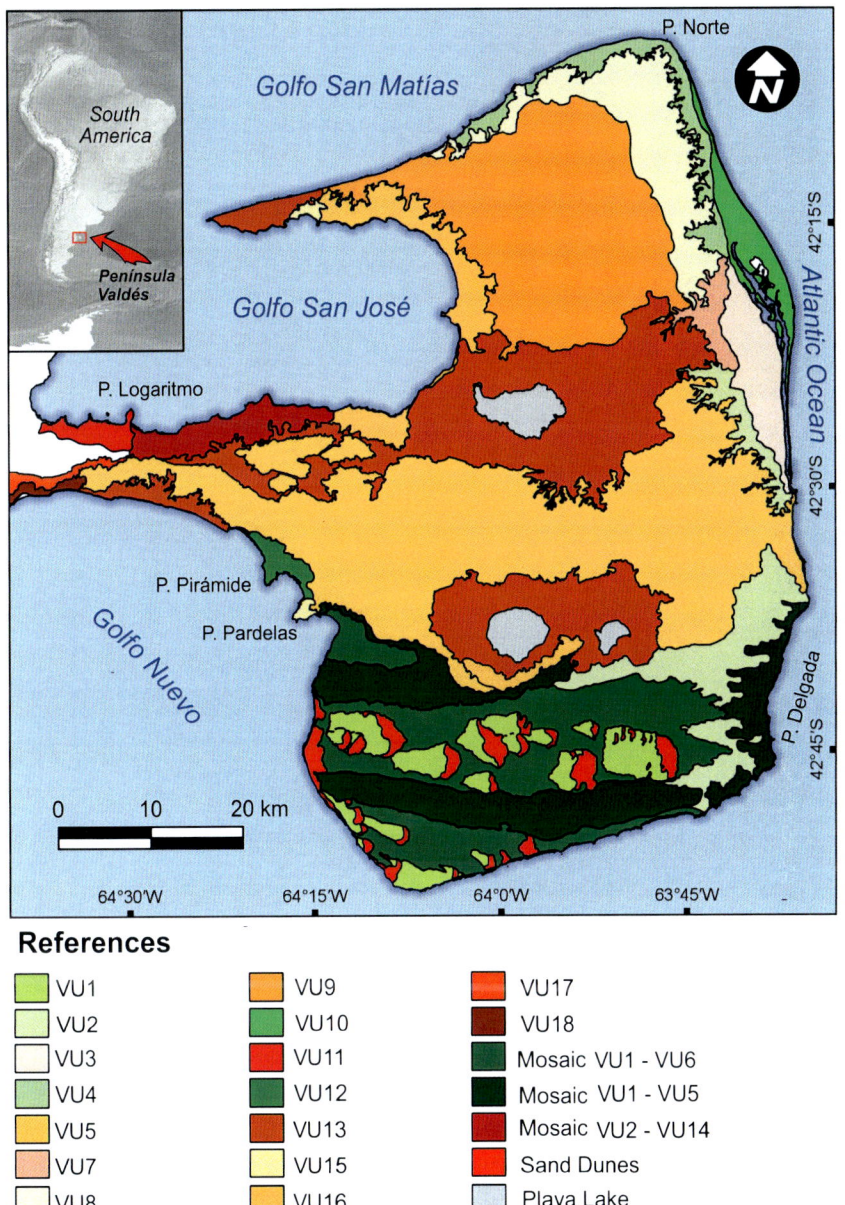

Fig. 1 Location of Península Valdés and Vegetation Units redrawn from Bertiller et al. (1981)

◄ **Fig. 2** Views of the main Vegetation Units of Península Valdés. **a** Shrub steppe of *Chuquiraga erinacea* ssp. *hystrix* and *Chuquiraga avellanedae* with perennial grasses (VU13); **b** Perennial grass steppe of *Piptochaetium napostaense*, *Nassella tenuis* and *Plantago patagonica* (VU2); **c** Mosaic VU2–VU5 (Shrub-perennial grass steppe of *Chuquiraga avellanedae* and *Nassella tenuis*; **d** Shrub-perennial grass steppe of *Chuquiraga avellanedae* and *Nassella tenuis* (VU5); **e** Shrub-perennial grass steppe of *Chuquiraga avellanedae* and *Nassella tenuis* (VU5); **f** Perennial grass steppe of *Sporobolus rigens* and *Nassella tenuis* (VU1); **g** Shrub steppe of *Hyalis argentea* (VU6); **h** Shrub steppe of *Chuquiraga erinacea* ssp. *hystrix*, *Cyclolepis genistoides*, and *Chuquiraga avellanedae* with perennial grasses (VU16); **i** Shrub steppe of *Chuquiraga avellanedae*, *Larrea divaricata* and *Nassauvia fuegiana* (VU14); **j** Shrub steppe of *Cyclolepis genistoides*, *Chuquiraga avellanedae*, and *Atriplex lampa* (VU15)

soil (*H. trifoliata*, *Baccharis darwinii*, *P. recurvata*, *Tetraglochin caespitosum*, and *A. seriphioides*). Perennial grass patches (15–20% cover, 5–10 cm tall) are dominated by *N. tenuis*, *Piptochaetium napostaense*, *P. speciosa*, *P. humilis*, with *J. neaei*, *P. ligularis*, with annual grasses (*S. barbatus* and *Bromus catharticus*) and the annual forb *Daucus pusillus*.

3.1.3 Vegetation of the Terrace Level III

The Terrace Level III is mainly covered by a two-phase mosaic (mosaic VU2–VU5, Fig. 1) consisting of the perennial grass steppe of *P. napostaense*, *N. tenuis*, and *Plantago patagonica* (VU2) and the shrub-perennial grass steppe of *C. avellanedae* and *N. tenuis* (VU5) characteristic of Terrace Level IV (see below). Also, pure patches of VU5 and a mosaic (VU1–VU6, Fig. 1) composed of the perennial grass steppe of *Sporobolus rigens* and *N. tenuis* (VU1) and the shrub steppe of *H. argentea*. (VU6), characteristic of the Stabilized aeolian fields (see Sect. 3.1.7), may cover in small patches Terrace Level III.

Perennial grass steppe of P. napostaense and N. tenuis with P. patagonica and isolated shrubs of C. avellanedae (VU2, Figs. 1 and 2b)

This vegetation unit has a total plant cover of 60–70% and is dominated by a conspicuous herbaceous layer (40–50% cover, 20 cm tall) characterized by the perennial grasses *P. napostaense* and *N. tenuis* accompanied by *B. catharticus*, *S. barbatus*, and the annual herb *P. patagonica*. The shrubby layer (15% cover, 50–200 cm tall) is dominated by *C. avellanedae* along with *S. johnstonii*, *L. chilense*, and *Discaria americana*. There are also scattered dwarf shrubs (*Baccharis melanopotamica* and *T. caespitosum*) and herbaceous perennial herbs (*Paronychia chilensis* and *H. trifoliata*) covering 5% of the soil with a height ca. 10 cm. This vegetation unit is mostly patchily associated (mosaic VU2–VU5, Fig. 1) with the shrub-perennial grass steppe of *C. avellanedae* and *N. tenuis* (VU5), characteristic of the central portion of the Península Valdés (Fig. 2c).

3.1.4 Vegetation of Terrace Level IV

The dominant vegetation at Terrace Level IV consists of shrub-grass steppes and shrubby steppes with perennial grasses. The most conspicuous vegetation unit occupying these terraces in central Península Valdés is the shrub-perennial grass steppe of *C. avellanedae* and *N. tenuis* (VU5) while in the northern areas vegetation is represented by the shrub steppe of *C. avellanedae* and *C. microphylla* (VU9).

Shrub-perennial grass steppe of *C. avellanedae* and *N. tenuis* (*VU5*, Figs. 1 and 2d–e)

The total plant cover is ca. 50% with a patchy arrangement consisting of shrubby and perennial grass patches. Shrubby patches (25% cover, 50–60 cm tall) are dominated by *C. avellanedae* along with *S. johnstonii* and *L. chilense*. Perennial grass patches (20–25% cover, 10 cm tall) are dominated by *N. tenuis* and *P. napostaense* with *P. ligularis* and *Nassella longiglumis*. Scattered dwarf shrub and herbaceous species (5% cover, 10 cm tall) represented by *H. trifoliata*, *P. chilensis* and *P. recurvata* are immersed in the perennial grass patches.

Shrub steppe of *C. avellanedae* and *C. microphylla* (*VU9*, Fig. 1)

This vegetation unit has a plant cover of 50–60% with a presence of shrubby and grass layers. The shrubby layer (40% cover, 40–120 cm height) is dominated by *C. avellanedae*, *C. microphylla*, and *P. globosum* along with *B. megalanthus*, *L. chilense* and *S. johnstonii*. The grass layer (15% cover, 15 cm height) is represented by *N. tenuis*, *P. speciosa*, *J. neaei*, *P. napostaense*, *N. longiglumis* and *S. barbatus*.

3.1.5 Vegetation of Terrace Level V

This Terrace level is covered by the shrub steppe of *C. avellanedae* and *N. tenuis* (VU5), characteristic of the Terrace Level IV, and by a two-phase mosaic (mosaic of this vegetation unit and the perennial grass steppe of *P. napostaense* and *N. tenuis* with *P. patagonica* and isolated shrubs of *C. avellanedae* (VU2), characteristic of Terrace Level III (VU2–VU5, Fig. 1).

3.1.6 Vegetation of Terrace Level VI

The Terrace Level VI consists of shrub-grass steppes and shrubby steppes with perennial grasses. Three vegetation units colonize this terrace level: The shrub steppe of *C. avellanedae* and *C. microphylla* (VU9) also characteristic of Terrace Level IV (see Sect. 3.1.4), the shrub steppe of *C. avellanedae* and *C. erinacea* ssp *erinacea* (VU8), and the shrub steppe of *C. erinacea* ssp. *hystrix* and *C. avellanedae* with perennial grasses (VU13), characteristic of Terrace Level II.

Shrub steppe of C. avellanedae and C. erinacea ssp. erinacea (*VU8*, **Fig. 1**)

This vegetation unit is characteristic of terrace levels at the northeastern Península Valdés. The total plant cover is 60–80% with two shrubby and a grass layers. The tall shrubby layer (30–40% cover, 50–100 cm height) is dominated by *C. avellanedae* and *C. erinacea* ssp. *erinacea*, accompanied with *L. chilense, C. microphylla, P. globosum* and *S. johnstonii*. The dwarf shrub layer (5% cover, 5–15 cm height) is represented by *P. chilensis, H. trifoliata, B. darwinii* and *P. recurvata*. The grass stratum (15–30% cover, 5–20 cm height) is dominated by *N. tenuis, J. neaei, P. speciosa* and *P. patagonica*. Other grasses less represented in this layer are *P. humilis, S. barbatus, P. ligularis, Vulpia myuros f. megalura* and *P. napostaense*.

3.1.7 Vegetation of the Stabilized Aeolian Fields

The Stabilized aeolian fields are covered by two main vegetation units: the perennial grass steppe of *S. rigens* and *N. tenuis* (VU1) and the dwarf shrub steppe of *H. argentea* (VU6). These vegetation units are mostly spatially arranged in a two-phase mosaic.

Perennial grass steppe of S. rigens and N. tenuis (*VU1*, **Figs. 1** and **2f**)

The plant canopy covers between 70 and 80% and consists of a continuous perennial grass layer, 30 cm tall, covering 50–60% of the soil. This layer is dominated by *S. rigens* and *N. tenuis* with *P. napostaense, Panicum urvilleanum* and *Poa lanuginosa*. Scattered shrubs (*Lycium* spp., *Baccharis divaricata*) or perennial herbaceous plants (*P. chilensis*) are immersed in the perennial grass matrix covering between 10 and 30%.

Dwarf shrub steppe of H. argentea (*VU6*)

This vegetation unit is patchily arranged in a mosaic (V1–V6, Figs. 1 and 2g) with the perennial grass steppe of *S. rigens* and *N. tenuis* (VU1). The plant cover varies between 70 and 90% and is characterized by a conspicuous shrubby layer dominated by the small shrub *H. argentea* (65–85% cover, 50 cm tall). This species may be accompanied by other shrubs (*B. divaricata*) or by dwarf shrubs (*A. seriphioides*) covering not more than 5% of the soil. *S. rigens, S. barbatus* and *P. lanuginosa* are the most frequent components of the herbaceous layer (5–10% cover, 5–25 cm tall).

3.2 *Vegetation of System B: Great Endorheic Basins*

3.2.1 Vegetation of the Piedmont Pediment and Bajada

Vegetation of Piedmont pediments and bajadas is mostly represented by the shrub steppe of *C. erinacea* ssp. *hystrix* and *C. avellanedae* with perennial grasses

(VU13), characteristic of the Terrace Level II (Fig. 2e) and small patches of the shrub steppe of *C. erinacea* ssp. *hystrix*, *C. genistoides* and *C. avellanedae* with perennial grasses (VU16).

Shrub steppe of *C. erinacea* ssp. *hystrix*, *C. genistoides*, and *C. avellanedae* with perennial grasses (*VU16*, Figs. 1 and 2h)

The total plant cover is 40–60% and consists of a shrubby stratum (30–40% cover, 60–100 cm height) dominated by *C. avellanedae* and *Mulinum spinosum* along with *C. erinacea* ssp. *hystrix* and *C. genistoides*. The dwarf shrub layer (5% cover, 5–10 cm tall) is represented by *A. seriphioides* and *P. recurvata*. The grass layer (5–10 cm tall) covering 5% of the soil is formed by the perennials *P. humilis*, *J. neaei*, *N. tenuis*, and *P. speciosa* accompanied by the annual *S. barbatus*.

3.3 Vegetation of the System C: Coastal Zone

3.3.1 Vegetation of the Coastal Piedmont Pediment

The vegetation of the Coastal piedmont pediment is highly heterogeneous consisting of shrub-perennial grass steppes of *C. erinacea* ssp. *erinacea* and *N. tenuis* (VU4) and shrub steppes with highly variable species composition: Shrub steppe of *C. avellanedae*, *Larrea divaricata* and *Nassauvia fuegiana* (VU14), shrub steppe of *Senecio filaginoides* and *M. spinosum* (VU12), Shrub steppe of *C. genistoides*, *C. avellanedae*, and *Atriplex lampa* (VU15), Shrub steppe of *C. avellanedae* and *M. spinosum* (VU11), and Shrub steppe of *L. divaricata*, *C. avellanedae* and *P. globosum* (VU18), the shrub steppe of *C. avellanedae* and *C. microphylla* (VU9), characteristic of Terrace Level VI, the shrubby steppe of *C. microphylla* and *L. chilense* (VU7), and the shrub steppe of *C. erinacea* ssp. *hystrix* and *C. avellanedae* with perennial grasses (VU13), characteristic of the Terrace Level II. Also, small patches of the perennial grass steppe of *P. napostaense*, and *N. tenuis* with *P. patagonica* and isolated shrubs of *C. avellanedae* (VU2) cover this landform.

Shrub-perennial grass steppe of *C. erinacea* ssp. *erinacea* and *N. tenuis* (*VU4*, Fig. 1)

This vegetation unit covers between 40 and 70% of the soil on narrow terrace flanks at northeastern Península Valdés. Plant canopy consists of three layers dominated by medium shrubs, perennial grasses, and dwarf shrubs, respectively. The medium shrub layer (40–70% cover, 80 cm height) is dominated *C. erinacea* along with *S. johnstonii*. The perennial grass layer (10–30% cover, 10 cm height) is dominated by *N. tenuis*, *N. longiglumis* and *P. napostaense* along with *B. catharticus*, *P. ligularis* and the annual herb *P. patagonica*. The dwarf shrub layer (5% cover, 10 cm tall) is composed by *A. seriphioides*, *B. darwinii* and *B. melanopotamica* accompanied by the perennial herbs *H. trifoliata* and *B. anthemoides*.

Shrub steppe of *C. avellanedae, L. divaricata* and *N. fuegiana* (**VU14**, **Figs.** 1 and 2i)

This vegetation unit is characteristic of the Golfo San José (San José Gulf) coast and appears patchily arranged in a two-phase mosaic (VU2–VU14, Fig. 1) with the perennial grass steppe of *P. napostaense* and *N. tenuis* with *P. patagonica* and isolated shrubs of *C. avellanedae* (VU2). The total plant cover is ca. 30% with two discontinuous shrubby layers and scattered grasses. The tallest shrubby layer (30% cover, 60 cm tall) is dominated by *C. avellanedae* along with *L. divaricata*, *C. microphylla* and *S. johnstonii*. The dwarf shrub stratum (4% cover, 20 cm height) is dominated by *N. fuegiana* and *B. darwinii*. *N. tenuis* and *S. barbatus* are the most common grasses covering ca. 1% of the soil.

Shrub steppe of *S. filaginoides* and *M. spinosum* (*VU12*, **Fig.** 1)

This vegetation unit occupies coastal piedmont pediments with aeolian deposits in the Golfo Nuevo. The total plant cover is 50% distributed in three layers. The shrubby layer (35% cover, 70–110 cm tall) is dominated by the medium shrubs *M. spinosum* and *S. filaginoides*, and the tall shrubs *L. chilense* and *S. johnstonii*. The dwarf shrub layer (10% cover, 30 cm tall) is formed by *B. darwinii* and *B. divaricata*. The lowest stratum (5% cover, 30 cm tall) is composed mostly of perennial grasses. *S. rigens* and *P. lanuginosa* dominate this layer along with *J. neaei*, *P. humilis*, and the annual grass *S. barbatus*.

Shrub steppe of *C. genistoides, C. avellanedae,* and *A. lampa* (*VU15*, **Figs.** 1 and 2j)

This vegetation unit occupies coastal patches at the Golfo San José and Golfo Nuevo. The total plant cover is 50–70% with three plant layers. The shrubby layer (30–50% cover, 60–100 cm tall) is dominated by the medium shrub *C. avellanedae* and the tall shrub *C. genistoides*. The dwarf shrub layer (10–20% cover, 5–10 cm tall) consists of *A. seriphioides* and *B. darwinii* along with *H. trifoliata* and *Gutierrezia solbrigii*. Finally, the herbaceous layer (10–20% cover, 10 cm tall) is dominated by *P. speciosa, P. humilis* and *J. neaei*.

Shrub steppe of *C. avellanedae,* and *M. spinosum* (*VU11*, **Fig.** 1)

This vegetation unit covers about 50% of the soil at coastal areas of the San José Gulf. This unit presents three plant layers (shrub, dwarf shrub, and herbaceous layers). The shrub layer (30–35% cover, 70–120 cm tall) is dominated by *C. avellanedae, M. spinosum,* along with *L. chilense* and *C. microphylla*. The dwarf shrub stratum (10–15% cover, 10 cm tall) is formed by *G. solbrigii* and *A. seriphioides*. The herbaceous layer (10–20% cover, 10 cm tall) consists of perennial grasses *N. tenuis, P. humilis, P. speciosa,* along with the annual grasses *B. catharticus* and *S. barbatus*.

Shrub steppe of L. divaricata, C. avellanedae, and P. globosum (*VU18*, **Fig.** 1)

This vegetation unit covers about 50–60% of the soil at coastal areas of the Golfo Nuevo. This unit consists of three plant layers (shrub, dwarf shrub, and herbaceous layers). The shrub layer (50–60% cover, 50–120 cm tall) is dominated by *L. divaricata*, *C. avellanedae*, *P. globosum* and *B. megalanthus* along with *C. microphylla*, and *Junellia* spp. The dwarf shrub stratum (10–15% cover, 5–10 cm tall) is formed by *B. darwinii*, *G. solbrigii*, *A. seriphioides* and *H. trifoliata*. The herbaceous layer (ca. 5% cover, 5–10 cm tall) consists of perennial grasses: *N. tenuis*, *P. humilis* and *J. neaei*.

Shrub steppe of C. microphylla and Lycium **spp**. (*VU7*, **Fig.** 1)

This vegetation unit covers about 60% of the soil in the east coast of Península Valdés. The plant canopy consists of two layers (shrubby and herbaceous layers). The shrubby layer (30% cover, 60–120 cm height) is dominated *C. microphylla*, *L. chilense*, and *C. avellanedae* accompanied by *Lycium gilliesianum* and *S. johnstonii*. The herbaceous layer (30% cover, 20 cm height) is dominated by *N. tenuis*, *J. neaei*, *P. humilis*, *P. napostaense*, *P. patagonica*, *N. longiglumis* and *S. barbatus*.

3.3.2 Vegetation of the Beach Ridges I–IV

Perennial grass steppe of N. tenuis and N. longiglumis with shrubs of C. avellanedae (*VU3*, **Fig.** 1)

This vegetation unit covers about 85% of the soil at flat coast plains in Caleta Valdés and consists of three plant layers (perennial grass, dwarf shrub, and medium shrub layers). The perennial grass layer (40% cover, 20 cm tall) dominated by *N. tenuis*, *N. longiglumis*, *P. napostaense* with, *B. catharticus*, *P. ligularis* and the annual herb *P. patagonica*. The dwarf shrub layer (5% cover, 10 cm tall) consists of scattered dwarf shrubs (*B. melanopotamica*) accompanied by perennial herbs (*P. chilensis* and *H. trifoliata*). The medium shrub layer (40% cover, 30 cm tall) is dominated *C. avellanedae* along with *Lycium tenuispinosum*.

3.3.3 Vegetation of the Beach Ridges V

Shrub steppe of S. johnstonii and L. chilense (*VU10*, **Fig.** 1)

This vegetation unit is characteristic of a narrow coastal plain at Caleta Valdés covering about 40% of the soil with two strata (shrubby and herbaceous) equally represented. The shrubby layer (20% cover, 50–100 cm height) is dominated by *S. johnstonii* and *L. chilense*. The herbaceous stratum (20% cover, 5–20 cm height) is dominated by *N. tenuis*, *J. neaei*, *P. speciosa*, *P. humilis* along with *P. ligularis* and *P. patagonica*.

4 Priority Sites for Conservation

Within the Península Valdés four priority sites for conservation (resolution scale 1:1) may be identified: Salt marshes, Uplands and Plains, The Great Endorheic Basin and Stabilized Aeolian field and Active Dunefields. Among them salt marshes deserve particular attention since they have unique and important ecological functions such as production and transport of nutrients and organic matter, and constitute specific areas for feeding, sheltering and/or nesting of a large number of marine and terrestrial organisms. In addition, inland sites of Península such as Uplands, Endorheic Basins, and Central rangelands deserve also attention since they are mostly submitted to disturbance processes triggered by grazing and water and wind erosion affecting not only plant canopy structure but also soils and ecosystem processes and services (see Chapter "Soil Degradation in Peninsula Valdes: Causes, Factors, Processes, and Assessment Methods").

4.1 Salt Marshes

The *salt marshes* are intertidal environments developed, in general, in estuarine or marine coasts, in which the slow movement of tidal water favours the accumulation of fine sediments (Fig. 1). They are colonized by halophytic herbs, grasses or low shrubs. While the salt marshes remain exposed to air most of the day, they are subject to periodic flooding product of fluctuations in the level of the adjacent water bodies, so these halophytes are tolerant both to immersion and hypersalinity (Adam 1990; Bortolus 2010). Because of their hydrological conditions and their location between the marine and terrestrial environments (see Chapter "Groundwater Resources of Península Valdés"), the salt marshes have unique features, and they are inhabited by both marine and terrestrial organisms (Mitsch and Gosselink 2000). However, they should not be considered *ecotones* (i.e. transition zone between two different ecosystems), because although they are composed of species from the surrounding environments (land or sea; Bortolus 2010) the salt marshes present an assemblage of species that characterizes and defines them.

4.1.1 Vegetation Communities of Salt Marshes

Plant community zonation is probably the most conspicuous feature characterizing salt marshes at the landscape scale (Adam 1990). Most salt marsh communities plants zonate in bands parallel to the coastline with specific species compositions (Fig. 3); usually changing with the relative elevation, the distance from the seashore, and the global position (Adam 1990). Few terrestrial plant species are able to survive in the low marsh, where tidal amplitudes and inundation frequency are high. However, as substrate elevation increases, tidal amplitude decreases and inundation becomes less frequent. The highest marsh levels are commonly characterized by more complex plant communities with a large variety of ecological interactions going on (Adam 1990). Along the Atlantic coast of South America, the salt marshes show a particular geographic pattern (Bortolus et al. 2009; Idaszkin and Bortolus

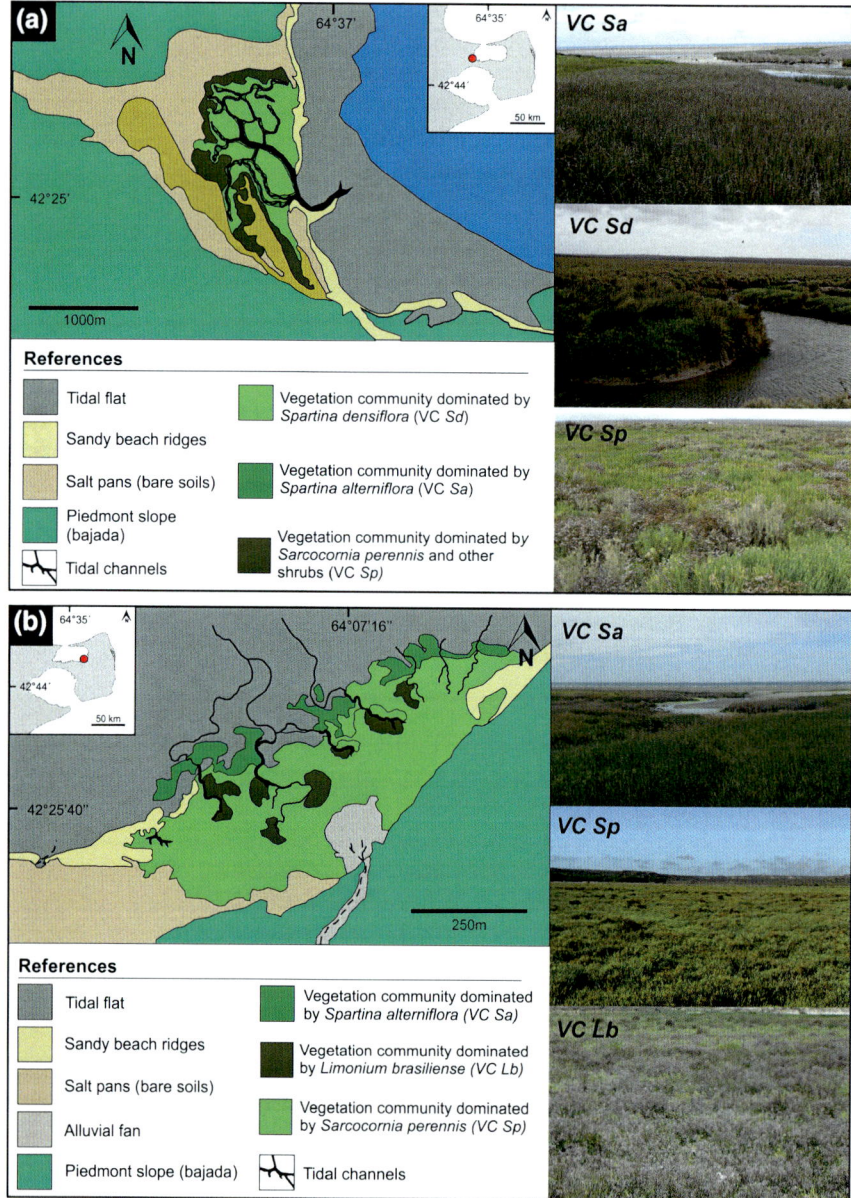

Fig. 3 Location and main vegetation communities of Riacho (**a**) and Fracasso (**b**) salt marshes

2011; Idaszkin et al. 2011). The northern salt marshes (i.e. located at the parallel 42° S or lower northern) are characterized by vegetation communities that dominated by the perennial cordgrasses *Spartina alterniflora* and *Spartina densiflora* (*Spartina*-marshes), while in the southern salt marshes (at latitudes greater than 43° S) plant communities are dominated by the succulent shrub *Sarcocornia perennis* with a rare or nil presence of *Spartina* species across the intertidal frame (*Sarcocornia*-salt marshes). In the Península Valdés region, between parallels 42° S and 43° S, these two types of salt marsh vegetation community overlap their geographic distribution. A feature shared by both vegetation communities is that *S. alterniflora*, occupies the lowest marsh level, and *S. perennis* and *S. densiflora* the highest one. While there are several salt marshes in Península Valdés, Riacho (42° 25' S, 64° 37' W; Fig. 3a) and Fracasso (42° 25' S, 64° 07' W; Fig. 3b) are the greatest salt marshes from the Península Valdés (Bortolus et al. 2009; Idaszkin et al. 2011).

In particular, Riacho is a *Spartina* marsh community, where *S. alterniflora* inhabits the low marsh and decreases its abundance towards higher marsh levels. These higher levels are commonly dominated by *S. densiflora*, accompanied by the shrubs *Limonium brasiliense*, *S. perennis* and *Atriplex vulgatissima* (Fig. 3a). On the other hand, the Fracasso salt marsh community is a *Sarcocornia*-marsh, where the presence of *S. alterniflora* has being increasing in the last decade in the low marsh. The high marsh is dominated by *S. perennis*, accompanied with *S. densiflora*, *Suaeda* spp. and *L. brasiliense* forming isolated patches in the more elevated spots (Fig. 3b).

4.1.2 Conservation Concerns

Salt marshes are widely recognized by the unique, and important ecological functions they provide, such as the high primary productivity on and the transport of nutrients and organic matter from and to the sea (Mitsch and Gosselink 2000). Currently, they are considered as one of the most productive environments in the world, and this high production is often essential for sustaining estuarine and coastal food chains (Weinstein and Kreeger 2000). Furthermore, salt marshes are critical to the maintenance of the regional integrity for both terrestrial and marine communities, because they constitute specific areas for feeding, sheltering and/or nesting of a large number of marine and terrestrial organisms (Mitsch and Gosselink 2000; Weinstein and Kreeger 2000; Adam 2002). Migratory and endemic birds, fish, mammals and invertebrates of great ecological and economic importance (e.g. mussels, clams and snails) are examples of organisms that depend on the existence and integrity of the salt marshes to ensure their survival (Bortolus 2006; Adam 2002). Both Fracasso and Riacho salt marshes are included in the RAMSAR site of Península Valdés wetlands.

4.2 Uplands and Plains

The Península Valdés region, as other areas of the extra-andean Patagonia, has been grazed by sheep (*Ovis aries*) since the beginning of the last century (Ares et al. 2003). There is evidence that sheep grazing in Patagonia have had negative effects on plant communities like the reduction in total plant cover, alteration of the spatial

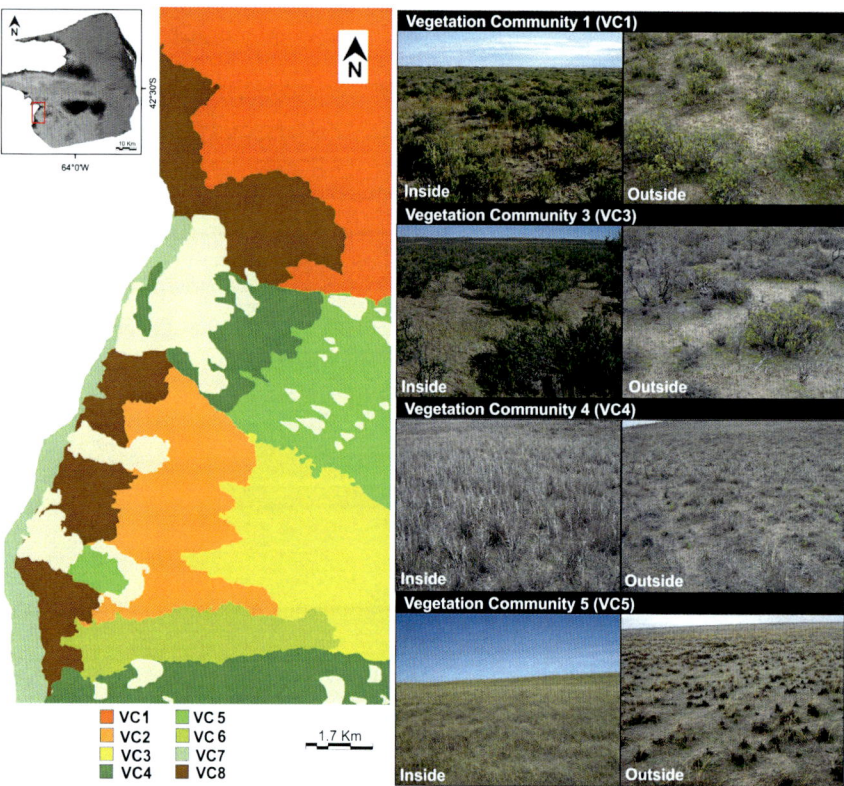

Fig. 4 Location and main vegetation communities of San Pablo de Valdés. VC1. Medium shrub steppe of *Chuquiraga avellanedae*, *Schinus johnstonii*, *Lycium ameghinoi*, *Menodora robusta* and *Acantholippia seriphioides;* VC2. Shrub-grass steppe of *Chuquiraga avellanedae*, *Nassella tenuis* and *Piptochaetium napostaense;* VC3. Tall shrub-grass steppe of *Chuquiraga erinacea* ssp. *hystrix*, *Chuquiraga avellanedae* and *Acantholippia seriphioides* with *Nassella tenuis*, *Piptochetium napostaense* and *Pappostipa speciosa;* VC4. Dwarf shrub steppe of *Hyalis argentea;* VC5. Grass steppe of *Sporobolus rigens* and *Nassella tenuis;* VC6. Grass-shrub steppe of *Sporobolus rigens*, *Nassella tenuis* and *Piptochaetium napostaense* with sparse shrub patches of *Chuquiraga erinacea* ssp. *hystrix*, *Acantholippia seriphioides*, *Mulinum spinosum* and *Baccharis divaricata;* VC7. Dwarf shrub steppe of *Hyalis argentea*, *Sporobolus rigens* and *Baccharis divaricata;* VC8. Mosaic of shrub, shrub-grass, and grass steppes. White areas are Active Dune Fields without vegetation. Photographs illustrate vegetation communities VC1, VC3, VC4 and VC5 inside San Pablo (*left*) and in adjacent sheep ranches (*right*)

structure of vegetation, soil degradation, and the reduction in the size of soil seed banks of herbivore-preferred species (Laycock 1995; Bertiller et al. 2002; Cipriotti and Aguiar 2005; Pazos and Bertiller 2008; Chartier et al. 2011). Despite the relative scarcity of specific studies, some of these impacts were also observed in the Península Valdés (Elissalde and Miravalles 1983; Blanco et al. 2008; Cheli 2009; Burgi et al. 2012). This raises important conservation concerns for these ecosystems considering the UNESCO World Heritage Site status of the area (Nabte et al. 2013).

A very important action taken for the conservation of representative terrestrial plant communities of the Península Valdés was the creation of strict reserves. This is the case of San Pablo de Valdés (henceforth San Pablo), a typical ranch dedicated to wool production that was converted into a wildlife reserve in 2005 by the local NGO "Fundación Vida Silvestre Argentina" (Fig. 4). This reserve added 7360 ha to the scarce 5% of protected lands of the arid Patagonia (Nabte et al. 2013). The immediate management actions taken were the removal of all domestic herbivores (ca. 3500 sheep), internal fences, and all structures related to grazing management.

4.2.1 Vegetation Communities of Uplands and Plains

San Pablo encloses a unique mosaic of plant communities representing the most extended vegetation units of Terrace Levels and Stabilized Aeolian field deposits in southern Península Valdés (Fig. 4) (Codesido et al. 2005). The main vegetation communities are

Medium shrub steppe dominated by the shrubs C. avellanedae, S. johnstonii, Lycium ameghinoi, Menodora robusta, and A. seriphioides (**VC1, Fig. 4**)

This community is established on Terrace Levels III. The total canopy cover ranges from 40 to 60%. The inconspicuous herbaceous layer is dominated by *N. tenuis*, *P. ligularis* and *P. lanuginosa*.

Shrub-grass steppe of C. avellanedae, N. tenuis and P. napostaense (**VC2, Fig. 4**)

This community is established on Stabilized Aeolian field deposits. The total canopy cover ranges from 45 to 60%. *C. erinacea* ssp. *hystrix* codominates the shrub layer.

Tall shrub-grass steppe of C. erinacea ssp. hystrix, C. avellanedae, and A. seriphioides with N. tenuis, P. napostaense and P. speciosa (**VC3, Fig. 4**)

This community is established on Terrace Levels and Stabilized Aeolian field deposits. The total plant cover ranges from 50 to 60%.

Dwarf shrub steppe of H. argentea (**VC4, Fig. 4**)

This community is established on undulated Active Dune Fields and Stabilized Aeolian field deposits. The total plant cover ranges from 80 to 90%, mostly represented by *H. argentea* with an incipient herbaceous layer dominated by *P. lanuginosa* and *N. tenuis*.

Grass steppe of S. rigens and N. tenuis (**VC5, Fig. 4**)

This community is established on undulated dune fields and Stabilized Aeolian field deposits. The total canopy cover ranges from 70 to 90%. *P. lanuginosa* and *P. urvilleanum* codominate the herbaceous layer. Also sparse shrub patches of *B. divaricata* are immersed in the grass matrix.

Grass-shrub steppe of S. rigens, N. tenuis and P. napostaense with sparse shrub patches of C. erinacea ssp. hystrix, A. seriphioides, M. spinosum and B. divaricata (**VC6, Fig. 4**)

This community is established on undulated dune fields and Stabilized Aeolian field deposits forming small patch mosaics with VC2 and VC5 and represents a degraded state of the grass steppe VC2.

Dwarf shrub steppe of H. argentea, S. rigens and B. divaricata (**VC7, Fig. 4**)

This community is established on coastal undulated Active Dune Fields and Stabilized Aeolian field deposits. The total plant cover ranges from 80 to 90% with presence of sparse blowouts.

Mosaic of shrub, shrub-grass and grass steppes (**VC8, Fig. 4**)

This plant community occupies heterogeneous landscapes associated with coastal Stabilized Aeolian field deposits and Small–Medium Closed Basins. It consists of a heterogeneous mosaic constituted by the plant communities described above. The conserved (San Pablo) and degraded (adjacent ranches) vegetation states of four of the above-mentioned plant communities are presented in Fig. 4. Additionally, Active Dune Fields without vegetation crossing eastwardly are distinctive components of the San Pablo landscape.

4.2.2 Conservation Concerns

San Pablo constitutes a unique opportunity in Península Valdés to describe the trajectory of plant communities under both livestock exclusion and grazing by native herbivores. A first vegetation survey carried out at the time of the reserve creation indicated that all these communities showed clear signs of degradation by livestock grazing (Codesido et al. 2005). Burgi et al. (2012) compared the structure and composition of plant communities VC1, VC3, VC4 and VC5 between San Pablo and adjacent ranches with sheep grazing production finding higher total plant cover, higher perennial-grass cover, and higher diversity of perennial grasses in San Pablo than in the adjacent grazed ranches (Fig. 4). Remarkably, all these changes were simultaneous with a steady increase of the guanaco density inside San Pablo (Marino et al. 2016). These results provide evidence on the need to conserve and protect terrestrial plant communities at relevant spatial scales serving as a basis for integrated management plans oriented to achieve ecological and economic sustainability under the long-standing scenarios of land degradation in Patagonia (Nabte et al. 2013; Marino et al. 2016).

4.3 The Great Endorheic Basin

A pilot area of ca. 400 km^2 in the centre of Península Valdés (see site location in Fig. 5) was selected where soil erosion by water is prevalent (Fig. 5). The geomorphology of the area comprises both the Uplands and Plains as the Great Endorheic Basin Systems. It is conspicuous that the presence of accelerated erosion indicators such as pedestals, rills and bare soil on the Terrace Levels and gullies on slopes of the piedmont pediments and bajadas. On the east part, there is a large dried playa lake called Gran Salitral and central to the area there is a burnt area due to a fire occurred in February 2004. Extensive, continuous sheep grazing for wool production is the main land use of these rangelands. Six vegetation communities are dominant in the area.

Fig. 5 Location and main vegetation communities of the Central Rangelands: VC1. Shrub-grass steppe of *Chuquiraga avellanedae*, *Condalia microphylla*, and *Nassella tenuis*, VC2. Shrub steppe of *Chuquiraga avellanedae* with desert pavement, VC3. Shrub-grass steppe of *Chuquiraga avellanedae*, *C. hystrix* and *Jarava* and *Pappostipa* species, VC4. Shrub-grass steppe of *Chuquiraga avellanedae*, *C. hystrix*, *Prosopidastrum globosum* and *Pappostipa* and *Jarava* species, VC5. Sandy grassland of *Nassella tenuis* and *Piptochaetium napostaense*, VC6. Grassland of *Nassella tenuis* and *Piptochaetium napostaense* with shrubs of *Chuquiraga avellanedae*

4.3.1 Vegetation Communities of the Great Endorheic Basin

Shrub-grass steppe of C. avellanedae, C. microphylla, and N. tenuis (VC1, Fig. 5)

This community occurs along with alluvial bajadas dominated by Calciargid soils (see Chapter "Soil–Geomorphology Relationships and Pedogenic Processes in Península Valdés"). Vegetation cover is about 60–80%, where *C. avellanedae* and *C. microphylla* accounts for up to 50–60% of plant cover and occur along with *B. megalanthus, L. chilense* and *P. globosum*. Dominant grasses, *N. tenuis, J. neaei, P. speciosa and P. napostaense*, cover around 10–20% of the soil.

Shrub steppe of C. avellanedae with desert pavement (VC2, Fig. 5)

This community forms extensive shrublands on Terrace Levels. It is dominated by the shrub species *C. avellanedae*, surrounded by *desert pavements*. Dominant soils are Natrargids (see Chapter "Soil–Geomorphology Relationships and Pedogenic Processes in Península Valdés"). Vegetation cover ranges from 30 to over 40%; shrub cover is around 25–35%, while grasses cover 5–10%. *S. johnstonii* and *P. globosum* are secondary shrub species in these communities. Grasses such as *N. tenuis, P. napostaense*, and less frequently, *P. humilis* and *P. ligularis may* occur scattered in the bare soil matrix.

Shrub-grass steppe of C. avellanedae, C. erinacea ssp. hystrix and Jarava and Pappostipa species (VC3, Fig. 5)

This community occurs along the Piedmont pediments, characterized by Torriorthents soils and the presence of gullies and erosion escarpments of sandstone of the Puerto Madryn Formation (see Chapter "Geology of Península Valdés"). Vegetation cover ranges from 60 to over 70%; shrub cover is around 45–55%, while grasses cover 10–20%. The most conspicuous shrubs are *C. avellanedae* and *C. erinacea* ssp. *hystrix*, which occur with *L. chilense and S. johnstonii*. Grasses such as *N. tenuis, J. neaei, P. speciosa* accompanied by, less frequently, *P. humilis* and *N. longiglumis*, make up the herbaceous stratum.

Shrub-grass steppe of C. avellanedae, C. erinacea ssp. hystrix, P. globosum and Pappostipa and Jarava species (VC4, Fig. 5)

This community occurs along the Terrace Levels, dominated by Haplocalcid soils and with a vegetation cover from 65 to over 85%. Shrub vegetation cover is about 50–60%, where *C. avellanedae, Chuquiraga hystrix*, and *P. globosum* account for up to 50% of plant cover and occurs along with grasses (20–30%) such as *N. tenuis, J. neaei*, and *P. speciosa* accompanied by, less frequently, *P. ligularis*.

Sandy grassland of N. tenuis and P. napostaense (VC5, Fig. 5).

This community forms extensive grasslands on thin sandsheets on Terrace Levels with Haplocalcid soils. Vegetation cover ranges from 40 to over 50%; grasses (20–30%) such as *N. tenuis, P. napostaense, P. speciosa, and N. longiglumis* make up the herbaceous stratum. Shrub cover is around 15–25%. The most conspicuous shrub is *C. avellanedae*, which occurs with *P. globosum* and other shrubs such as *S. johnstonii* and *L. chilense*.

Grassland of *N. tenuis* and *P. napostaense* with shrubs of *C. avellanedae* (*VC6*, Fig. 5)

This community principally occurs on sandy sediment that filled the Holocene stream valleys that dissected the Coastal Piedmont Pediments with Haplocalcid soils. Vegetation cover is about 40–60%, where *N. tenuis* and *P. napostaense* accounts for up to 40–50% of grass cover and occur along with *P. speciosa* and *Nasella longiglumis*. Some shrubs (5–10%) *of C. avellanedae* and less frequently, *P. globosum* are present.

4.4 Vegetation Communities of Stabilized Aeolian Field and Active Dunefields

A pilot area is located in grazed lands in the southern portion of the Península Valdés dunefields (see site location in Fig. 6). The source of sediment for the aeolian landforms are the western sandy beaches of Golfo Nuevo where a continued supply of loose, sand-sized sediment is available to be transported inland by the prevailing

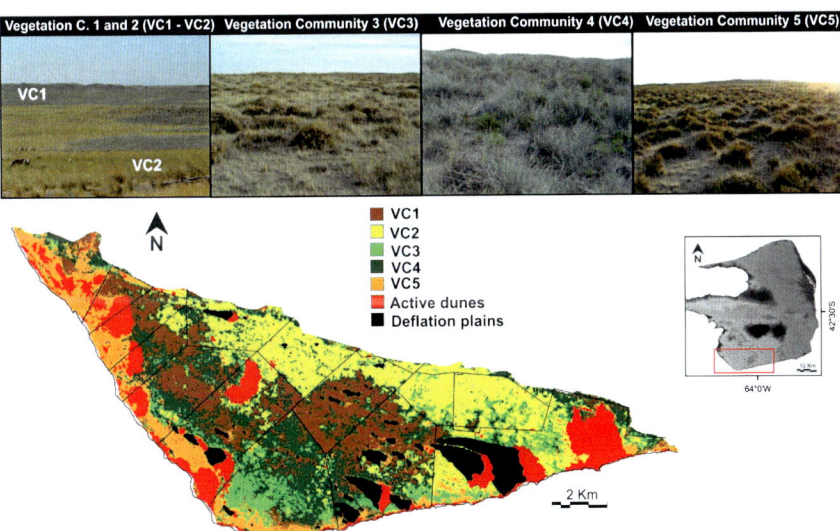

Fig. 6 Location and main vegetation communities of Dunefields: VC1. Grassland of *Sporobolus rigens* and *Aristida spegazzinii*, VC2. Subshrub steppe of *Hyalis argentea*, VC3. Shrub-grass steppe of *Brachyclados megalanthus*, *Mulinum spinosum* and *Nassella tenuis*, VC4. Grass-shrub steppe of *Nassella tenuis* and *Mulinum spinosum*. VC5. Shrub-grass steppe of *Mulinum spinosum* and *Hyalis argentea*

westerly winds (see Chapter "Late Cenozoic Landforms and Landscape Evolution of Península Valdés"). General features in the topography of dunefields are Stabilized aeolian field (relict aeolian landforms) and mega-patches of Active sand dunes with deflation plains. Relict aeolian landforms would include sand sheets and longitudinal dunes, which nowadays are mostly stabilized by psammophile plant species. The presence of blowouts on these areas is evidence of current erosive processes in the dunefield. Five vegetation communities dominate the area.

4.4.1 Vegetation Communities of Stabilized Aeolian Field and Active Dunefields

Dwarf shrub steppe of H. argentea (VC1, Fig. 6)

Vegetation cover is about 90%; the dominant plant is *H. argentea* forming big dense patches. Other grass species associated are *P. urvilleanum*, *P. lanuginosa* and *S. rigens*.

Grassland of S. rigens and Aristida spegazzinii (VC2, Fig. 6)

This grassland has around 75% of vegetation cover. *S. rigens* and *A. spegazzinii* account for up to 60–70% of cover and occur along with *P. urvilleanum*, *P. lanuginosa*, *N. tenuis* and *P. napostaense*. *Maihueniopsis darwinii* and *Marrubium vulgare* occurs on degraded areas.

Shrub-grass steppe of B. megalanthus, M. spinosum and N. tenuis (VC3, Fig. 6)

This community is dominated by the shrub species *B. megalanthus* and *M. spinosum*, with *N. tenuis* (perennial grass) dominating the herbaceous stratum. The vegetation cover is about 65%.

Shrub-grass steppe of M. spinosum and H. argentea (VC4, Fig. 6)

This community is dominated by the subshrubs species *M. spinosum* and *H. argentea*, with herbaceous secondary species such as *S. rigens*, *P. urvilleanum*, *P. lanuginosa* and *N. tenuis*. Some clumps of *B. divaricata* are found. Vegetation cover in this community is about 65%.

Grass-shrub steppe of N. tenuis and M. spinosum (VC5, Fig. 6)

This community (with around 70% plant cover) is dominated by the grass species *N. tenuis* and the shrub *M. spinosum*. Grasses such as *S. rigens*, *P. urvilleanum* and *P. lanuginosa* are codominant species.

Vegetation of Península Valdés: Priority Sites for Conservation

5 Perspectives and Future Work

This chapter provides a synthesis of the state of knowledge on the vegetation of Península Valdés and identifies priority sites for conservation programs. Future work should be aimed to intensify studies on the vegetation dynamics of these priority sites and also to recognize new sites which could be sensitive to degradation due to human activities.

Acknowledgements We are grateful to the Project of Ecology and Regional Development of Arid and Semiarid Regions (OEA-CONICET-INTA) and to the Centro Nacional Patagónico (CENPAT-CONICET) for supporting field research for the description of the Vegetation Units and to the Comisión Nacional de Actividades Espaciales (CONAE) from Argentina that supplied the satellite images. We also thank Fundación Vida Silvestre Argentina and landowners for granting the access to San Pablo de Valdés and the adjacent ranches, respectively. GEP acknowledges Victoria Rodríguez and Andrea Marino for important input in the San Pablo de Valdés work. GEP was partially supported by PIP 11220120100369CO. Research permissions were granted by Secretaría de Turismo y Áreas Naturales Protegidas and Dirección de Fauna y Flora Silvestre de Chubut. We also thank G. Bernardelo and Marta Collantes for their helpful comments in the revision of this chapter.

Glossary

Adaptation	The process of adjustment of an individual organism to environmental stress
Ecotone	Transitional zone between adjacent plant communities or biomes
Floristic composition	A list of plant species of a given area, habitat, or association
Floristic element	In phytogeography, a convenient term for any group of plants sharing a common feature of importance
Halophytic	Plant tolerating saline conditions
Life form	The characteristic structural traits of a plant species
Patchy	Contagious distribution
Perennial	Plants that persist for several years with a growth period each year
Physiognomic	Appearance of a plant community or vegetation
Phytogeographical Provinces	Geographical divisions characterized by floristic composition
Stratum	Horizontal layer of vegetation

Appendix: Floristic List

Family	Species and authors
Ephedraceae	*Ephedra ochreata* Miers
Juncaginaceae	*Triglochin concinna* Burtt Davy
Poaceae	*Amelichloa ambigua* (Speg.) Arriaga and Barkworth
	Aristida spegazzinii Arechav.
	Avena sativa L.
	Bromus catharticus Vahl
	Bromus unioloides Humboldt, Bonpland et Kunth
	Distichlis scoparia (Kunth) Arechav.
	Distichlis spicata L.
	Eremium erianthum (Phil.) Seberg and Lindle-Laursen
	Hordeum comosum J. Presl
	Hordeum euclaston Steud.
	Hordeum murinum L.
	Jarava neaei (*Nees ex Steud.*) Peñalillo
	Koeleria mendocinensis (Hauman) C.E. Calderón and Nicora
	Nassella longiglumis (Phil.) Barkworth
	Nassella tenuis (Phil.) Barkworth
	Panicum urvilleanum Kunth
	Pappostipa chysophylla (E. Desv.) Romasch.
	Pappostipa humilis (Cav.) Romasch.
	Pappostipa speciosa (Trin. and Rupr.) Romasch.
	Piptochaetium napostaense (Speg.) Hack.
	Poa lanuginosa Poir.
	Poa ligularis Nees ex Steud.
	Polipogon monspeliensis (L.) Desf.
	Schismus barbatus (L.) Thell.
	Spartina alterniflora Loisel
	Spartina densiflora Brongn.
	Sporobolus rigens (Trin.) E. Desv.
	Vulpia myuros (L.) C.C. Gmel. *megalura* Phil.
Amaryllidaceae	*Rhodophiala mendocina* (Phil.) Ravenna
Oleaceae	*Menodora robusta* (Benth.) A. Gray
Schoepfiaceae	*Arjona tuberosa* Cav.
Polygonaceae	*Polygonum brasiliense* K. Koch
Chenopodiaceae	*Atriplex lampa* (Moq.) D. Dietrich
	Atriplex sagittifolia Speg.
	Dysphania ambrosioides (L.) Mosyakin and Clements
	Sarcocornia perennis (Mill.) A.J. Scott.
	Suaeda argentinensis A. Soriano

(continued)

(continued)

Family	Species and authors
	Suaeda divaricata Moq.
Nictaginaceae	*Bougainvillea spinosa* (Cav.) Heimerl
Aizoaceae	*Mesembryanthemum crystallinum* L.
Caryophyllaceae	*Cardionema ramosissima* (Weinm.) A. Nelson and J.F. Macbr.
	Cerastium arvense L.
	Cerastium glomeratum Thuill.
	Cerastium junceum Möschl
	Heniaria cinerea DC.
	Paronychia chilensis DC.
Capparaceae	*Capparis atamisquea* Kuntze
Rosaceae	*Tetraglochin caespitosum* Phil.
	Tetraglochin ameghinoi (Speg.) Speg.
Fabaceae	*Adesmia candida* Hook. f.
	Adesmia af. acuta Burkart
	Anarthrophyllum rigidum (Gillies ex Hook. and Arn.) Hieron.
	Hoffmannseggia trifoliata Cav.
	Prosopidastrum globosum (Gillies ex Hook. and Arn.) Burkart
	Prosopis alpataco Phil.
	Prosopis denudans Benth.
	Vicia pampicola Burkart *burkartii* Giangualani
Geraniaceae	*Erodium cicutarium* (L.) L'Hér. ex Aiton
Zygophyllaceae	*Larrea divaricata* Cav.
	Larrea nitida Cav.
Euphorbiaceae	*Euphorbia portulacoides* L.
	Euphorbia serpens Kunth
Anacardiaceae	*Schinus johnstonii* F.A. Barkley
Rhamnaceae	*Condalia microphylla* Cav.
	Discaria americana Gilles and Hook.
Malvaceae	*Malvela leprosa* (Ortega) Krapov.
Frankeniaceae	*Frankenia patagonica* Speg.
	Frankenia pulverulenta L.
Loasaceae	*Loasa bergii* Hieron.
Cactaceae	*Maihuenia patagonica* (Phil.) Britton and Rose
	Maihueniopsis darwinii (Hensl.) Ritter
Onagraceae	*Camissonia dentata* (Cav.) Reiche
	Oenothera versicolor Lehm
	Oenothera stricta Ledeb. ex Link *altissima* W. Dietr.
Apiaceae	*Bowlesia incana* Ruiz and Pav.
	Daucus pusillus Michx.
	Eryngium chubutense Neger ex Dusén
	Mulinum spinosum (Cav.) Pers.
Plumbaginaceae	*Limonium brasiliense* (Boiss.) Kuntze

(continued)

(continued)

Family	Species and authors
Apocyanceae	*Phillibertia candolleana* (Hook. and Arn.) Goyder
Convolvulaceae	*Dichondra microcalyx* (Haller f.) Fabris
Polemoniaceae	*Gilia crassifolia* Benth.
Boraginaceae	*Amsinckia calycina* (Moris) Chater
	Lappula redowskii (Hornem.) Greene
	Pectocarya linearis (Ruiz and Pav.) D.C.
Verbenaceae	*Acantholippia seriphioides* (A Gray) Moldenke
	Glandularia aurantiaca (Speg.)Botta *aurantiaca*
	Mulguraea ligustrina (Lag.) N. O'Leary and P. Peralta var. *lorentzii* (Niederl. ex Hieron.) N. O'Leary and P. Peralta
Lamiaceae	*Marrubium vulgare* L.
Solanaceae	*Lycium ameghinoi* Speg.
	Lycium chilense Miers ex Bertero
	Lycium gilliesianum Miers
	Lycium tenuispinosum Miers
Plantaginaceae	*Plantago myosuros* Lam.
	Plantago patagonica Jacq.
Rubiaceae	*Galium richardianum* (Gilles ex Hook. and Arn.) Endl.ex Walp
Calyceraceae	*Boopis anthemoides* Juss.
Asteraceae	*Baccharis crispa* Spreng.
	Baccharis darwinii Hook. et Arn.
	Baccharis divaricata Hauman
	Baccharis gilliesii A. Gray
	Baccharis melanopotamica Speg.
	Baccharis spartioides (Hook. et Arn. Ex DC.) J. Remy
	Baccharis tenella Hook. et Arn.
	Baccharis triangularis Hauman
	Brachyclados megalanthus Speg.
	Chuquiraga aurea Skottsb.
	Chuquiraga avellanedae Lorentz
	Chuquiraga erinacea D.Don ssp. *erinacea*
	Chuquiraga erinacea D.Don ssp. *hystrix*(Don) C. Ezcurra
	Cyclolepis genistoides D.Don
	Gamochaeta chamissonis (DC.) Cabrera
	Grindelia chiloensis (Cornel.) Cabrera
	Gutierrezia solbrigii Cabrera
	Hyalis argentea D. Don ex Hook. and Arn. var. *latisquama* Cabrera
	Hypochaeris radicata L.
	Hysterionica jasionoides Willd.

(continued)

(continued)

Family	Species and authors
	Nassauvia fuegiana (Speg.)Cabrera
	Nassauvia ulicina (Hook. f.) Macloskie
	Perezia recurvata (Vahl) Less.ssp. *recurvata*
	Noticastrum sericeum (Less.) Less. ex Phil.
	Senecio chrysocomoides Hook. et Arn.
	Senecio filaginoides DC.
	Sonchus asper (L.) Hill.

References

Adam P (1990) Saltmarsh ecology. Cambridge University Press, Cambridge
Adam P (2002) Saltmarshes in a time of change. Environ Conserv 29:39–61
Ares JO, Beeskow AM, Bertiller MB, Rostagno CM, Irisarri MP, Anchorena J, Defossé GE, Merino CA (1990) Structural and dynamic characteristics of overgrazed grasslands of northern Patagonia. In: Breymeyer A (ed) Managed grasslands. Regional studies. Elsevier, Amsterdam, pp 149–175
Ares JO, del Valle HF, Bisigato AJ (2003) Detection of process-related changes in plant patterns at extended spatial scales during early dryland desertification. Glob Change Biol 9:1643–1659
Barros V, Rivero M (1982) Mapas de probabilidad de precipitación de la Provincia del Chubut. Monografía 54. Centro Nacional Patagónico, Puerto Madryn, 12 p
Bertiller MB, Beeskow AM, Irisarri MP (1981) Caracteres fisonómicos y florísticos de la vegetación del Chubut. 2. La Península Valdés y el Istmo Ameghino. SECyT, CONICET. CENPAT. Contribución N° 41, 20p
Bertiller MB, Beeskow AM, Coronato F (1991) Seasonal environmental and plant phenology in arid Patagonia (Argentina). J Arid Environ 21:1–11
Bertiller MB, Ares JO, Bisigato AJ (2002) Multiscale indicators of land degradation in the Patagonian Monte, Argentina. Environ Manage 30:704–715
Bertiller MB et al (2006) Leaf strategies and soil N across a regional humidity gradient in Patagonia. Oecologia 148:612–624
Blanco PD et al (2008) Grazing impacts in vegetated dune fields: predictions from spatial pattern analysis. Rangeland Ecol Manage 61:194–203
Bortolus A (2006) The austral cordgrass *Spartina densiflora* Brong.: its taxonomy, biogeography and natural history. J Biogeogr 33:158–168
Bortolus A (2010) Marismas Patagónicas: las últimas de Sudamérica. Ciencia Hoy 19:10–15
Bortolus A et al (2009) A characterization of Patagonian salt marshes. Wetlands 29:772–780
Burgi MV et al (2012) Response of guanacos to changes in land management in Península Valdés, Argentine Patagonia. Conservation implications. Oryx 46:99–105
Cabrera AL (1976) Regiones Fitogeográficas Argentinas. In: Kugler WF (ed) Enciclopedia Argentina de Agricultura y Jardinería. Buenos Aires, pp 1–85
Campanella MV, Bertiller MB (2008) Plant phenology, leaf traits, and leaf litterfall of contrasting life forms in arid Patagonian Monte, Argentina. J Veg Sci 19:75–85
Castellanos A, Pérez Moreau R (1944) Los tipos de vegetación de la República Argentina. Monografías del Instituto de Estudios Geográficos de Tucumán 4:1–154

Chartier M, Rostagno CM, Pazos GE (2011) Effects of soil degradation on infiltration rates in grazed semiarid rangelands of northeastern Patagonia, Argentina. J Arid Environ 75:656–661

Cheli GH (2009) Efectos del disturbio por pastoreo ovino sobre la comunidad de artrópodos epígeos en Península Valdés (Chubut, Argentina). Universidad Nacional del Comahue, Bariloche

Cipriotti PA, Aguiar MR (2005) Effects of grazing on patch structure in a semi-arid two-phase vegetation mosaic. J Veg Sci 16:57–66

Codesido MA et al (2005) Relevamiento ambiental de la "Reserva de Vida Silvestre San Pablo de Valdés". Caracterización ecológica y evaluación de su condición como unidad de conservación y manejo. General technical report N-1. Programa "Refugios de Vida Silvestre", Sistema de Relevamientos Ecológicos Rápidos. Fundación Vida Silvestre Argentina, Buenos Aires, Argentina

Correa MN (1969) Flora Patagónica. Tomo VIII. Parte II. Typhaceae a Orchidaceae (excepto Gramineae). Colección Científica de INTA. Buenos Aires, 219 pp

Correa MN (1971) Flora Patagónica. Tomo VIII Parte VII. Compositae. Colección Científica de INTA. Buenos Aires, 451 pp

Correa MN (1978) Flora Patagónica. Tomo VIII. Parte III. Gramineae. Colección Científica de INTA. Buenos Aires, 563 pp

Correa MN (1984a) Flora Patagónica. Tomo VIII. Parte IV a. Dicotyledones Dialipétalas (Salicaceae a Cruciferae). Colección Científica de INTA. Buenos Aires, 556 p

Correa MN (1984b) Flora Patagónica. Tomo VIII. Parte IV b. Dicotyledones Dialipétalas (Droseraceae a Leguminosae). Colección Científica de INTA. Buenos Aires, 309 p

Correa MN (1988) Flora Patagónica. Tomo VIII. Parte V. Dicotyledones Dialipétalas (Oxalidaceae a Cornaceae). Colección Científica de INTA. Buenos Aires, 381 p

Correa MN (1998) Flora Patagónica. Tomo VIII. Parte I. Colección Científica de INTA. Buenos Aires

Correa MN (1999) Flora Patagónica. Tomo VIII. Parte VI Flora Dicotyledones Gamopétalas (Ericaceae a Calyceraceae). Colección Científica de INTA. Buenos Aires

Elissalde NO, Miravalles H (1983) Evaluación de los campos de pastoreo de Península Valdés. 70, Centro Nacional Patagónico (CONICET), Puerto Madryn, Argentina

Frenguelli J (1940) Rasgos principales de la Fitogeografía argentina. Revista del Museo de La Plata (nueva serie). Botánica 3:65–181

Golluscio RA, Sala OE (1993) Plant functional types and ecological strategies in Patagonian forbs. J Veg Sci 4:839–846

Hauman L (1920) Ganadería y Geobotánica en la Argentina. Revista del Centro de Estudios Agronómicos y Veterinarios (Buenos Aires) 102:45–65

Hauman L (1926) Etude phytogéographique de la Patagonie. Bull Soc R Bot Belg 58:105–180

Hauman L (1931) Esquisse phytogéographique de l'Argentine subtropicale et de ses relations avec la Geobotanique sudaméricaine. Bull Soc R Bot Belg 64: 20–64

Hauman L (1947) Provincia del Monte. In: Hauman L, Burkart A, Parodi LR, Cabrera AL (eds) La Vegetación de la Argentina, Geografía de la República Argentina. Sociedad Argentina de Estudios Geográficos, GAEA, Buenos Aires, pp 208–249

Holmberg EL (1898) La flora de la República Argentina. Segundo Censo de la República Argentina 1:385–474

Idaszkin YL, Bortolus A (2011) Does low temperature prevent *Spartina alterniflora* from expanding toward the austral-most salt marshes? Plant Ecol 212:553–561

Idaszkin YL, Bortolus A, Bouza PJ (2011) Ecological processes shaping Central Patagonian salt marsh landscapes. Austral Ecol 36:59–67

Kühn F (1922) Fundamentos de Fisiografía Argentina. Biblioteca del Oficial, Buenos Aires

Kühn F (1930) Geografía de la Argentina. Ed. Labor, Buenos Aires

Laycock WA (1995) New perspectives on ecological condition of rangelands: can state-and-transition or other models better define condition and diversity? In: Proceedings of the International Workshop on plant genetic resources, desertification, and sustainability, INTA-EEA Rio Gallegos, Argentina, pp 140–164

León RJC et al (1998) Grandes unidades de vegetació´n de la Patagonia extraandina. Ecol Austral 8:125–144
Lorentz P (1876) Cuadro de la Vegetación de la República Argentina. In: Napp R (ed) La República Argentina. Buenos Aires, pp 77–136
Marino A, Rodríguez MV, Pazos GE (2016) Resource-defense polygyny and self-limitation of population density in free-ranging guanacos. Behav Ecol 27:757–765
Mitsch WJ, Gosselink JG (2000) Wetlands. Wiley
Morello J (1958) La Provincia Fitogeográfica del Monte. Opera Lilloana 2:5–115
Mueller-Dombois D, Ellenberg H (1974) Aims and methods of vegetation ecology. Wiley, New York, 547 pp
Nabte MJ et al (2013) Range management affects native ungulate populations in Península Valdés, a World Natural Heritage. PLoS ONE 8(2):e55655
Parodi LR (1934) Las plantas indígenas no alimenticias cultivadas en la Argentina. Rev Argent Agron 1:165–212
Parodi LR (1945) Las regiones fitogeográficas argentinas y sus relaciones con la industria forestal. Plants and Plant Science in Latin America, pp 127–132
Parodi LR (1951) Las regiones fitogeográficas argentinas y sus relaciones con la industria forestal. Rev Uruguaya Geogr 2:89–100
Pazos GE, Bertiller MB (2008) Spatial patterns of the germinable soil seed bank of coexisting perennial-grass species in grazed shrublands of the Patagonian Monte. Plant Ecol 198:111–120
Roig FA, Martínez Carretero E (1998). La vegetación Puneña en la Provincia de Mendoza, Argentina. Phytoecoenologia 28:565–608
Sala OE et al (1989) Resource partitioning between shrubs and grasses in the Patagonian steppe. Oecologia 81:501–505
Soriano A (1950) La vegetación del Chubut. Rev Argent Agron 17:30–36
Soriano A (1956) Los distritos florísticos de la Provincia Patagónica. Serie Fitogeográfica, RIA. Tomo X. N° 4. INTA
Weinstein MP, Kreeger DA (eds) (2000) Concepts and controversies in tidal marsh ecology. Kluwer Academic Publishers

Soil–Geomorphology Relationships and Pedogenic Processes in Península Valdés

Pablo Bouza, Ileana Ríos, César Mario Rostagno and Claudia Saín

Abstract The soil landscape of the Penínusla Valdés region is review based in soil genesis analysis and soil–geomorphic relationships. The main soil types of the study area are grouped into two Soil Orders: Aridisols and Entisols. The oldest geomorphic surface corresponds to a relict terrace level from the Rodados Patagónicos lithostratigraphic unit, with Xeric Petrocalcids—Xeric Haplocalcids soil complex. Pleistocene landforms as youngest terrace levels, piedmont pediments (endorheic basins and coastal zone), and paleo-beach ridge units, an intricate soil distribution pattern occurs. This soil distribution is registered by a soil complex constituted by Natrargids, Natrigypsids, Calciargids and Haplocalcids, all of them with xeric regime soil moisture. In playa lakes of the endorheic basins, the soils were classified as Calcic Aquisalids. The Entisols are developed on Holocene geomorphic surfaces; Xeric Torripsamments in stabilized aeolian fields, and Typic Torriorthents in *bajadas* (coalescing alluvial fans associated to piedmont pediments). In salt marshes, geomorphic elements, anoxia degree, and vegetation communities are associated to soil type; Haplic Sulfaquents, Sodic Endoaquents and Sodic Psammaquents, are related to the low salt marsh, while Typic Fluvaquents and Sodic Hydraquents are developed in the high salt marshes. The main pedogenic processes registered in the Península Valdés soils are clay illuviation, calcification, gypsification, and sulfide production—sulfidization. As the calcretization process progresses, a transformation

P. Bouza (✉) · I. Ríos · C.M. Rostagno · C. Saín
Instituto Patagónico para el Estudio de los Ecosistemas Continentales (IPEEC),
Consejo Nacional de Investigaciones Científicas y Técnicas (CONICET)—CCT
Centro Nacional Patagónico (CENPAT), Boulevard Brown 2915,
ZC: U9120ACD Puerto Madryn, Chubut, Argentina
e-mail: bouza@cenpat-conicet.gob.ar

I. Ríos
e-mail: irios@cenpat-conicet.gob.ar

C.M. Rostagno
e-mail: rostagno@cenpat-conicet.gob.ar

C. Saín
e-mail: sain@cenpat-conicet.gob.ar

and neoformation of clay minerals occurs in the following sequence: smectite—palygorskite—sepiolite. The isotopic composition of $\delta^{13}C$ y $\delta^{18}O$ in pedogenic carbonate could be used as paleoecological and plaeoclimate proxy indicators, respectively.

Keywords Argillic horizon · Pedogenic calcretes · Isotope composition · Framboidal pyrite

1 Introduction

The Peninsula Valdés area presents a climate that varies from arid to semiarid, with mediterranean features. The scarcity of available water seems to be the limiting factor not only for plants, but also for soil formation. Chemical weathering, clay formation, and translocation of particles and ionic solutions should not occur under these conditions, and only weakly developed Entisols (Soil Survey Staff 1999) would be finding in the area.

However, strongly differentiated soil profiles appear in Peninsula Valdés. These would be relict features of colder or wetter climates of the past (Nettleton and Peterson 1983) or they could be attributed to the seasonally marked extreme rainfall years occurring in the present thus producing deep leaching (Buol et al. 1990). A common attribute, as in most Patagonian *Aridisols*, is the *polygenetic* nature of soils resulting from the alternation of morphogenesis periods with pedogenic periods under wetter climatic conditions than at present (Laya and Pazos 1976; Súnico 1996; Súnico et al. 1996; Bouza et al. 2005).

Besides, soils influenced by a local dominant factor over the normal effect of climate and vegetation (intrazonal soils) occur in restricted areas, as it is the case of hydromorphic soils of the coastal salt marshes (Aquents). A general discussion on soil parent materials and soil processes is presented in this chapter. In this sense, the geomorphological setting offered in Chapter "Late Cenozoic Landforms and Landscape Evolution of Península Valdés" is used to characterize the soil–geomorphic relationships (Fig. 1). In this sense, the geomorphic surface concept is addressed here to characterize the landscape in terms of genetic landform components, geologic age, and related pedogenic features (Gile et al. 1981).

The objectives of this chapter are: (1) characterize the soils of the Península Valdés region through the presentation of soil-forming factors and pedogenetic processes, and (2) discuss the soil–geomorphology relationship and the influence of geomorphic processes on soil evolution. In this chapter, information about the soil properties is uneven because to different data sources.

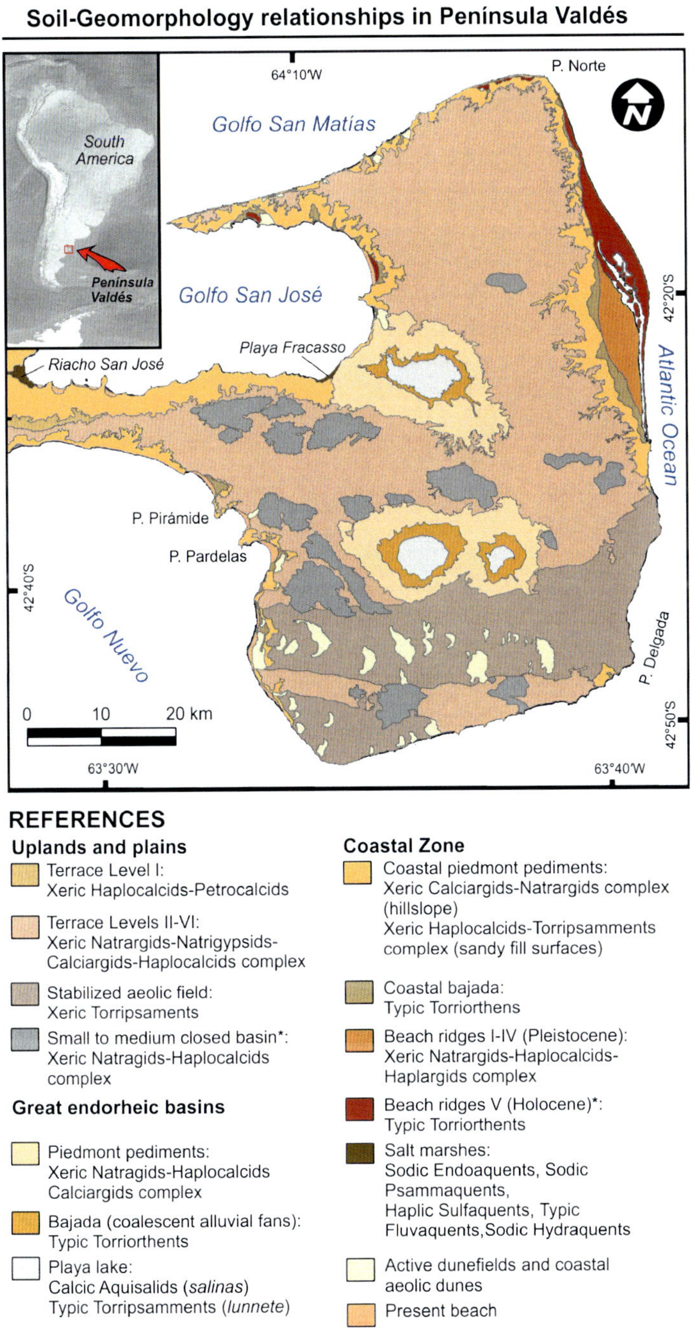
Soil-Geomorphology relationships in Península Valdés
64°10'W
P. Norte
Golfo San Matías
South America
Península Valdés
Golfo San José
42°20'S
Playa Fracasso
Riacho San José
Atlantic Ocean
P. Pirámide
P. Pardelas
42°40'S
Golfo Nuevo
P. Delgada
0
10
20 km
42°50'S
63°30'W
63°40'W
REFERENCES
Uplands and plains
Terrace Level I:
Xeric Haplocalcids-Petrocalcids
Terrace Levels II-VI:
Xeric Natrargids-Natrigypsids-
Calciargids-Haplocalcids complex
Stabilized aeolic field:
Xeric Torripsaments
Small to medium closed basin*:
Xeric Natragids-Haplocalcids
complex
Great endorheic basins
Piedmont pediments:
Xeric Natragids-Haplocalcids
Calciargids complex
Bajada (coalescent alluvial fans):
Typic Torriorthents
Playa lake:
Calcic Aquisalids (salinas)
Typic Torripsamments (lunnete)
Coastal Zone
Coastal piedmont pediments:
Xeric Calciargids-Natrargids complex
(hillslope)
Xeric Haplocalcids-Torripsamments
complex (sandy fill surfaces)
Coastal bajada:
Typic Torriorthens
Beach ridges I-IV (Pleistocene):
Xeric Natrargids-Haplocalcids-
Haplargids complex
Beach ridges V (Holocene)*:
Typic Torriorthents
Salt marshes:
Sodic Endoaquents, Sodic
Psammaquents,
Haplic Sulfaquents, Typic
Fluvaquents, Sodic Hydraquents
Active dunefields and coastal
aeolic dunes
Present beach

Fig. 1 Soil–Geomorphology relationships in the Península Valdés. *Small to medium closed basin and Beach ridges V, the soils were classified tentatively according to field controls and soil profile descriptions

2 Soil–Geomorphic Relationship: Origin, Soil Parent Materials, and Pedogenic Features on Geomorphic Surfaces

2.1 Uplands and Plains System

Uplands and plains system is described in Chapter "Late Cenozoic Landforms and Landscape Evolution of Península Valdés", and it corresponds to stepped sequences of old fluvial terraces of the *Rodados Patagónicos lithostratigraphic unit* (see Chapter "Geology of Península Valdés"). On this regional landscape other geomorphological units such as small to medium closed basins, stabilized aeolian fields and active dunefields are superimposed.

2.1.1 Relict Terrace Levels

The parent materials of relict terrace levels are the Rodados Patagónicos deposits (see Chapters Geology of Península Valdés and Late Cenozoic Landforms and Landscape Evolution of Península Valdés; Fig. 2a). However, while all the terraces have similarities in their parent materials (gravel and sand deposits, free of carbonates, gypsum and soluble salts) important differences in their morphological, physical and chemical properties are clearly observed (Table 1).

The soil profile on Terrace Level I was described and analyzed out of the Península Valdés region. In this oldest geomorphic surface, Petrocalcids (Fig. 2b) are the dominant soils, associated with Haplocalcids. The upper part of the Rodados Patagónicos deposits has carbonate accumulation, partially cementing the matrix of the gravel deposit (3Ck horizon). This surface was presumably eroded and later covered by a deposit in which the intense carbonate leaching filled pores and formed a petrocalcic horizon (2Bkm horizon), composed of a hard massive layer, in which platy laminar crusts of 2–3 cm thick were found to be intercalated. Geomorphic processes would have truncated the soil surface, exhuming the thick and hard petrocalcic horizon, which was later buried by sandy material of presumably aeolian origin, in which an A–C1–C2 horizon sequence developed. On Terrace Levels II–VI, an intricate soil distribution pattern constituted by Natrargids–Natrigypsids–Calciargids–Haplocalcids *complex* occurs (Fig. 2c, d, e; Table 1), all of them with *xeric moisture regime* (Rostagno 1980, 1981; Bouza et al. 2007). The morphological soil descriptions showed lithologic discontinuities, which delimit successive depositional units in turn affected by pedogenetic processes (Bouza et al. 2005, 2007).

The relief of these terrace levels, as well as of the majority land surfaces of gently sloping, is mildly undulated, with mounds colonized by shrubs (*Chuquiraga avellanedae*) and grasses (*Nassella tenuis*), and inter-mounds characterized by desert pavements and *vesicular horizons* (Av). The altitudinal differences between

Table 1 Soil properties of selected soil profiles of the Uplands and Plains system

Profile[a] horizon	Depth (cm)	Color (dry)	Structure[b]	Boundary[b]	Gravels (%)	Sand (%)	Silt (%)	Clay (%)	CaCO$_3$ (%)	SOC[c] (%)	pH (1:2.5)	EC[c] (dS/m)	ESP[c] (%)	CEC[c] (cmol$_c$ kg^{-1})
TL I	Xeric Petrocalcid													
A	0–20	10YR6/3	gr, vf, 1	as	9	78.8	15.6	5.7	0.89	0.82	8.36	0.71	0.6	14.3
C1	20–41	10YR6/3	sg	gi	15	78.6	14.8	6.6	4.78	0.47	8.60	0.59	2.4	13.5
C2	41–68	10YR6/3	sg	ai	13	78.7	14.8	6.5	2.46	0.45	9.02	1.23	11.3	13.5
2Bkmb	68–102	10YR8/2	pl, m, 3	aw	4	51.1	11.7	37.2	60.36	0.29	9.39	2.80	30.4	6.0
3Ck	>102	10YR6/3	m	–	78	63.3	14.6	22.1	12.82	0.19	9.45	4.24	42.4	12.0
TL II-1	Xeric Natrigypsid													
Av	0–2	10YR5/3	gr, vf, 1	as	0	85.3	11.8	2.9	0.13	0.72	7.76	0.99	1.6	8.1
2Btn	2–15	10YR3/4	pr, f, 3	aw	0	45.9	8.4	45.6	0.17	0.86	8.56	1.50	17.6	22.8
2Btkn	15–40	10YR5/3	sbk, m, 2	as	1	59.2	3.8	37.0	13.02	0.20	8.74	6.97	43.9	14.3
2Ck	40–65	10YR7/3	m	as	1	61.9	5.2	32.9	11.36	0.16	8.44	8.44	42.5	14.3
3Cky1	65–91	10YR8/3	m	as	0	41.3	6.1	52.6	17.73	0.14	7.87	7.88	34.7	6.6
3Cky2	91–124	10YR8/3	m	gi	15	47.7	14.7	37.6	16.40	0.13	7.83	7.40	34.1	8.5
3Cky3	124–240	10YR8/3	m	–	55	–	–	–	21.69	–	–		–	–
3C	>240	10YR6/3			68	–	–	–	0.84	–	–		–	–
TL II-2	Xeric Haplocalcid													
A	0–2	10YR5/3	pl, m, 1	as	9	71.7	20.6	7.7	nd	0.72	7.58	0.67	1.2	10.8
C1	2–22	10YR6/3	sg	gs	6	75.9	15.7	8.4	nd	0.40	7.88	0.49	1.4	12.0
C2	22–42	10YR6/3	sg	gs	8	76.3	16.4	7.3	nd	0.36	8.35	0.58	3.2	12.0
C3	42–62	10YR6/3	sg	as	8	75.6	16.7	7.7	nd	0.33	8.65	0.56	4.5	12.0
2Ck1	62–120	10YR8/3	m	gw	3	43.5	12.5	44.0	30.4	0.42	8.68	0.67	4.8	11.6
2Ck2	120–150	10YR8/3	m	gw	7	44.6	16.9	38.5	22.5	–	–	–	–	–
3Ck3	>150	10YR8/3	m	–	46	40.3	18.4	41.3	14.4	–	–	–	–	–

(continued)

Table 1 (continued)

Profile[a] horizon	Depth (cm)	Color (dry)	Structure[b]	Boundary[b]	Gravels (%)	Sand (%)	Silt (%)	Clay (%)	CaCO$_3$ (%)	SOC[c] (%)	pH (1:2.5)	EC[c] (dS/m)	ESP[c] (%)	CEC[c] (cmol$_c$ kg^{-1})
TL VI-1	Xeric Natrargid													
A	0–7	10YR6/3	gr, vf, 1	as	8.9	72.0	15.0	13.0	nd	0.48	8.48	1.76	2.5	10.0
2Btn	7–22	7.5YR4/4	pr, f, 3	aw	2.3	54.4	7.2	38.4	nd	0.65	8.55	6.24	35.6	20.8
3Btkn	22–31	7.5YR6/4	sbk, m, 2	as	7.7	53.7	6.3	40.0	5.3	0.44	9.00	7.02	38.3	17.0
3Ck1	31–45	7.5YR8/2	m	as	14.2	41.6	15.9	42.5	20.1	0.26	9.25	8.10	37.4	11.3
4Ck2	>45	7.5YR8/2	m	–	64.2	51.6	19.0	29.4	15.7	–	–		–	–
TL VI-2	Xeric Calciargid													
A	0–10	10YR6/3	gr, vf, 1	aw	2	74.6	14.2	11.0	0.77	1.07	8.02	0.90	0.6	13.9
C	10–29	10YR6/3	sg	aw	4	71.2	15.9	12.9	0.03	0.57	8.29	0.66	2.0	11.6
2Bt	29–38	7.5YR5/4	sbk, m, 2	gi	1	57.2	19.2	23.7	2.97	0.49	8.35	1.43	5.3	18.9
2Btk	38–52	7.5YR7/4	sbk, m, 2	gw	3	55.0	7.0	38.0	14.25	0.36	8.58	1.32	6.6	13.5
2Bk	52–66	7.5YR8/2	m	ai	13	59.3	15.2	25.5	20.36	0.34	8.77	1.32	8.2	13.6
3Ck1	66–74	7.5YR8/2	m	gi	75	53.0	5.3	41.7	26.12	0.31	8.67	1.54	10.2	12.3
3Ck2	74–103	7.5YR8/2	m	gi	65	–	–	–	30.23	–	–		–	–
3C	>103	10YR6/3	sg	–	22	–	–	–	0.97	–	–		–	–
TL VI-3	Xeric Haplocalcid													
A1	0–5	10YR 5/3	gr, vf, 1	as	0	89.1	10.1	0.8	0.48	1.09	7.53	0.66	0.0	–
A2	5–25	10YR 5/3	gs	aw	3	79.6	16.7	3.7	0.35	0.48	7.47	0.49	0.0	–
C	25–65	10YR 6/3	gs	gi	4	79.0	17.0	4.0	0.41	0.38	7.74	0.30	0.0	–
2Bkb1	65–100	10YR 8/3	sbk. m, 1	gi	2	61.0	12.3	26.7	27.13	0.29	9.34	0.78	2.5	–
2Bkb2	100–135	10YR 8/2	sbk. m, 1	gi	1	60.4	12.1	27.5	27.29	0.15	9.36	0.58	2.3	–
2Bkb3	135–170	10YR 6/4	sbk. m, 1	gi	2	61.3	14.4	24.3	25.67	0.08	9.45	0.98	3.2	–

(continued)

Table 1 (continued)

Profile[a] horizon	Depth (cm)	Color (dry)	Structure[b]	Boundary[b]	Gravels (%)	Sand (%)	Silt (%)	Clay (%)	CaCO$_3$ (%)	SOC[c] (%)	pH (1:2.5)	EC[c] (dS/m)	ESP[c] (%)	CEC[c] (cmol$_c$ kg^{-1})
2Bkb4	170–205	10YR 6/4	sbk, m, 1	gi	1	61.1	19.9	19.0	27.34	0.15	9.35	0.74	3.2	–
2Cky1	205–268	10YR 8/1	m	aw	7	72.7	7.7	19.6	8.89	0.17	8.04	9.80	4.3	–
3Cky2	268–280	10YR 8/1	m	aw	24	75.8	8.7	15.5	6.89	0.14	7.81	10.00	4.0	–
3Cy	>280	10YR 7/4	sg	–	51	82.5	12.6	4.9	2.09	0.05	7.53	3.79	0.6	

[a]Abbreviations for soil profile designation; *TL* Terrace Level; Roman numerals indicate chronosequence levels (geomorphic surfaces)
[b]Abbreviations for morphological description are from Schoeneberger et al. (2002); structure: *gr* granular, *sg* single grain, *m* masive, *pl* platy, *pr* prismatic, *sbk* subangular blocky; size: *vf* very fine, *f* fine, *m* medium; grade: *1* weak, *2* moderated, *3* strong; boundary, distinctness: *a* abrupt, *g* gradual; topography: *w* wavy, *s* smooth, *i* irregular
[c]Soil properties: *EC* electrical conductivity, *SOC* soil organic carbon, *ESP* exchangeable sodium percentage, *CEC* cation exchange capacity
Data from Rostagno (1981), Bouza, et al. (2005, 2007) and Bouza (2012)

Fig. 2 Soils in the Uplands and Plains System; **a** A relict Terrace level developed in deposits of the Rodados Patagónicos (Pliocene—Early Pleistocene geomorphic surfaces); Note the calcretized zone on top of the deposit and the fossil cryogenic features on the bottom; **b** Terrace level I, Xeric Petrocalcid; **c** Terrace level II, Xeric Natrigypsid; **d** Terrace level VI, Xeric Calciargid–Natrargid complex; **e** Terrace level VI, Xeric Haplocalcid; **f** Stabilized Aeolian Fields, Xeric Torripsamments

the mounds and inter-mounds were minor (20–25 cm). These relationships between soil, vegetation and micro-relief explain the mentioned pedocomplex.

2.1.2 Stabilized Aeolian Fields

The soil development in stabilized aeolian fields landforms corresponds to Xeric Torripsamments (Rostagno 1981), that support psammophile plant species as Sporobolus rigens and Hyalis argentea (see Chapter "Vegetation of Penĺnsula Valdés: Priority Sites for Conservation"). The weak soil development is demonstrated by the horizons sequence AC–2C1–2C2 (Fig. 2f). The pH value (1:2.5 soil–water relationships) increases in depth, and ranges from 7.2 to 8.5, presumably due to the higher organic carbon content (organic acids) in the upper horizon (Rostagno 1981).

2.2 Great Endorheic Basins System

The Great Endorheic Basin system is characterized by a typical centripetal drainage network. This drainage dissects the Piedmont Pediments units, give rise to

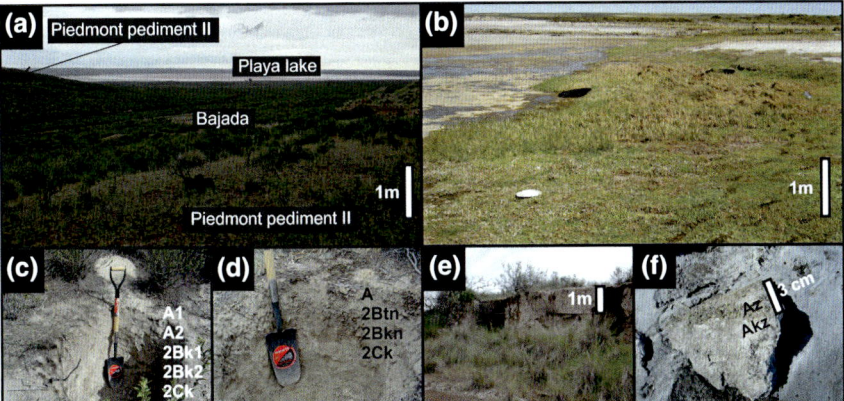

Fig. 3 Soils in the Great Endorheic Basins System. **a** Piedmont pediments and Bajadas. **b** Playa lakes. **c** Piedmont pediment, Xeric Haplocalcid. **d** Piedmont pediment, Xeric Natrargid. **e** Bajada, Typic Torriorthent. **f** Playa lake, Calcic Aquisalid

alluvial-fluvial fans that form the *Bajada* unit, and ends at the bottom of the basin in the Playa Lake unit (Fig. 3a, b).

2.2.1 Piedmont Pediments

In the Piedmont Pediment unit, two levels (geomorphic surfaces) were identified (i.e., Pediment I and Pediment II; see Chapter "Late Cenozoic Landforms and Landscape Evolution of Península Valdés"). The soil parent materials comprise sandstone from the Miocene Puerto Madryn Formation and Quaternary sediments of alluvial-colluvial origin. The Miocene sandstone constituted a *paralithic material* at <50 cm depth (*paralithic contact*) and this is a difference with Aridisols of Terrace levels. On the Puerto Madryn Formation, soils with calcic and argillic horizons have been developed (2Bt–2Bk, 2Btn–2Bkn horizon sequences). Surface runoff processes eroded and then buried the exhumed 2Btn or 2Bk horizons by alluvial-colluvial deposits, which evolved to superficial soil horizons (A, A–C horizons). The soils on Pediment I surface are mainly made of Xeric Haplocalcids–Xeric Natrargids complex (Fig. 3c–d), while on Pediment II Xeric Calciargids are the main soil types (Table 2; Bouza et al. 2008). The calcic horizons were identified by pedogenic carbonate accumulation (nodules and soft micritic masses), which surpassed in all cases to 10 by volume percent. Vegetation comprises *C. avellanedae* and *Chuquiraga hystrix* as dominant species, whit *N. tenuis* and *Stipa sp.* (see Chapter "Vegetation of PenÍnsula Valdés. Priority Sites for Conservation").

Table 2 Soil properties of selected soil profiles of the Great Endorheic Basin system

Profile[a] horizon	Depth (cm)	Color (dry)	Structure[b]	boundary[b]	Gravels (%)	Sand (%)	Silt (%)	Clay (%)	SOC (%)	pH (1:2.5)	EC (dS/m)	ESP (%)
PP I-1	Xeric Haplocalcid											
A1	0–7	10YR 7/2	gr, f,1	as	0	66.4	27.6	6.0	1.18	7.80	0.60	0.0
A2	7–13	10YR 7/3	sg	as	13	69.1	23.9	7.0	0.69	7.86	0.34	0.0
2Bk1	13–28	10YR 7/4	sbk, m, 2	gs	5	73.6	21.4	5.0	0.65	7.84	0.54	1.5
2Bk2	28–45	10YR 8/3	sbk, f, 2	as	0	74.1	21.9	4.0	0.52	7.36	1.17	8.1
2Ck	>45	2.5YR 7/2	m	–	0	75.0	22.6	2.4	0.31	7.11	9.18	22.5
PP I-2	Xeric Natrargid											
A	0–8	10YR 7/3	gr, f,1	aw	0	86.6	11.4	2.0	0.48	7.61	0.51	2.5
2Btn	8–23	10YR 3/4	pr, f, 2	gw	15	66.6	25.4	8.0	0.70	7.36	1.10	17.7
2Bkn	23–43	10YR 7/4	sbk, m, 2	gs	5	66.0	21.8	12.2	0.54	7.53	5.52	30.4
2Ck	>43	2.5YR 7/2	m	–	0	71.2	24.8	4.0	0.23	7.55	16.99	37.4
PP II	Xeric Calciargid											
A	0–13	10YR 7/3	gr, f, 3	aw	10	73.6	19.4	7.0	0.56	7.30	0.30	0.9
2Bt	13–35	10YR 3/4	pr, f, 1	gw	0	55.0	27.8	17.2	0.48	7.92	1.28	8.3
2Bk	>35	10YR 8/3	sbk, m, 2		0	57.4	17.6	25.0	0.36	7.52	0.55	8.2
PL D	Typic Torripsamment		Lunnete									
A	0–20	10YR 5/3	gr, f, 2	gs	5	92.4	6.6	1.0	0.32	7.72	0.38	0.3
C1	20–40	10YR6/3	sg	gs	0	91.4	7.1	1.5	0.38	7.17	0.52	2.2
C2	40–60	10YR6/3	sg	gs	0	88.6	10.4	1.0	0.18	7.25	0.83	1.9
C3	60–80	10YR6/3	sg	–	0	86.9	12.1	1.0	0.15	7.82	0.61	3.1

[a] Abbreviations for soil profile designation; *PP* Piedmont Pediment, *PL D* Playa Lake, Dune, *CPP* Coastal Piedmont Pediment, *PBR* Pleistocene Beach Ridge, *SM* salt marshes; Roman numerals indicate chronosequence levels (geomorphic surfaces)

[b] Abbreviations for morphological description are from Schoeneberger et al. (2002); structure: *gr* granular, *sg* single grain, *m* masive, *pl* platy, *pr* prismatic, *sbk* subangular Blocky; size: *vf* very fine, *f* fine, *m* medium; grade: *1* weak, *2* moderated, *3* strong; boundary, distinctness: *a* abrupt, *g* gradual; topography: *w* wavy, *s* smooth, *i* irregular

[c] Soil properties: *EC* electrical conductivity, *SOC* soil organic carbon, *ESP* exchangeable sodium percentage, *CEC* cation exchange capacity, *nd* not detected

Data from Bouza et al. (2008)

2.2.2 Bajadas

The soil parent materials of the *Bajada* unit (coalescent alluvial fans) are made of Quaternary sandy gravel deposits of alluvial-colluvial origin. The soils have a weak development pedogenic degree and are tentatively classified as Typic Torriorthents (Fig. 3e; Bouza et al. 2008). The vegetationcomprises a shrub steppe of *C. hystrix* and *C. avellanedae* with perennial grasses.

2.2.3 Playa Lakes

In the Great Endorheic Basin system, both the radial and centripetal drainage network as the perched water table in the Puerto Madryn Formation generated a perimeter zone in playa lakes where the water table fluctuation is registered by redoximorphic features within 100 cm in deep. In the Gran Salitral playa lake, the parent material of the soils is sandy alluvium and the profile has an Az–Akz–Ckg horizon sequence (Fig. 3b, f; Rostagno 1981). The Az horizon (0–3 cm, 2.5Y 6/6 olive yellow) is a massive saline crust, with loamy sand texture, while the Akz horizon (3–40 cm, 2.5Y 5/4 light olive brown) is a *salic horizon* (electrical conductivity, EC 63 dS/m) with sandy loam texture and abundant carbonate nodules. The Ckg horizon (>40 cm to more of 100 cm, 2.5Y 4/0 dark gray) has a sandy loam texture and common carbonate nodules and the EC decreases to 12 dS/m. The vegetation comprises *Distichlis spicata, D. scoparia* and *Sarcocornia perennis*. The soils were tentatively classified as Calcic Aquisalids (Aquollic Salorthids; Rostagno 1981).

In the Playa Lake of the Gran Salitral (Fig. 1), a crescent-shaped lunette is found on the downwind side (leeward). The soils correspond to Typic Torripsamments (Table 2) with grass-steppe dominated by *N. tenuis* and *Piptochaetium napostaense*.

2.3 Coastal Zone System

In the same way that in the Great Endorheic Basins system, on the coastal piedmont area, pediments and associated *bajadas* were distinguished (see Chapter "Late Cenozoic Landforms and Landscape Evolution of Península Valdés").

On the Atlantic coast, there is a sequence of the Late Pleistocene beach ridges corresponding to the Caleta Valdés Formation, while that in all the coastal area of Península Valdés, a sea level rise during Middle Holocene is recorded by beach ridges, which correspond to the San Miguel Formation (see Chapters "Geology of Península Valdés and Late Cenozoic Landforms and Landscape Evolution of Península Valdés").

2.3.1 Coastal Piedmont Pediments

Piedmont pediments are well developed in the Coastal Zone System. Whereas in the eastern coastline of the Atlantic Ocean, only one pediment level was recognized (see Chapter "Late Cenozoic Landforms and Landscape Evolution of Península Valdés"), in the coastal zone of the Golfo Nuevo and Golfo San José, two piedmont pediment levels exist, as is observed in the Great Endhorheic Basin system.

The soils developed on these hillslopes compose a Xeric Calciargids–Xeric Natrargids complex (pedons CPP 1 and CPP 2, respectively; Table 3). The vegetation consists of a shrubby layer dominated by *Condalia microphylla*, *Lycium chilense*, and *C. avellanedae* that alternate with patches of grass layers dominated by *N. tenuis* (see Chapter "Vegetation of Península Valdés. Priority Sites for Conservation"). On the eastern coastline of the Atlantic Ocean, several Holocene stream valleys that dissected the Coastal Piedmont Pediments are filled with aeolian sandy sediments. These sandy fill surfaces were colonized by grassland of *N. tenuis* and *P. napostaense* with shrubs of *Chuquiraga hystrix* and *Ch. avellanedae* (see Chapter "Vegetation of Península Valdés. Priority Sites for Conservation"). According to the occurrence of calcic horizon depth, the soils were classified as Xeric Haplocalcids and/or Xeric Torripsamments (Table 3; Bouza et al. 2008).

2.3.2 Pleistocene Beach Ridges

This unit is located at the west part of the Penísnula Valdés region near the Atlantic coastline (Fig. 1). It is made of Middle–Late Pleistocene gravel deposits of the Caleta Valdés Formation (see Chapter "Geology of Península Valdés"). The unit includes four geomorphic surfaces identified by Roman numbers (i.e. PBR I–IV) in descending height order and age (see Chapter "Late Cenozoic Landforms and Landscape Evolution of Península Valdés"). In general, the dominant soil developed on these geomorphic surfaces was classified as Xeric Natrargid. In this chapter are referred the properties of Xeric Haplargids, Xeric Haplocalcids and Xeric Natrargids developed on the Pleistocene beach ridges I, II and IV, respectively (Table 3).

The complete surficial stratigraphy in PBR IV was observed where three depositional units were alternatively modified by pedogenetic processes (Fig. 4a; Table 3). The most ancient reddish-brown paleosol (3Btkb) was formed at the top of the gravelly deposits of the Caleta Valdes Formation (*Marine Isotope Stage 5e*; Rostami et al. 2000; Brückner et al. 2007). In the lower limit of this buried horizon a 3Bkmb (petrocalcic crust) was formed. Geomorphic processes would have truncated the soil surface exhuming the paleosol, which was buried by other soil parent material (second depositional unit), presumably related with periglacial processes of *Llanquihue glaciations* (Bouza 2014). During a lapse of surficial stability, this soil parent material evolved to a 2Btb–2Btkb–2Bkb–2Ck horizon

Table 3 Soil properties of selected soil profiles in the Coastal Zone system

Profile[a]/horizon	Depth (cm)	Color (dry)	Structure[b]	boundary[b]	Gravels (%)	Sand (%)	Silt (%)	Clay (%)	SOC (%)	pH (1:2.5)	EC (dS/m)	ESP (%)
CPP 1	Xeric Calciargid											
A	0–17		gr, f, 3		15	81.6	16.0	2.4	0.75	6.49	0.29	0.1
2Bt	17–24		pr, m, 3		5	58.0	20.4	21.6	0.67	8.37	1.15	9.7
2Bk	24–30		sbk, m, 2		0	60.4	29.2	10.4	0.90	9.15	1.07	9.2
2Ck	>30		sbk, m, 2		0	69.2	26.0	4.8	0.19	9.65	1.96	17.2
CPP 2	Xeric Natrargid											
Av	0–3		pl, m, 1	as	40	74.6	15.8	9.6	0.48	7.69	0.94	8.3
2Btn	3–18		pr, f, 3	gw	5	57.6	12.8	29.6	0.93	8.14	5.07	25.6
2Bkn	>18		sbk, m, 2	–	0	73.6	14.4	12.0	0.45	9.28	9.93	35.4
CPP 3	Xeric Haplocalcid	Cañadón										
A	0–20	10YR6/3	gr, f, 3	gs	10	81.3	14.9	3.8	0.69	8.19	0.30	0.0
C	20–32	10YR6/3	sg	aw	0	81.3	16.9	1.8	0.70	8.42	0.37	0.0
2Bk	>32	10YR 8/3	sbk, m, 2	–	0	52.6	35.8	11.6	0.83	8.48	0.52	0.1
CPP 4	Typic Torripsamment	Cañadón										
A	0–2	10YR5/3	gr, f, 3	as	7	79.0	18.2	2.8	0.82	6.44	0.35	0.3
C1	2–20	10YR6/3	sg	gs	0	81.0	16.2	2.8	0.62	7.38	0.44	0.0
C2	20–48	10YR6/3	sg	gs	0	77.8	20.2	2.0	0.39	7.79	0.34	0.2
C3	>48	10YR6/3	sg	–	0	82.8	15.4	1.8	0.28	8.98	0.36	1.9
PBR 1	Xeric Haplargids											
A	0–6	10YR6/3	gr, f, 2	aw	5	48.4	34.7	16.9	1.89	8.08	1.94	–
2Btb	6–16	7.5YR4/4	pr, f, 3	aw	5	40.8	20.8	38.4	0.95	8.66	0.97	–
2Btkb1	16–30	10YR7/4	pr, m, 3	gw	24	47.6	6.3	46.2	0.50	9.24	0.68	–
2Btkb2	30–53	10YR7/4	sbk, f, 3	gi	64	43.0	22.6	34.4	0.70	9.04	1.09	–
3Bkyb1	53–71	10YR7/4	sbk, m, 3	gw	43	49.2	44.7	6.1	0.19	8.02	2.08	–
3Bkyb2	71–89	10YR7/4	sbk, m, 2	aw	82	86.2	10.8	3.0	0.18	7.88	3.10	–

(continued)

Table 3 (continued)

Profile[a] horizon	Depth (cm)	Color (dry)	Structure[b]	boundary[b]	Gravels (%)	Sand (%)	Silt (%)	Clay (%)	SOC (%)	pH (1:2.5)	EC (dS/m)	ESP (%)
3Ck	89–93	10YR7/4	sg	as	76	81.5	17.9	0.6	0.19	8.07	–	–
3C	93–116	10YR7/4	sg	as	89	73.5	25.5	1.0	0.21	8.06	–	–
4Ck	>116	10YR7/4	m	–	53	84.8	14.3	0.9	0.10	8.16	2.38	–
PBR II	Xeric Haplocalcid											
A1	0–7	10YR6/3	gr, vf, 1	as	6	59.9	40.1	0.0	0.58	7.02	1.02	–
A2	7–31	10YR6/3	sbk, m, 2	gi	12	60.9	21.2	17.8	0.67	7.78	0.73	–
C	31–63	10YR6/3	m	aw	9	61.4	26.3	12.3	0.62	8.46	1.60	–
2Bkyb	63–87	10YR7/4	sbk, m, 2	gi	20	17.0	78.7	4.3	0.24	8.09	2.84	–
2Cky	87–131	10YR7/4	sbk, m, 2	gi	27	10.8	88.2	1.0	0.29	8.05	3.05	–
3Btkyb	>131	5YR5/4	sbk, f, 3		72	8.0	85.8	6.2	0.19	8.02	7.52	–
PBR IV	Xeric Natrargid											
A	0–5	10YR6/3	gr, vf, 1	as	4	69.2	11.5	19.3	0.94	8.32	1.49	4.5
C	5–13	10YR6/3	m	aw	22	75.3	12.6	12.1	0.62	8.39	0.67	9.1
2Btn	13–26	7.5YR4/4	pr, f, 3	gw	0	48.5	7.2	44.3	0.70	8.50	3.57	18.7
2Btkn	26–46	10YR7/4	sbk, m, 1	gi	2	39.7	14.4	45.6	0.25	8.76	12.50	43.4
2Bkn	46–67	10YR7/4	sbk, m, 1	aw	4	39.0	13.7	47.4	0.21	8.32	44.00	50.1
2Ck	67–87	10YR7/4	m	aw	82	50.2	14.8	35.0	0.16	8.57	35.20	46.1
3Btkb	87–235	5YR5/4	sbk, f, 3	gi	73	26.7	28.2	45.1	0.22	7.98	38.20	42.3
3Bkmb	235–240	5YR8/2	pl, m, 3	as	10	–	–	–	–	–	–	–
3C	>240	10YR6/3	sg	–	80	–	–	–	–	–	–	–

[a]Abbreviations for soil profile designation: *CPP* Coastal Piedmont Pediment, *PBR* Pleistocene Beach Ridge, *SM* salt marshes; Roman numerals indicate chronosequence levels (geomorphic surfaces)

[b]Abbreviations for morphological description are from Schoeneberger et al. (2002); structure: *gr* granular, *sg* single grain, *m* massive, *pl* platy, *pr* prismatic, *sbk* subangular Blocky; size: *vf* very fine, *f* fine, *m* medium; grade: *1* weak, *2* moderated, *3* strong; boundary, distinctness: *a* abrupt, *g* gradual; topography: *w* wavy, *s* smooth, *i* irregular

[c]Soil properties: *EC* electrical conductivity, *SOC* soil organic carbon, *ESP* exchangeable sodium percentage, *CEC* cation exchange capacity, *nd* not detected
Data from Bouza et al. (2008) and Bouza (2014)

Fig. 4 Soils in the Coastal Zone System. **a** Horizon sequence, depositional units and paleosol of the Pleistocene Beach Ridges IV (Caleta Valdés Formation, Upper Pleistocene), Xeric Natrargid. **b** Riacho San José salt marshes, Haplic Sulfaquents (*S. alterniflora*). **c** Playa Fracasso salt marsh, Typic Fluvaquent (*S. perennis*). **d** Playa Fracasso salt marsh, Typic Fluvaquent (*L. brasiliense*)

sequence. Finally, erosion processes truncated the argillic horizon, which then was buried by new parent materials, presumably of aeolian origin. These materials evolved to an A–C horizons sequence.

2.3.3 Tidal Salt Marshes

In the Península Valdés the tidal salt marshes are mainly developed in Golfo San José on the Istmo Carlos Ameghino coast, and in the Caleta Valdés coastal lagoon (e.g., Riacho San José and Playa Fracasso). Tidal salt marshes are areas located in the upper part of the intertidal plain, where a muddy substratum supports a varied and dense stands of halophytic plants (Allen and Pye 1992). These coastal landforms were formed by Holocene sea-level changes and are frequently flooded by tides and subjected both to marine erosion–deposition processes, as runoff and continental sedimentation. For this reason, the most of tidal salt marshes have little or no evidence of development of pedogenic horizons, the soils thus classifying as *Entisols*. Since the salt marsh soils are continuously saturated with water and periodically affected by tidal flooding (and waterlogging), they are commonly included in the Suborder of *Aquents* (Soil Survey Staff 1999).

In this chapter, the soils of Riacho San José and Playa Fracasso sites are presented (Fig. 1). The soils were characterized according their *landform elements*, physiographic positions (low and high marshes), and vegetation dominance (Table 5; Ríos 2015; see Chap. 6).

In Playa Fracasso, the soils in the low marsh and with *Spartina alterniflora* were classified as *Sodic Endoaquents* and *Sodic Psammaquents,* which were associated with inter-sandbars and sandbars, respectively. Also, *Sodic Psammanquents* are formed in sandy point-bars from tidal channels. In Riacho San José, the soil in the low marsh was classified as *Haplic Sulfaquents* (Fig. 4b) due to the occurrence of sulfidic materials. The soils from the high marshes were classified as *Typic Fluvaquents* and *Sodic Hydraquents*, with *Limonium brasiliense* and *S. perennis* as dominant species (Fig. 4c, d). In Playa Fracasso *L. brasiliense* vegetation unit is associated with tidal-channel levee deposits. In high salt marshes the water table level fluctuating and the root-zone are identified by reddish-brown redoximorphic features (hue 5YR) mainly around macropores (e.g., roots and rhizomes), stratified sediments, and nodules, presumably composed of ferrihydrite ($5Fe_2O_3.9H_2O$).

3 Pedogenic Processes

3.1 Illuviation

In Bt horizons, with medium subangular blocky structure, common thin clay coatings ($\sim$10 to 30 μm thickness), both on the mineral grain surfaces and on void walls, are the main evidences of pedogenic illuviation in Calciargids (Fig. 5a). On the other hand, the Btn horizons in Natrargids, with fine prismatic and columnar structures, abundant and well preserver thick clay coatings ($\sim$40–80 μm thickness) occurs also on grains and voids walls (Fig. 5b). This greater clay coating development is attributed to the high content of exchangeable sodium (Tables 1, 2, 3 and 4), which favors the dispersion of clay particles and promotes the formation of *natric horizon*.

Fig. 5 Illuviation, micromorphological features in the argillic horizons of the Terrace level VI soils. **a** Thin clay coatings of the grains and void walls (2Bt horizon, TL VI-2 Xeric Calciargid). **b** thick clay coatings on the grains and the voids walls (2Btn horizon, TL VI-1 Xeric Natrargid)

The argillic horizons in arid regions constitute a tool for studying Quaternary paleoclimates (Birkeland et al. 1991). The formation of these horizons is associated with surfaces that have remained stable since at least the Late Pleistocene. Palynological and sedimentological analyses in northeastern Patagonia indicate a wetter episode between 19 and 15 ka (Galloway et al. 1988), followed by arid climatic conditions with aeolian morphodynamic processes during the Middle Holocene, and finally semiarid conditions with a higher rainfall frequency in the Late Holocene (Schäbitz 1994). These climatic changes were used to explain the spatial variations and polygenetic character of the Argids of the Península Valdés region (Bouza et al. 2005).

3.2 Calcification

Since the terrace levels and Pleistocene beach ridge deposits are free of carbonates and since the Ca^{2+} released during weathering is not adequate to explain the $CaCO_3$ content in the soils, this mineral must have an allochthonous origin, mainly by aeolian influx (Bouza 2012, 2014); these carbonates were likely redistributed in the soils.

Generally, the soils developed on soil parent materials free of carbonates have a deep carbonate accumulation that affected the upper part of the gravelly deposits. It is because the downward water movement is restricted when fine-textured materials are underlain by gravel layers as occurs in the generalized limit 2Bk–3Ck horizons (Table 1). The pedogenic carbonates are deposited at or near the top of the sand or gravel layers as water are removed by evapotranspiration at these interfaces (Stuart and Dixson 1972), so that a asymmetric profile of carbonate accumulation occurs (Fig. 6a). This carbonate content increases with the age of the geomorphic surfaces (terraces), constituting a tool to correlate soils and *paleosols* (*chronosequence*; Bouza 2012; Bouza and del Valle 2014).

Also, the pedogenic origin is revealed by the presence of typical morphological features as: (1) calcitic coatings on gravel surfaces of *pendant* type, (2) *matrix nodules* and *floating coarse mineral components*, (3) micritic calcite nodules surrounded by *circumgranular cracks*, (4) lenticular gypsum aggregates intergrown in the micritic groundmass, (5) *pellets* and *ooids*, and (6) rhombohedral micrite. (Fig. 6a–g; Bouza et al. 2007; Bouza 2012). These characteristics indicate an *alpha-type* microfabric (physicochemical origin) and displacive micritic calcite crystallization by evaporation from supersaturated soil solution (Wright 1990).

The beta-type microfabric (biological origin; Fig. 6h–k) is made of *calcified filaments* (remains of fungal hyphae; Klappa 1979), *calcispheres* (fungal spores; Jones 2011) and *needle-fiber calcite* (bacterial activity during organic matter decomposition and fungal hyphae break; Cailleau et al. 2009).

Table 4 Soil properties of selected soil profiles of salt marshes in the Coastal Zone system

Profile[a] horizon	Depth (cm)	Color (wet)	Structure[b]	Boundary[b]	Gravels (%)	Sand (%)	Silt (%)	Clay (%)	CaCO$_3$ (%)	SOC[c] (%)	pH (se)	pHi[c] (se)	EC[c] (dS/m)	ESP[c] (%)	CEC[c] (cmol$_c$ kg^{-1})
RSM Sa	Haplic Sulfaquent														
Ag	0–20	5Y 3/2	sg	gs	1	3.8	79.6	16.6	0.7	12.90	7.50	4.90	90.2	77.2	52.2
Cg1	20–44	5Y 3/2	sg	gs	0	4.3	79.2	16.5	0.6	14.40	7.20	3.50	92.0	72.0	52.2
Cg2	44–64	5Y 3/2	sg	gs	0	2.3	79.7	18.0	0.5	14.70	7.30	5.40	117.1	82.1	52.2
Cg3	64–138	5Y 3/1	sg	gs	0	3.0	82.0	15.0	0.7	8.80	7.30	3.90	52.7	66.4	51.3
2Cg4	>138	5Y 4/1	sg	–	0	76.7	15.8	7.6	0.9	1.30	7.70	4.30	17.4	35.0	11.7
FSM Sp	Typic Fluvaquent														
A	0–12	7.5YR 5/2	gr	gs	0	11.0	87.1	1.8	1.4	2.60	8.2	8.1	13.9	41.2	47.6
C	12–27	7.5YR 5/2	sg	gs	0	41.6	57.6	0.9	0.4	1.50	7.95	7.3	13.1	38.5	29.2
2Cg1	27–75	10YR 6/2	sg	gs	0	75.6	22.5	1.9	0.2	0.40	7.25	7.1	10.2	36.5	7.0
2Cg2	>75	7.5YR 3/0	sg	–	0	88.2	10.3	1.5	0.1	0.40	3.99	3.7	9.1	26.9	6.1
FSM Lb	Typic Fluvaquent														
A1	0–10	10YR 5/3	sg	aw	0	14.6	78.6	6.8	0.4	4.20	7.73	7.5	10.6	35.7	45.3
2A2	10–35	10YR 6/3	sg	as	0	18.4	68.8	12.8	0.4	2.20	7.22	7.1	14.7	38.2	39.2
3C	35–75	10YR 5/2	sg	gs	0	90.4	8.1	1.5	1.0	0.30	7.05	7.2	6.5	24.7	4.9
3Cg	>75	7.5YR 3/0	sg	–	0	92.9	5.7	1.5	0.3	0.2	4.19	4.2	6.6	20.0	3.1

[a] Abbreviations for soil profile designation: *RSM* Riacho salt marshes, *FSM* Fracasso Sal Marsh, *Sa* S. *alterniflora*, *Sp* *Sarcocornia perennis*, *Lb* L. *brasiliense*
[b] Abbreviations for morphological description are from Schoeneberger et al. (2002); structure: *gr* granular, *sg* single grain, *m* massive, *pl* platy, *pr* prismatic, *sbk* subangular Blocky; size: *vf* very fine, *f* fine, *m* medium; grade: *1* weak, *2* moderated, *3* strong; boundary, distinctness: *a* abrupt, *g* gradual; topography: *w* wavy, *s* smooth, *i* irregular
[c] Soil properties: *EC* electrical conductivity, *SOC* soil organic carbon, *se* saturation extract, *pHi* incubation pH, *ESP* exchangeable sodium percentage, *CEC* cation exchange capacity, *nd* not detected
Data from Ríos (2015)

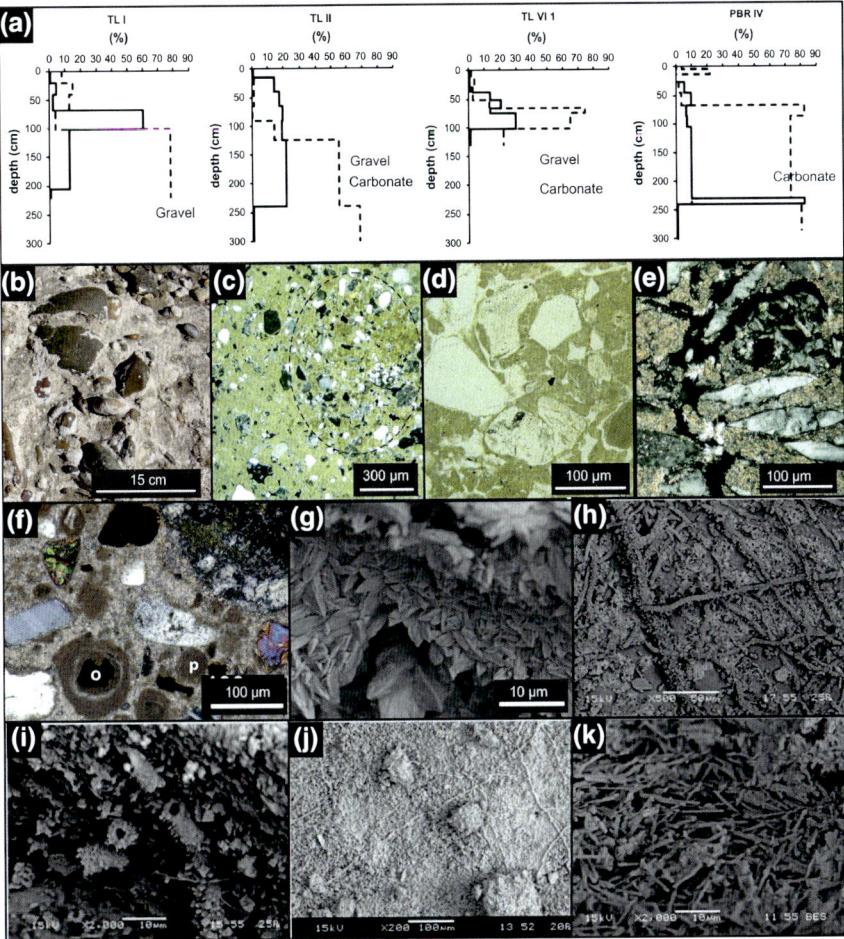

Fig. 6 Calcification. **a** Depth distribution of carbonate (*solid line*) and gravel (*dashed line*) contents of the selected soils of the Península Valdes region. **b** Calcic horizon from terrace level soil, calcite coatings (pendants) with evidence of gravel displacement by growth of authigenic calcite; **c** matrix nodules and floating coarse mineral components in the groundmass (LPP). **d** Calcite nodules surrounded by circumgranular cracks (LPP, TL VI soil). **e** Pedogenic gypsum crystal aggregates of few millimeters in diameter (LPX, TL II soil). **f** o, spherical ooids, p, spherical pellet (LPP, TL I). **g** Rhombohedral micrite (PBR IV soil). **h** and **i** Calcified filaments (PBR IV soil). **j** Calcispheres (PBR IV soil). **k** Needle-fiber calcite (PBR IV soil)

3.3 Gypsification

Another pedogenic feature observed in many soils of the Penísnula Valdés region is the typical lenticular crystals of pedogenic gypsum in the Natrigypsid of TL II soils (Fig. 6e). The lenticular habit of gypsum also indicates high Ca^{2+}/SO_4^{2-} ratios of the soil solution and relatively high temperatures during its growth (Porta 1998).

The multiple lenticular crystallitic aggregates intergrown in the micritic groundmass may be the result of rapid evaporation of the solution migrating through the soil (Watson 1985).

The occurrence of gypsum in the lower part of the profile and below the maximum carbonate accumulation constitutes an evidence of *per descensum*-leaching model (pedogenic origin).

3.4 Redox Processes and Redoximorphic Features in Salt Marshes

Due to the aquic conditions of many soils, the Fe-cations are found in different states of oxidation and conferring colors that may vary from brown, yellowish brown and dark yellowish brown (hues 10YR and 7.5YR). Thus presumably indicating the occurrence of goethite [α-FeO(OH)] in the well drained upper parts of the profiles, to olive, olive gray and dark gray (hue 5Y), presumably indicating the occurrence of green rust [Fe(OH)$_2$] in the strongly hydromorphic soils.

The redoximorphic features (see Sect. 2.3.3) like redox concentration around macropores (as roots and rhizome or rhizo-concretions) (see Sect. 2.3.3), stratified sediments, and nodules showed reddish-brown colors (hue 5YR) presumably indicating the occurrence of ferrihydrite (5Fe$_2$O$_3$.9H$_2$O; Fitzpatrick and Shand 2008). On the other hand, same soils with black horizons emanate a smell of rotten eggs, indicating the occurrence of sulfides, and therefore strongly reduced conditions (Fig. 7a). In sediments with a highly homogeneous texture, an increase in

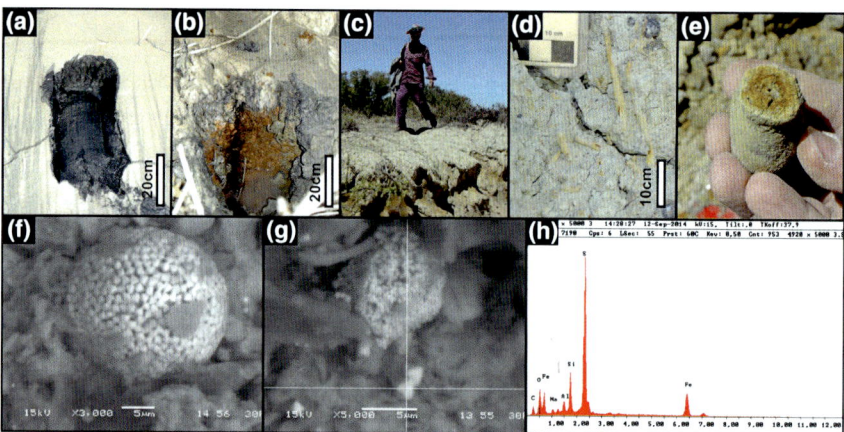

Fig. 7 Redoximorphic pedofeature. **a** Sulfidic materials. **b** Alternating anoxic/oxic condition in lithological discontinuity of the soil profile. **c, d** and **e** Salt marsh paleosol, rhizo-concretions (redox concentrations) in soil matrix (redox depletions). **f** SEM detail of octahedral pyrite in framboid. **g** Microcrystals of octahedral pyrite (twinned crystal). **h** EDS-spectrum on octahedral pyrite

darkness is commonly and positively associated with depth and anoxia intensity (Patrick and Delaune 1972). The soil surveys show that when marshes grew atop Holocene gravel beach ridges, the typical pattern of rapidly increasing anoxic conditions with depth is inverted. Instead, anoxic conditions near the surface of the marsh substratum occur above the oxidized layers (Fig. 7b). Oxygenated and nutrient-rich seawater easily flows horizontally across the gravel deposits of the beach ridges at considerable depth under the soil surface in other marine environments along the Patagonian coast (Esteves and Varela 1991). Thus in the Penísnula Valdés region this seems to be the case for some soil anoxia pattern. Some of this redoximorphic features as rhizo-concretions can be seen preserved in fossil salt marshes associated to beach ridges of Middle Holocene age (Fig. 7c–e).

The high salinity and reducing conditions present in this kind of soils are propitious for the identification of sulfidic materials (Great Group of Sulfaquents) originated by the biological reduction of the sulfates dissolved in the seawater and of Fe^{3+} oxides from sediments (Eqs. 1 and 2). Anaerobic bacteria use the organic matter as their source of energy (Van Breemen 1982).

$$SO_4^{2-} + 2CH_2O + 2H^+ \rightarrow H_2S + 2CO_2 + H_2O \qquad (1)$$

$$2Fe_2O_3 + CH_2O + 8H^+ \rightarrow Fe^{2+} + CO_2 + 5H_2O \qquad (2)$$

The main end product is pyrite (Eq. 3) although iron monosulfides and hydrogen sulfide gas can be formed as precursors.

$$Fe_2O_3(s) + 4SO_4^{2-} + 8CH_2O + 1/2\ O_2 \rightarrow 2FeS_2(pyrite) + 8HCO_3^- + 4H_2O \qquad (3)$$

Scanning electron microscopy (SEM) analysis in Sulfaquents of Riacho San José marsh (Fig. 1) reveals the occurrence of *framboidal pyrite* (Fig. 7f), varying in size from 3 to 35 µm in diameter. Also, numerous individual, twinned and intergrown octahedral microcrystals were also observed (Fig. 7g, h; Rios 2015). Similar observations on pyrite framboids were reported by Wilkin et al. (1996), who were able to calculate that framboids consist of between 102 and 105 discrete pyrite microcrystal. Their formation is related to rapid nucleation from supersaturated solution with respect to FeS_2 (Sawlowicz 1993). When these soils are drained, the sulfidic materials are oxidized forming sulfuric acid producing a drop of the pH below 3 and an increase of Fe and Al in soil solutions to toxic levels (Reddy et al. 1995).

3.5 Genesis of Soil Clay Minerals

Clay mineral types and their genesis in soils could be summarized by describing Terrace Levels I, II, VI, and Pleistocene Beach Ridges IV (see Sects. 2.1.1 and 2.3.2;

Table 5). The clay fraction has a mixture of minerals formed by: illite, interstratified illite–smectite, smectite, kaolinite, palygorskite, and sepiolite (Bouza et al. 2007; Bouza 2014). Illite was identified by X-ray diffraction analysis (XRD) from its characteristic peaks at 10.0 Å (001) and 5.0 Å (002). This mineral predominates in A, C and 2Btn horizons. In general, the 2Bt–2Btk–2Ck sequence horizons, including the 3Btkb–3Bkmb horizons of paleosol of PBR IV, have well-defined peaks in the 14–15 Å (001) region, indicating the predominance of smectites. The XRD patterns of the calcic and petrocalcic deeper horizons of the Terrace level soils, showed smectite and strong peaks at 10.6 Å (110) as well as moderate lines in the 6.4 Å (200) region, indicating the presence of palygorskite. In the 2Bkm horizon of TL I soil and in fragments of the petrocalcic crusts included in the 3Ck horizon of the TL VI-1 soil, sepiolite was identified as the dominant clay mineral together with palygorskite. This conclusion was based on the presence of strong peaks at 12.2 Å (110) and moderate peaks at 4.4 Å (131). Also, in 2Bkm horizon of TL I soil profile; fluorite was identified by the three peaks at 3.15, 1.93, and 1.65 Å (Bouza et al. 2007). The presence of fibrous clay minerals (palygorskite and sepiolite) in the soils studied was confirmed by their typical morphology observed through the scanning electron microscopy (SEM), where a chaotic occurrence of palygorskite fibers were detected in separated clay fraction (Fig. 8a, b). Sepiolite was observed in planar aggregates, the fibers of which had curved and frayed ends (Fig. 8c, d). The occurrence of different types of clay minerals is related to the age of the pedogenetic periods that affected the formation of geomorphic surfaces and to calcretization processes. Pedogenetic carbonate was qualified through X-ray powder diffraction analysis as low-Mg calcite, indicating that during its precipitation, the Mg^{2+} activity increased in the soil solution, favoring the smectite → palygorskite transformation. The petrocalcic horizons (and fragments of them), represent the oldest pedogenetic period, where sepiolite was the dominant clay mineral. During the calcretization processes, the sepiolite was precipitated from the soil solution following the formation of palygorskite.

3.6 Stable Isotopic Composition of C and O from Pedogenic Carbonate

The isotopic composition of $\delta^{13}C$ and $\delta^{18}O$ in pedogenic carbonate was used as indicator of climate influence on soil genesis (Cerling 1984). The $\delta^{13}C$ isotope in pedogenic carbonates is related with $\delta^{13}C$ of biomass from local vegetation communities (C4 vs. C3, vegetation photosynthetic pathways), while that of $\delta^{18}O$ in pedogenic carbonate present a good correlation with the $\delta^{18}O$ of the local water precipitation, which is determined by temperature (Cerling and Quade 1993).

In this section the isotopic composition of $\delta^{13}C$ y $\delta^{18}O$ in soils from Terrace Levels I, II and VI (Rodados Patagónicos) and Pleistocene Beach Ridges IV (Caleta Valdés Formation) is referred (Table 6). The carbon and oxygen isotopic

Table 5 Mineral distribution of the selected soils

Profile Horizon	Bulk samples (powder XRD)							Clay fraction (oriented aggregates)							
	Q	Pl	Fk	Phyll	Calc	Gy	Fl	Sp	P	S	I	K	Q	Pl	Fl
TL I															
A	+++	++	nd	nd	+	nd	nd	nd	nd	++	++	+	++	+	nd
C1	++	+++	nd	++	+	nd	nd	nd	nd	++	++	+	++	+	nd
C2	++	++	nd	++	+	nd	nd	nd	nd	++	++	+	++	++	nd
2Bkmb	++	(+)	nd	+	+++	nd	(+)	+++	++	+	(+)	nd	(+)	+	+
3Ck	++	++	+	++	++	nd	nd	nd	++	++	++	(+)	+	++	nd
TL II-1															
A	++	+++	nd	nd	nd	nd		nd	nd	+	++	+	++	++	nd
2Btn	++	++	+	+++	nd	nd		nd	nd	+	+++	+	+	+	nd
2Btkn	++	+	+	++	++	nd		nd	nd	+++	++	+	+	+	nd
2Ck	++	++	nd	++	++	nd		nd	nd	+++	++	+	+	+	nd
3Cky1	+	+	nd	++	++	++		nd	+++	+	(+)	(+)	+	+	nd
3Cky2	+	+	nd	++	++	++		nd	+++	++	(+)	(+)	+	+	nd
TL VI-1															
A	++	+++	+	++	nd	nd		nd	nd	+	+++	(+)	++	+	nd
C	++	+++	+	++	nd	nd		nd	nd	+	+++	+	+	+	nd
2Bt	++	++	+	+++	+	nd		nd	nd	++	++	+	+	+	nd
2Btk	++	+	+	++	++	nd		nd	nd	++	++	+	+	+	nd
2Bk	++	++	+	++	++	nd		nd	nd	++	++	+	+	+	nd
3Ck	++	++	+	++	++	nd		nd	+++	+	++	+	+	+	nd
CF								+++	++	(+)	+	(+)	(+)	(+)	nd
TL VI-3															
A2	−	−	−	−	−	−		nd	nd	+	+++	+	++	+	nd
C	−	−	−	−	−	−		nd	nd	+	+++	+	++	+	nd
2Bk1	−	−	−	−	−	−		nd	nd	+++	+	+	+	+	nd
2Bk2	−	−	−	−	−	−		nd	++	++	+	+	+	+	nd
2Bk3	−	−	−	−	−	−		nd	+++	++	+	+	+	+	nd
2Bk4	−	−	−	−	−	−		nd	+++	++	+	+	+	+	nd
2Cky1	−	−	−	−	−	−		nd	+++	++	+	+	+	+	nd
3Cky2	−	−	−	−	−	−		nd	++	+++	+	+	+	+	nd

(continued)

Table 5 (continued)

Profile Horizon	Bulk samples (powder XRD)							Clay fraction (oriented aggregates)						
	Q	Pl	Fk	Phyll	Calc	Gy	Fl	Sp	P	S	I	K	Q	Pl	Fl
PBR IV															
A	+++	++	nd	nd	nd	nd		nd	nd	++	++	+	++	+	nd
C	+++	++	nd	nd	nd	nd		nd	nd	+	+++	+	+	+	nd
2Btn	++	+	+	+++	nd	nd		nd	nd	++	+++	+	++	+	nd
2Btkn	–	–	–	–	–	–		nd	nd	++	++	++	+	+	nd
2Bkn	++	++	+	+++	+	nd		–	–	–	–	–	–	–	nd
2Ck	–	–	–	–	–			nd	nd	+	++	++	++	+	nd
3Btkb	++	+	+	+++	+	nd		nd	nd	++	++	+	+	++	nd
3Bkmb	–	–	–	–	–			nd	nd	+++	++	+	+	+	nd

Q quartz, *Fk* potassic feldspar, *Pl* plagioclase, *Phyll* phyllosilicates, *Calc* calcite, *Gy* gypsum, *Fl* fluorite, *Sp* sepiolite, *P* palygorskite, *S* smectite, *I* illite, *K* kaolinite, *nd* not detected, (+) traces, + 5–10%, ++ 15–40%, +++ > 40%, *CF* calcrete fragment
Data from Bouza (2012)

Fig. 8 Morphologies of fibrous clay minerals, taken from Bouza et al. (2007) with modifications. **a** Palygorskite fibers grouping from 2Bk4b horizon, TL VI-3 soil. **b** Chaotic arrangement of palygorskite microfibers from 3Ck2 horizon, TL VI-1 soil. **c** Sepiolite fibers grouping from 2Bkmb horizon, TL I soil. **d** Details of c-scanning electron micrograph: planar aggregate of microfibers, with curved ends and spherules of microcrystal aggregates of fluorite and opal CT (*arrows*)

Table 6 Isotope composition $\delta^{13}C$ and $\delta^{18}O$ of pedogenic carbonates from calcic and petrocalcic horizons

Profile horizon	Morphology	Depth (cm)	$\delta^{13}C‰$ (VPDB) (± 0.1)	$\delta^{18}O‰$ (VPDB) (± 0.1)	C4 (%)	Paleo-Temperature* (°C)
TL I						
2Bkm	Petrocalcic crust	68–102	−3.8	−5.3	58	16.9
3Ck	Bulk sample	>102	−4.7	−6.0	51	14.2
TL II-1						
2Btkn	Calcitic nodules	15–40	−5.2	−4.2	48	21.2
2Bkn	Calcitic nodules	40–65	−5.0	−4.7	49	19.4
3Cky1	Bulk sample	65–91	−5.5	−5.2	46	17.6
3Cky2	Bulk sample	91–124	−5.1	−5.3	49	16.9
3Cky3	Bulk sample	124–240	−4.5	−4.8	53	18.7
TL IV-1						
2Btk	Calcitic nodules	38–52	−3.9	−1.9	57	29.2
2Bk	Calcitic nodules	52–66	−3.4	−2.4	61	27.5
	Petrocalcic fragment		−2.9	−3.1	64	25.1
	Petrocalcic fragment		−3.2	−3.6	62	23.4
3Ck	Bulk sample	66–74	−3.7	−2.8	59	26.2
3Cky	Bulk sample	74–103	−4	−3.4	56	24.1
PBR IV						
2Btkn	Bulk sample	26–46	−6.2	−3.5	41	23.7
	Calcitic nodules		−6.8	−3.4	36	23.9
2Bkn	Calcitic nodules	46–67	−5.9	−4.6	43	19.3
	Calcitic nodules		−5.3	−3.9	47	22.0
2Ck	Fillings between gravels	67–87	−8.3	−5.6	26	15.8
	Pendants		−9.7	−6.7	16	11.0
3Btkb	Fillings between gravels	87–235	−7.8	−6.0	29	14.6
	Pendants		−7.9	−6.3	29	13.1
	Pendants		−7.7	−5.8	30	14.9
3Bkmb	Petrocalcic crust	235–240	−7.8	−5.3	29	17.2

*Paleotemperatures estimated according to Dworkin et al. (2005)

compositions of carbonates (Table 6) showed an acceptable positive correlation (r^2 = 0.5, p < 0.001; Bouza 2012, 2014) indicating a correspondence between assemblages of plants with higher proportion of C4, which are tolerant to water stress and high temperatures (values more positive of $\delta^{13}C$ and $\delta^{18}O$ respectively).

The paleosols (3Btkb–3Bkmb), formed in littoral gravelly deposits of the Pleistocene Beach Ridge (Caleta Valdés Formation), present a dominance of smectite clays (see Sect. 2.5) and lower $\delta^{13}C$ and $\delta^{18}O$ contents, indicating a higher proportion of C3 plants and a seasonally contrasting, Mediterranean type climate (xeric soil moisture regime). Considering that these littoral deposits could have been formed during MIS 5e (see Chapter "Late Cenozoic Landforms and Landscape Evolution of Península Valdés"), the paleosol might have been developed at the end of this interglacial isotopic stage (MIS 5a–c). This paleosol was probably eroded and covered by sedimentary episodes, probably associated to periglacial processes of isotope stages 4–2 (*Llanquihue Glaciation*; Mercer 1976). These deposits, along with the eolian deposits, were the parent materials of a new soil that evolved to a 2Btb–2Btkb–2Bkb–2Ck horizon sequence during a period of environmental stability. This pedogenetic event is probably associated to a more humid period, with a marked seasonality in the precipitations that occurred during the Late Glacial—Early Holocene, followed by a drier period and a temperature increase that was recorded in the region during the mid Holocene. The third pedogenetic event is represented by the A–C horizon sequences. Besides, the oldest pedogenic carbonates in Terrace levels of Rodados Patagónicos, have a higher proportion of C4 plants and they are associated with typical fibrous clay minerals (palygorskite and sepiolite) of arid and hyper-arid regions. This pedogenic carbonates would have developed in a warmer and drier interglacial period, presumably during MIS 11, with estimated maximum mean annual temperature about 22 °C (Table 6).

4 Perspectives and Future Work

As has been discussed in this chapter, studies on soil–geomorphology relationships constitute an important tool for Quaternary geology and paleosols (paleoecology, paleoclimate), geoecology (distribution patterns of vegetation and soil–plant relationships), and soil degradation studies. However, studies in detail micro-scale will be needed to elucidate both the polygenesis and the pedological evolution of the landscapes of Peninsula Valdes. Besides, studies more accurately on geochemistry will be necessary to elucidate investigative aspects about acid generation in salt marshes and isotope composition on pedogenic carbonate as paleoecological indicators.

Acknowledgements The authors would like to thank the helpful reviews of Profs. Perla Imbellone, Héctor Morrás and Augusto Varela which improved the final version of this manuscript. This research has been funded by the CONICET (PIP 2014 00190 CO) and Agencia Nacional de Promoción Científica y Tecnológica (PICT 2013-1876).

Glossary

Chronosequence	A sequence of related soils that differ, one from the other, in certain properties primarily as a result of time as a soil-forming factor
Framboidal pyrite	From French word: framboise, raspberry patterns; spheroidal aggregates of octahedral pyrite microcrystals
Geomorphic surface	A mappable area of the earth's surface that has a common history; the area is of similar age and is formed by a set of processes during an episode of landscape evolution. A geomorphic surface can be erosional, constructional or both
Lithostratigraphic unit	A body of rocks that is defined and recognized on the basis of its lithology or combination of lithologic properties and stratigraphic relations and not by their inferred age, the time span they represent, inferred geologic history, or manner of formation
Llanquihue glaciation	Sequence of glacier fluctuation for the last glaciation interval registered in the Chilean Lake District from 70 to about 14 ka (marine Oxygen Isotope Stages 4–2)
Marine Isotope Stage 5e (MIS 5e)	Stage in the geologic temperature record determined through the benthic oxygen isotope records from almost all areas of the ocean, between 80 and 130 ka ago. Sub-stage MIS 5e covers the last major interglacial period before the Holocene
Matrix nodules	(micromorphology) Matrix pedofeature unrelated to surfaces. Matrix pedofeatures result from changes in composition of the groundmass; this implies that several characteristics of the groundmass remain visible in the pedofeature. Most commonly the nature and distribution pattern of the coarse fraction are preserved
Micritic calcite	Term used to indicate individual crystals of fine calcite (<4 µm)
Pendants	Coating on the lower surface of a free or embedded grain or aggregate (e.g., gypsum or calcite bear below a pebble)
Ooids	Carbonate grains which commonly range in diameter from 0.25 to 2.00 mm. They are spherical to ellipsoidal with a nucleus covered by one or more concentric coatings that under microscope show a series of concentric laminations

Paleosols	A soil that formed on a landscape in the past with distinctive morphological features resulting from a soil-forming environment that no longer exists at the site. The former pedogenic process was either altered because of external environmental change or interrupted by burial
Pellets	It is a grain (particles that have formed by chemical or biochemical precipitation) composed entirely of micrite. They are elongated with size range between 0.03 and 0.3 mm and generally structureless
Polygenetic soils	A soil that has been formed by two or more different and contrasting processes so that all the horizons are not genetically related
Psammophile plant	A psammophile (meaning "sand loving") is a plant that lives on sand or sandy soils
Soils complex	A soil complex consists of areas of two or more soils, so intricately mixed or so small in size that they cannot be shown separately on the soil map

References

Allen J, Pye K (1992) Coastal saltmarshes: their nature and importance. In: Allen J, Pye K (eds) Saltmarshes: morphodynamics, conservation and engineering significance. Cambridge University Press, Cambridge, pp 1–18

Birkeland P, Machette M, Haller K (1991) Soils as tool for applied Quaternary Geology. Utah Geological and Mineral Survey. Utah Department of Natural Resources, publications 91-3, p 64

Bouza PJ (2012) Génesis de las acumulaciones de carbonatos en Aridisoles Nordpatagónicos: su significado paleopedológico. Revista de la Asociación Geológica Argentina 69(2):298–313

Bouza PJ (2014) Paleosuelos en cordones litorales de la Formación Caleta Valdés, Pleistoceno superior, NE del Chubut. Revista de la Asociación Geológica Argentina 71(1):1–10

Bouza PJ, del Valle H (2014) Capítulo 8: Propiedades y génesis de las acumulaciones de carbonatos en Aridisoles del centro-este del Chubut. In: Imbellone P (ed) Suelos con acumulaciones calcáreas y yesíficas. Asociación Argentina de la Ciencia del Suelo, Buenos Aires, Argentina, pp 199–219

Bouza PJ, Simón M, Aguilar J, Rostagno M, del Valle H (2005) Chapter 1: genesis of some selected soils in the Valdés Peninsula, NE Patagonia, Argentina. In: Faz Cano A, Ortiz R, Mermut A (eds) Genesis, classification and cartography of soils. Adv Geoecol Catena Verlag, Reiskirchen, pp 1–12

Bouza PJ, Simón M, Aguilar J, del Valle H, Rostagno CM (2007) Fibrous-clay mineral formation and soil evolution in Aridisols of northeastern Patagonia, Argentina. Geoderma 139:38–50

Bouza PJ, Blanco PD, Álvarez P, del Valle H (2008) Estudio de caso Chubut: Centro-Norte Península Valdés. In: Cantú M, Becker A, Bedano C (eds) Evaluación de la sustentabilidad ambiental en sistemas agropecuarios. Fundación Universidad Nacional de Río Cuarto, Río Cuarto, pp 165–181

Brückner H, Schellmann G, Daut G, Mäusbacher R, Schnack E, Schneider H (2007) Erste Befunde zu Veränderungen des holozänen Meeresspiegels und zur Größenordnung holozäner 14C-Reservoireffekte im Bereich des Golfo San José (Península Valdés, Argentinien). Bamberger Geographische Schriften 22:93–111

Buol SW, Hole FD, Mccracken RJ (1990). Génesis y clasificación de suelos (2nd ed) Trillas, México, p 417

Cailleau G, Verrecchia E, Braissant O, Emmanuel L (2009) The biogenic origin of needle fibre calcite. Sedimentology 56:1858–1875

Cerling TE (1984) The stable isotopic composition of modern soil carbonate and its relationship to climate. Earth Planet Sci Lett 71(2):229–240

Cerling TE, Quade J (1993) Stable carbon and oxygen isotopes in soil carbonates. In: Swart PK, Lohmann KC, McKenzie J, Savin S (eds) Climate change in continental isotopic records. Geophys Monogr 78:217–231

Dworkin S, Nordt L, Atchley S (2005) Determining terrestrial paleotemperatures using the oxygen isotopic composition of pedogenic carbonate. Earth Planet Sci Lett 237:56–68

Esteves JL, Varela DE (1991) Nutrient dynamic of the Valdés Bay-Punta Cero pond system (Península Valdés, Patagonia) Argentine. Oceanol Acta 14(1):51–58

Fitzpatrick R, Shand P (2008) Inland acid sulfate soils: overview and conceptual models. In: Fitzpatrick R, Shand P (eds) Inland acid sulfate soil systems across Australia. CRC LEME Open File Report 249, Perth, Australia, pp 6–74

Galloway R, Markgraf V, Brudbury J (1988) Dating shorelines of lakes in Patagonia, Argentina. J S Am Earth Sci 1(2):195–198

Gile L, Hawley H, Grossman R (1981) Soil and geomorphology in the basin and range area of southern New Mexico. Guidebook to the Desert Proyect, Memoir 39. New Mexico Bureau of Mines and Mineral Resoursers, Mexico, p 222

Jones B (2011) Biogenicity of terrestrial oncoids formed in soil pockets, Cayman Brac, British West Indies. Sediment Geol 236:95–108

Klappa CF (1979) Calcified filaments in Quaternary calcretes: organo-mineral interactions in the subaerial vadose environment. J Sediment Petrol 49:955–968

Laya HA, Pazos MS (1976) Horizontes B2t discontinuos, estudio micromorfológico. IDIA, suplemento N° 3. Séptima Reunión Argentina de la Ciencia del Suelo, diciembre 1976. Bahía Blanca, pp 728–733

Mercer JH (1976) Glacial history of southernmost South America. Quatern Res 6:125–166

Nettleton WD, Peterson F (1983) Chapter 5, Aridisols. In: Wilding L, Smeck N, Hall G (eds) Pedogenesis and soil taxonomy II, the soil orders. Developments in soil science 11B. Elsevier, Amsterdam, pp 165–215

Patrick WH, Delaune RD (1972) Characterization of the oxidized and reduced zones in flooded soil. Soil Sc Soc Am J 36(4):573–576

Porta J (1998) Methodologies for the analysis and characterization of gypsum in soils: a review. Geoderma 87:31–46

Reddy KJ, Wang L, Gloss SP (1995) Potential solid phases controlling dissolved aluminium and iron concentration in acidic soils. In: Date RA et al (eds) Plant and soil interaction at Low pH. Kluwer Academic Publishers, Netherlands, pp 35–40

Ríos I (2015) Relaciones edafo-geomorfológicas y geo-ecología de plantas vasculares en marismas patagónicas: propiedades morfológicas, físicas, químicas y biogeoquímicas. Doctoral Thesis. Facultad de Ciencias Exactas, Físicas y Naturales, Universidad Nacional de Córdoba, Argentina, p 170

Rostagno CM (1980) Propiedades, génesis de dos suelos de Península Valdés. IX Reunión Argentina de las Ciencias del Suelo, Paraná, pp 955–963

Rostagno CM (1981) Reconocimiento de suelos de Península Valdés. SECyT. Serie contribución n° 44. Centro Nacional Patagónico (CENPAT-CONICET), Puerto Madryn, Argentina, p 24

Rostami K, Peltier W, Mangini A (2000) Quaternary marine terraces, sea-level changes and uplift history of Patagonia, Argentina: comparisons with predictions of the ICE-4G (VM2) model of the global process of glacial isostatic adjustment. Quat Sci Rev 19:1495–1525

Sawlowicz Z (1993) Pyrite framboids and their development: a new conceptual mechanism. Geol Rundsch 82:148–156

Schäbitz F (1994) Holocene climatic variations in northern Patagonia, Argentina. Palaeogeogr Palaeoclimatol Palaeoecol 109:287–294

Schoeneberger PJ, Wysocki DA, Benham EC, Broderson WD (2002) Field book for describing and sampling soils, Version 2.0. Natural Resources Conservation Service, National Soil Survey Center. Lincoln, Nebraska, p 298

Soil Survey Staff (1999) Soil taxonomy. A basic system of soil classification for making and interpreting soil surveys; 2nd edition. Agricultural Handbook 436; Natural Resources Conservation Service, USDA, Washington, p 869

Stuart DM, Dixson RM (1972) Water movement and caliche formation in layered arid and semiarid soils. Soil Sc Soc Am Proc 37:323–324

Súnico A (1996) Geología del Cuaternario y Ciencia del Suelo: relaciones geomórficas-estratigráficas con suelos y paleosuelos. Doctoral Thesis, Facultad de Ciencias Exactas y Naturales, Departamento de Graduados, Universidad Nacional de Buenos Aires, p 227

Súnico A, Bouza P, del Valle H (1996) Erosion of subsurface horizons in Northeastern Patagonia, Argentina. Arid Soil Res Rehab 10:359–378

Van Breemen N (1982) Genesis, morphology and classification of acid sulphate soils in coastal plains. In: Kittrick JA, Fanning DS, Hossner LR (eds) Acid sulphate weathering. SSSA Special Publication 10. Soil Sc Soc Am J, Madison, Wisconsin, pp 95–108

Watson A (1985) Structure, chemistry and origin of gypsum in southern Tunisia and in central Namib Desert. Sedimentology 32:855–875

Wilkin RT, Barnes HL, Brantley SL (1996) The size distribution of framboidal pyrite in modern sediments: an indicator of redox conditions. Geochim Cosmochim Ac 60:3897–3912

Wright VP (1990) A micromorphological classification of fossil and recent calcic and petrocalcic microstructures. In: Douglas LA (ed) Soil micromorphology: a basic and applied science. Developments in Soil Science. Elsevier, Amsterdam, pp 401–407

Soil Degradation in Peninsula Valdes: Causes, Factors, Processes, and Assessment Methods

Paula D. Blanco, Leonardo A. Hardtke, Cesar M. Rostagno,
Hector F. del Valle and Gabriela I. Metternicht

Abstract In semiarid rangelands where the anthropogenic impact is currently increasing, as occurs in the rangelands of the Península Valdés, the detrimental impacts of soil degradation on land resources became really dramatic. This chapter presents a review on the current knowledge of soil degradation in the Península Valdés rangelands. Section 1 introduces the chapter, Sect. 2 focuses on soil degradation main processes, factors and causes, and Sect. 3 presents a review of soil degradation assessment methods and several soil degradation studies carried out since 1990 in the Península Valdés region. Water and wind erosion are the degradation processes that are most strongly evidenced. Major causes of soil degradation are attributed to a combination of climatic and *anthropic* factors, with overgrazing being perceived to be a major factor. Four key causes associated with overgrazing in the Península Valdés region rangelands are described: (1) Poor range management with respect to flock distribution and overstocking, (2) Limited access to information, (3) Top-down and largely ineffective government policy, and (4) Overdependence on grazing systems for sustained livelihoods. Assessment methods for assessing soil degradation include: expert judgment, remote sensing,

P.D. Blanco (✉) · L.A. Hardtke · C.M. Rostagno · H.F. del Valle
Instituto Patagónico para el Estudio de los Ecosistemas Continentales (IPEEC), Consejo Nacional de Investigaciones Científicas y Técnicas (CONICET)—CCT Centro Nacional Patagónico (CENPAT), Boulevard Brown 2915, ZC: U9120ACD, Puerto Madryn, Chubut, Argentina
e-mail: blanco@cenpat-conicet.gob.ar

L.A. Hardtke
e-mail: hardtke@cenpat-conicet.gob.ar

C.M. Rostagno
e-mail: rostagno@cenpat-conicet.gob.ar

H.F. del Valle
e-mail: delvalle@cenpat-conicet.gob.ar

G.I. Metternicht
Institute of Environmental Studies (IES), University of New South Wales (UNSW), Sydney, Australia
e-mail: g.metternicht@unsw.edu.au

productivity changes, field monitoring, pilot studies at farm level based on field criteria and expert opinion, and modeling.

Keywords Rangelands · Overgrazing · Soil erosion · Soil degradation assessment · Remote sensing

1 Introduction

Soil degradation, either natural or induced by humans, is considered one of the main factors responsible for declining soil fertility potential and desertification (Bestelmeyer et al. 2006; Michaelides et al. 2009; Schlesinger et al. 1990). It is one of the most important environmental problems in the world and can irreversibly affect soil productivity. In arid and semiarid regions, soil degradation through topsoil loss by water or wind erosion, can have a high impact on soil quality, as soils are generally shallow. In these regions soil erosion has been considered a significant component of the desertification processes (Rostagno 1989; Schlesinger et al. 1990; Ravi et al. 2010).

Drylands of Peninsula Valdes are fragile ecosystems, dominated by rangelands exposed to increasing anthropogenic impacts, with detrimental effects on the state of land resources. Soil formation in these soils, dominated by Entisols and Aridisols with poor concentration of organic matter and coarse textured, is a slow process, often taking thousands of years to form a few inches of topsoil, which under mismanaged or overgrazed systems can be lost on the order of months to years due to accelerated rates of wind and water erosion (Chartier et al. 2009).

This chapter reviews current knowledge of soil degradation in the PV rangelands. Section 2 focuses on causes, factors and processes leading to soil degradation in this geographic area. A review of soil degradation assessment methods follows (Sect. 3), including several soil degradation studies carried out since 1990 in Península Valdés.

2 Processes, Factors, and Causes of Soil Degradation in Península Valdés Rangelands

2.1 Soil Degradation Processes and Factors in Península Valdés Rangelands

The main soil degradation processes occurring in semiarid rangelands are compaction, wind and water erosion, deposition of sediments and loss of soil structure. The factors of soil degradation are the biophysical environmental characteristics that determine the degradation processes, including the following: soil properties (e.g., texture, clay minerals, structure), climate (e.g., rainfall amount and intensity, wind severity), terrain (e.g., slope gradient, slope length), and vegetation (e.g., percent ground cover, vegetation spatial pattern). Península Valdés experiences severe levels of soil degradation, largely due to the interaction between fragile soils of high erodibility, winds of high erosivity, low vegetation cover, and poor natural resource management. Degradation assessments carried out in several areas of Península Valdés describe environmental conditions and main degradation processes affecting different landscape units. The review of these works points to water and wind erosion as the mechanisms driving land degradation in the area.

This section describes these processes and their factors.

2.1.1 Water Erosion

As in others semiarid rangelands, soils of Península Valdés are very susceptible to water erosion. Scarce vegetation cover and increased soil compaction resulting from continued grazing of high stocking-density over the years, are exposed to torrential precipitation events, of short duration and high intensity. Rainfall frequency and magnitude, in combination with the soil geomorphic thresholds are main factors related to rainfall erosion potential (Bisigato et al. 2009). High rainfall intensity drives runoff (Renard et al. 1974). Península Valdés receives a high amount of rainfall over a short period of time in the year (see Chapter "The Climate of Península Valdés Within a Regional Frame"); these short-term, high-intensity events are more erosive than the total amount of precipitation. For example, an analysis of precipitation events in Península Valdés for the period 2006–2007 showed a total of 115 events, with precipitation amount varying from 1 to 26.9 mm (Blanco 2010); 65% of the rainfall events recorded less than 1 mm, and on average, only 6% of rainfall events were greater than 10 mm, with low probability of occurrence but with high potential consequences due to their erosivity. Overall, rainfall intensity recorded over 10 min varied from 0.98 to 87 mm/h. These conditions favor surface runoff, and soil detachment occurs through overland flow and sheet wash (sheet erosion) or concentrated flows (rills or gullies) (Blanco 2010), and is defined as the balance between *erosivity* and *erodibility*. Península Valdés soils are very erodible due to the relatively low clay and organic matter content. Native vegetation patterns, characterized by plant-cover patches called "islands of fertility", are very significant in the regulation of surface hydrological processes. Accelerated soil water erosion in these rangelands has created a mosaic of different vegetation types, each of them associated with a particular soil surface condition. Large tracts of the once dominant grass with scattered shrubs steppe have been transformed into *Chuquiraga avellanedae* dominated shrub steppes, characterized by patchy, discontinuous cover of *desert pavements*. Areas dominated by grasses exhibit soils with intact A horizon and high contents of coarse fragment in the soil matrix. A high correlation between surface gravel and the thickness of partially eroded A horizon evidences accelerated soil erosion has played an important role in the formation of desert pavements.

2.1.2 Wind Erosion

Wind erosion is widely evidenced throughout Península Valdés, favored by climatic characteristics of windy conditions (see Chapter "The Climate of Península Valdés Within a Regional Frame"), with annual mean wind speed >4.0 m/s. However, westerly winds are less predominant here, while north and northeast winds become more frequent given the stronger influence of the southwest Atlantic anticyclone (Prohaska 1976; Paruelo et al. 1998, see Chap. 4). Severe and frequent wind storms occur mainly during spring and summer (September–February), with wind speeds regularly exceeding 120 km/h. Wind storms can cause significant aeolian sediment transport especially on barelands. When the protective steppe vegetation cover is removed by human activities and sheep *overgrazing*, soils become exposed to wind action (Rostagno et al. 2004).

Landforms resulting from aeolian erosion and accumulation are present in the different landscape units of Península Valdés. In the *wind erosion* process three transport modes can be distinguished: *creep, saltation*, and *suspension* (Bagnold 1941). It is a process influenced by factors including the severity of the climate, soil susceptibility to erosion (erodibility) and the soil surface conditions (soil texture, soil roughness, and soil cover). The wind removes not only soil particles, but also organic matter and nutrients, affecting rangeland productivity through declining physical and chemical fertility of the topsoil.

Soil degradation by sand encroachment is present in Península Valdés, with evident progression of active sand dunes fronts, grouped in discrete megapatches (see chapter "Late Cenozoic Landforms and Landscape Evolution of Península Valdés"). Prevailing strong winds cause eastward sand migration and associated dune mobility (del Valle et al. 2008). The shape of the coastline influences air circulation; dunefields are related to the loose sediments of these windward coasts that are unique cases in the whole Patagonia according to the main orientation of the shoreline. Three dunefields located in the southern portion of Península Valdés cover approximately 590 km^2, and are the most remarkable aeolian feature of Península Valdés, evidencing the advance of numerous dune fronts (del Valle et al. 2000).

2.2 Causes of Soil Degradation in Península Valdés Rangelands

Major causes of soil degradation in semiarid rangelands have been attributed to a combination of climatic and *anthropic* factors, with *overgrazing* being perceived to be a major factor of soil degradation in Patagonia (Soriano and Movia 1986; Ares et al. 1990).

In the previews section rainfall and wind as triggering factors of soil degradation processes were analyzed. In addition, wildfires are one of the major disturbance agents in many ecosystems and they are one of the main environmental issues in Península Valdés (Villagra et al. 2009). Studies on the role of wildfires in rangelands show their strong impact on ecosystem functional processes, affecting bio-geo-chemical cycling, and determining the plant species structure and composition (Bowman 2009). Península Valdés is prone to wildfires during late spring and summer, when high winds and temperatures are coupled with low relative humidity (see Chapter "The Climate of Península Valdés Within a Regional Frame"). In this region, wildfires reduce plant biomass drastically; followed by a relatively quick recovery of the herbaceous strata and the much slower recovery of the shrubby strata thereafter (Rostagno et al. 2006).

Sheep grazing for wool production is the main landuse of Península Valdés rangelands, with extensive continuous grazing practices and stocking rates ranging from 0.34 to 0.95 heads/ha annum. The area caters for 47 established sheep farms ranging from 1000 to 10,000 ha, and 62% with a size between 2000 and 5000 ha. Around 160,000 sheep graze in Península Valdés with an average wool production of 1.83 kg/ha, ranging from 1.28 to 4.79 kg/ha. Hereafter main causes associated with *overgrazing* in Península Valdés rangelands are described.

Possible root causes of poor range management are grazers' limited knowledge and awareness of early signs and effects of soil degradation. Producers apply traditional management, and their perception of risk includes annual variation of productivity with rainfall and market variations. Slow processes such as grass replacement or topsoil loss that undermine the productivity of rangelands over decades fail to be noticed. Historical loss of range carrying capacity is explained in terms of climate changes with reduced rainfall.

Another cause is the low level of technology applied in livestock management that support uncontrolled grazing throughout the paddocks. Land suitability to range use is frequently ignored, leading to overestimation of the carrying capacity, and overstocking. Likewise are extension officers' recommendations on flock distribution, with the consequence that land has been subdivided in an orthogonal pattern, barely accounting for the distribution of watering holes. Information needed for technical projections and decision-making is incomplete, dispersed and does not reach producers and policy makers in an orderly way. Limited access to information and advice on timing of stocking rate variation increases the probability of losses due to climatic events (droughts, fires, high-intensity precipitation).

In addition to the aforementioned knowledge gaps, pastoral systems are vulnerable to external factors such as market prices and climatic fluctuations. Uncertainty reduces profitability of the system, increases poverty, induces migration to seek off-farm income and reduces the level of management available to prevent rangeland degradation. Since the years 1950–1960, this environmental context has been aggravated by the weakening of the sheep industry in Patagonia, comprising both wool and meat production, and negatively influenced by the vagaries of Argentine politics and by globalization, especially through the great variations for agricultural products in the international market. As result surviving family-based farms are exposed to greater vulnerability, especially the smaller ones, who had somehow found ways of adapting their management to mitigate a declining productivity.

Knowledge gaps also exist on viable alternative production options for economic diversification of ranches and non-pastoral opportunities. The lack of effective alternatives for economic diversification of the farm system, plus market prices that do not favor good practices over inappropriate ones, constitute root causes that contribute to economic vulnerability.

3 Soil Degradation Factors and Processes Assessment Studies Conducted in Península Valdés Rangelands

Common methods for assessing soil degradation as identified in the LADA-FAO approach (Koohafkan et al. 2003) include: expert judgment, remote sensing, productivity changes, field monitoring, pilot studies at farm level based on field criteria and expert opinion, and modeling. Table 1 shows the relationship between methods and survey scales. In practice, synergistic uses combining several of these

Table 1 Common methods for assessing land degradation

Method	Scale	Comments
Expert judgment	Global, small scale	Subjective, particularly in the definition of degradation classes: not degraded; slight, moderate, severe, very severe degradation Hardly reproducible Relatively low cost
Remote sensing	Global, regional, sub-regional and local	Acquisition on a repetitive basis enables monitoring Clear identification of indicators (direct or indirect) is needed Indicator must carry direct spectral absorption features or be correlated to a soil chromophera Cost varies with platform and sensor used
Field monitoring	Subnational, large scale	Costly, depending on the intensity of fieldwork Stratified sampling recommended Enables monitoring over time
Pilot studies at plot level	Local	Enables a grass roots view on the severity of degradation and its causes Relying on field indicators of degradation can be subjective Can be costly depending on area coverage
Modeling	Global to local	Uses established models for soil erosion by wind and water Enables the integration of biophysical with socioeconomic factors Prediction of degradation hazard
Productivity changes	National, local	Uses crop performance indicators, biomass production related to land degradation as an expression of lowered productivity. Productivity decline could be caused by factors other than land degradation Reliable data sources are required (e.g., national yield statistics, yield monitoring, etc.)

approaches are more common than the implementation of individual methods. Hereafter we analyze soil degradation factors and processes assessment studies conducted in Península Valdés rangelands.

Existing assessments of soil and land degradation have used a variety of geospatial technologies and approaches, depending on scale and soil degradation process to be analyzed. Field methods and leading practice in the synergistic use of remote sensing, field surveys, and GIS-based analysis modeling are described in the next sections. Remote sensing provides homogeneous data over large regions with a regular revisit capability, and can therefore greatly contribute to regional landscape erosion assessment and prediction (Vrieling 2006), including factors and drivers as described hereafter.

3.1 Desert Pavement Formation and Its Use for Soil Erosion Assessment

Desert pavements (the continuous soil cover of rounded or angular stones) are prominent features of many geomorphic surfaces in arid lands (Dregne 1977). In the semiarid soils of north-eastern Patagonia, gravel commonly covers interspace areas of shrub-dominated communities, in contrast with that of grass-dominated patches where gravel cover is either absent or negligible.

The cover and size of the coarse fragments may affect the dynamics of various hydrological and soil degradation processes (Poesen et al. 1998; Cerdà 2001). Desert pavements have been described in soils of different landscapes (Cooke et al. 1993), although no single set of processes seem to be uniquely responsible for their formation. Cooke et al. (1993) described three groups of particle concentration processes: (1) wind erosion, (2) surface runoff removal of fines, and (3) shrink-swell process of soils causing upward migration of coarse particles. In shrub-dominated communities, rainsplash erosion can contribute to desert pavement formation (Parsons et al. 1992), although Wainwright et al. (1995) considered that raindrop erosion on its own cannot account for the development of pavements suggesting that other mechanisms leading to the surface concentration of coarse particles must also operate. Additionally, the excavating activity of fossorail fauna (organisms adapted for digging such as rodents) causes upward transport of coarse fragments, which can also favor their concentration on the soil surface (Johnson et al. 1987).

Desert pavements are typically found in areas where soils containing coarse fragments dominate, and where plant cover is sparse and there is little impediment to wind and water erosion. Importantly, they are prominent in those areas where erosion has been accelerated, as is the case of many overgrazed rangelands (Simanton et al. 1984), and where some grasslands have been transformed into shrub lands (Herbel 1979). In some shrub-dominated rangelands, the surficial A-soil horizon between the shrubs has been eroded, leaving swales mantled by a gravel lag which forms a desert pavement (Abrahams et al. 1995).

In Punta Ninfas rangelands (the point that constitute the entrance to the Golfo Nuevo, in front of Península Valdés). Beeskow et al. (1995) described the changes from grass-shrub steppes into shrub steppes as a consequence of sheep grazing. Sheep grazing affects directly the cover of perennial grasses and indirectly the soil stability, accelerating soil degradation through the loss of soil organic matter, and topsoil crust development. Soil erosion in the shrub interspaces favors the formation of desert pavements and mounds associated to shrubs (Fig. 1).

Soil erosion can be estimated considering the surface gravel mass and the thickness of the remnant A horizon (depth to Bt-soil horizon). Rostagno and Degorgue (2011) determined the mean gravel content (fragments >2.0 mm) in well conserved Xeric Calciargid soils and then determined the gravel mass per unit area of land (i.e. kg/m^2) produced by the removal of a unit depth of soil (fractions <2.0 mm). Soil erosion was estimated as the ratio between the surface gravel mass and the mean gravel mass present in a unit depth of the well conserved soil for the same position. Soil erosion was also determined as the difference between mean thickness of the A horizon of the well conserved soil and that of the remnant A horizon of the eroded soil. The relationship between gravel cover and mass, and that between gravel mass and the depth to the Bt horizon (the A horizon thickness) can then be analyzed through regression analysis, establishing for a given soil, the relationship between gravel cover (desert pavement) and the eroded layer. Rostagno and Degorgue (2011) found a strong correlation between the surface gravel mass and the thickness of the remaining A horizon. Regression analysis indicated that depth to the Bt horizon explained 54, 84, and 63% of the variability in coarse fragments concentration on the soil surface for the upper, middle, and lower slope positions of a flank pediment, respectively. The mean soil erosion for the upper,

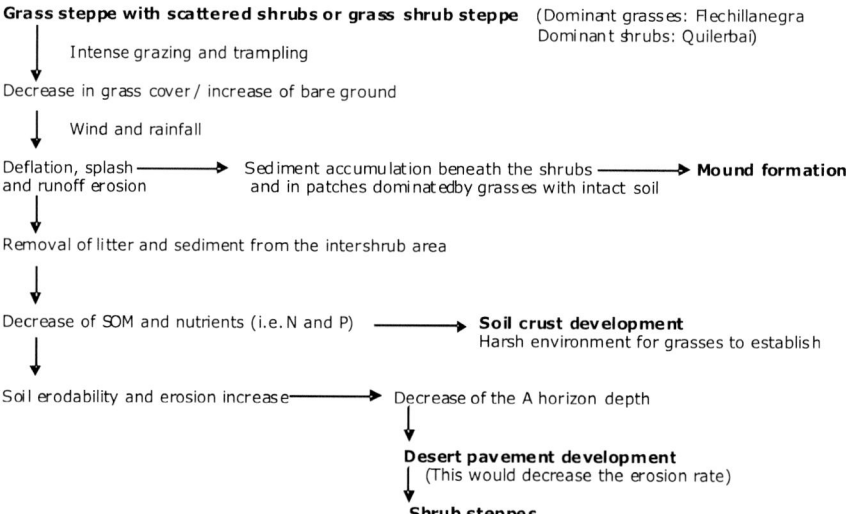

Fig. 1 Model of soil degradation and desert pavement formation in Xeric Calciargids of the central Península Valdés region

middle, and lower slope positions, as determined by the surface gravel mass, were 28.3, 27.0, and 33.1 mm, respectively. However, the mean soil loss for each of these positions, as determined by the difference between the intact and the remaining A horizon of the eroded areas, were 50.0, 52.5, and 82.0 mm, respectively. Where most of the A horizon was removed, the Bt was exhumed and a very pale brown (10 YR 7/3), vesicular (Av) horizon develops on top of it, showing a strong color and structural contrast with the very dark grayish brown (10 YR 4/2) Bt (argillic) horizon with a granular structure.

3.2 Mapping Water Erosion Using a Distributed Model and Remotely Sensed Data

Accelerated water erosion driven by *overgrazing* is regarded as one of the major forms of land degradation in Península Valdés. Factors such as low vegetation cover, intensive and erratic precipitation, shallow soils, rough topography, and extreme temperatures increase water erosion risk in these ecosystems (Holecheck et al. 2003). Regional-scale water erosion models that can identify where and when accelerated erosion occur, and quantify current and potential soil losses are a critical first step toward the development of policies and regulations to effective rangeland management.

EROSAR, an erosion model for semiarid rangelands was developed by Blanco et al. (2015) to produce maps of simulated soil loss for a test site in Península Valdés rangelands. The model predicts soil loss at regional scale based on the simulation of hydrologic and erosional processes unique to rangelands; it describes splash and sheet erosion as a process of soil detachment by raindrop impact and surface water flow, and sediment transport by overland flow.

EROSAR is structured around the concept of *ecological sites* (ES, the basic entity used in rangeland health assessment) and includes processes specific to rangelands, such as the effects of soil surface degradation (desert pavements, rocky outcrops, bare soil) and the impact of vegetation patches distribution on infiltration rate and runoff. The model also considers *Horton overland flow* resulting when precipitation intensity exceeds soil infiltration capacity, this pathway of runoff is the most common in rangelands (Wilcox et al. 2003).

Adopting a GIS-based modular approach, shown in Fig. 2a, the model is kept simple with limiting number of input factors and using available information from panchromatic, multispectral, and hyperspectral satellite images and digital elevation models (*DEMs*).

The spatial pattern of eroded areas as predicted by EROSAR over the period June 2006–Ocotber 2007 (115 rainfall events) is shown in Fig. 2. The median observed erosion rate was 22 g/m, ranging from 0 to 2840 g/m. Model outputs were reclassified into six erosion classes, using the natural breaks method for interval setting. Model performance was assessed by comparing the simulated erosion rates with field observed erosion, resulting in a global reliability value of 73.8%.

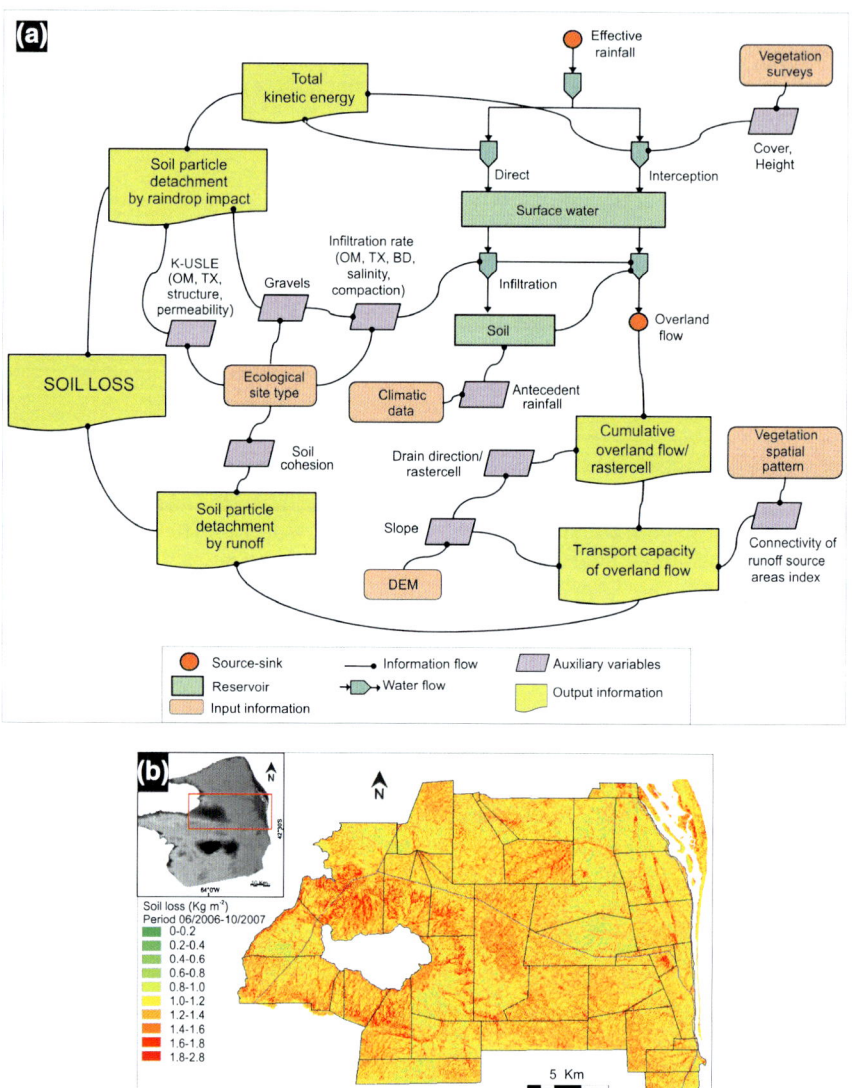

Fig. 2 **a** Flowchart of the EROSAR model. **b** Soil loss spatial distribution in the period June 2006–October 2007. *OM* organic matter, *TX* texture, *BD* bulk density

Spatial and temporal variability of overland flow and observed erosion were high, as compared to the characteristics of the ES and rainfall events. Model outputs were assessed in two ways: considering the spatial distribution of erosion severity in association with ES characteristics (soil, geomorphology, vegetation, surface soil condition, Table 2), and analyzing the temporal differences in erosion rates in connexion with each rainfall event (amount, kinetic energy, intensity, antecedent precipitation).

Table 2 Ecological site characteristics

Code	Physiognomy	Dominant species	Dominant soil	Landform
ES1	Shrub-grass steppe	ChA–CoM–StT	Calciargid	*Bajada*
ES2	Shrub steppe/desert pavement	ChA	Natrargid	*Terrace Levels*
ES3	Shrub-grass steppe	ChA–ChH–PrG–StT–StL	Torriorthents-Tertiary outcrops	Piedmont pediments
ES4	Shrub-grass steppe	ChA–ChH–ScP–LyC–StT–StL	Haplocalcid	*Terrace Levels*
ES5	Sandy grassland	StT–PiN–PoL	Haplocalcid	Stabilized aeolic field
ES6	Grassland	StS–StH	Haplocalcid	Coastal piedmont pediment, paleochannel fill (aeolic origin)

Cha *Chuquiraga avellanedae*, ChH *Chuquiraga hystrix*, CoM *Condalia microphylla*, PrG *Prosopidastrum globosum*, ScP *Schinus polygamus*, LyC *Lycium chilense*, StT *Stipa tenuis*, StL *Stipa longiglumis*, PiN *Piptochaetium napostaense*, PoL *Poa ligularis*, StS *Stipa speciosa*, StH *Stipa humilis*

Erosion rates were generally higher in ES characterized for long and steep slopes; desert pavements also showed high values (ES2); the presence of crusts and gravels reduced infiltration and encourage runoff and erosion. The lowest erosion rates were found in ES4; the higher grass cover intercepts rainfall, protecting the soil from raindrops direct impact, and dissipating the energy available to cause erosion. Ecological sites with high grass cover (ES6) showed higher erosion than expected erosion. These channels receive large volumes of runoff and sediment from alluvial 'bajadas' (ES1), partly transported, but largely are deposited.

Regarding the effect of temporal variation of rainfall events, the main factor influencing erosion was kinetic energy (KE); a KE of 1000 J/ha appeared to be a threshold level beyond which disintegration of aggregates was severe, and influence of KE on erosion processes and sediment sorting could change. These findings demonstrate the need for considering KE-influenced sediment transport when predicting erosion (Blanco et al. 2015).

3.3 Mapping of Soil Degradation Patterns with Radar Data

Wind-driven soil degradation negatively impacts on rangeland production and infrastructure in Península Valdés. Wind erosion patterns have been assessed at different scales in this region, but often with limited data. For more than four decades remote sensing images have been used to monitoring and understanding the desert biomes (Tueller and Lorain 1973; Barrett and Hamilton 1986; Cohen and Goward 2004; Mulder et al. 2011). Because of large data availability, optical images have been used both in research and operational basis. On the other hand, microwave remote sensing is of significantly useful due mostly to low soil moisture content and homogeneous and temporally stable distribution of dielectric constant. Thus, temporal variability of scattering and emission from targets are mainly a function of geometrical characteristics.

del Valle et al. (2013) analyzed soil degradation patterns that control radar backscatter in Península Valdés using Shannon Entropy mapping method on ALOS PALSAR polarimetric Synthetic Aperture Radar (PolSAR) images. The visibility of soil degradation features on radar images showed to be mostly a function of wavelength, polarization and incidence angle (Fig. 3). Stabilized sand deposits were clearly observed in the Shannon Entrpy image, with defined edges but also signals of ongoing wind erosion. One of the most conspicuous features corresponded to old track sand dunes, a mixture of active and inactive barchanoid ridges and parabolic dunes. The latter is a clear example of deactivation of migrating dunes as function of increased vegetation. Shannon Entropy image also

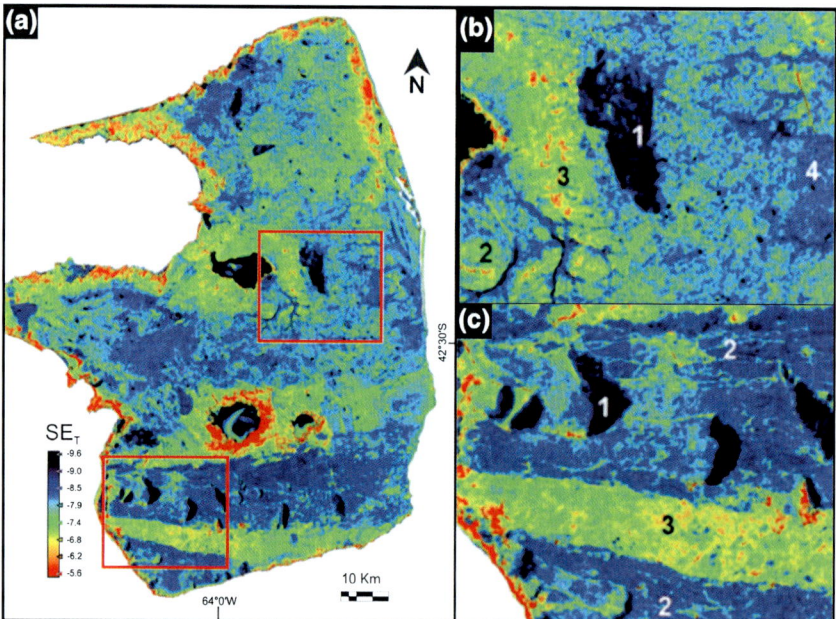

Fig. 3 a Shannon Entropy map. **b** Zoom to Península Valdés central area, *1*. Old burnt area, *2*. Water erosion (gullies, rills), *3*. Aeolian mantles stabilized (shrub with grass steppe), *4*. Overgrazing. **c** Zoom to Península Valdés dunefields. *1*. Active sand dune megapatches (colony of sand dunes), *2*. Old track of sand dunes stabilized, *3*. Aeolian mantles discontinuous stabilized on gravel plains

showed the conspicuous presence of water degradation features such as gullies and rills, *overgrazing* areas and an old burnt area. The results showed that the Shannon Entropy it is a feasible way to detect disturbance targets and to extract land degradation characteristics. Shannon Entropy of the Península Valdés confirms that land degradation is highly dispersed as the entropy. It indicates that the rate of land degradation is quite high and needs proper management to attain land sustainable development. Hence it is imperative that with reliable data, planning and managements should be monitored and managed in a sustainable way to protect these ecosystems.

3.4 Mapping Soil Erosion Indicators Related with Grazing Impacts

Livestock grazing can generate and accelerate soil erosion processes through intensive utilization of the rangeland (Pickup et al. 1994). Intensive grazing can be conditioned, among other factors, by the distribution of water points. In large grazed paddocks, with a single water point located at one corner, a direct trend predominates when gradients of several paddocks are pooled in a regional estimate: higher grazing pressure around water points, diminishing at increasing distance from it (Bastin et al. 1993). Lange (1969) coined the term *piosphere* for these utilization gradients.

The assessment grazing impact on soil erosion processes prevailing in arid rangelands requires the selection of adequate indicators, or relevant process-related parameters to quantify them. Blowouts have been indicated as the most common aeolian erosional landforms in dune landscapes in arid zones (Hesp and Hyde 1996). Blowouts are depressions or hollows formed by wind erosion on a preexisting sand deposit (Hesp 2002) and are common features in vegetation-stabilized dune fields (Livingstone and Warren 1996). Vegetated parabolic dunes are leeward extensions of blowouts. Plants colonize, and may eventually stabilize, the trailing arms, which are formed by the advancing dune apex (Melton 1940; Verstappen 1968; Pye 1982). Grazing, however, can contribute to destabilization of dunes (e.g., De Stoppelaire et al. 2004).

Blanco et al. (2008) studied the changes induced by grazing on soil erosion processes in the vegetated dunefields of Península Valdés, using the spatial pattern of blowouts as an indicator of erosion intensity. Distance-dependent spatial statistics were used to identify scales at which the spatial pattern of blowouts were significantly aggregated around water points (i.e., "critical scales"). It was hypothesized that large numbers of livestock concentrated at water points may destroy vegetation cover, resulting in dune reactivation and consequently generating blowouts. Accordingly, it was expected that the impact of livestock on soil erosion processes was greatest in sites near water points relative to those farther from water points (i.e., a utilization gradient).

The utilization gradient should result in aggregation of blowouts near water holes, though the spatial structure of the sites sensitive to erosion (i.e., dune crests) may confound a direct effect. Therefore, the research analyzed blowout densities at varying distances away from water points to assess whether they were higher than expected by a random distribution of blowouts over the dune crest (i.e., after removing potential spatial structures). It was expected the critical scale of aggregation should increase under grazing. Considering that vegetation cover varies with position on the dune, the research analyzed the effects of the spatial distribution of dune crests around water points on the spatial pattern of blowouts. It was hypothesized that the effect of livestock on soil erosion processes is influenced by the preexisting spatial pattern of dune crests. Thus, it was expected that the grazing disturbance was greatest in sites with a higher density of dune crests.

Blowouts, used as main indicators of aeolian erosion processes, as well as dune crests, which are susceptible to erosion, were mapped on aerial photographs and images from Landsat 7 Enhanced Thematic Mapper Plus, in eight paddocks under two grazing conditions: lightly (0.4 sheep/ha) and heavily grazed (0.8 sheep/ha) (Fig. 4). The cartography of water points, crests and blowouts was used to calculate a spatial statistic (O-ring statistic), which gives the expected intensity of blowouts within the area covered by crests as function of distance away from water points. To ascertain whether density of crests around water points influences density of blowouts, the intensity of dune crests in the neighborhood of water points was estimated and compared the densities of blowouts among water points with low, medium and high densities of crests.

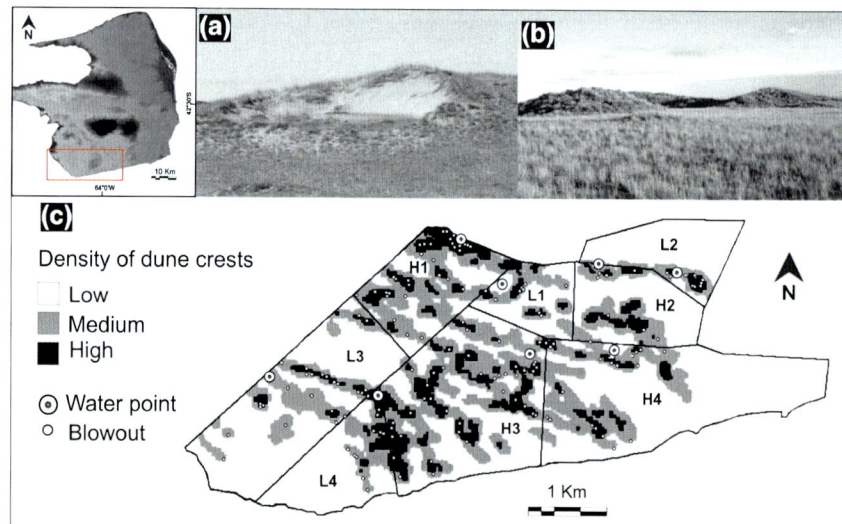

Fig. 4 a Blowout. b Dune crests stabilized by vegetation. c Blowouts spatial distribution. d Density of dune crests and points representing blowouts and water points. *H1–H4* paddocks heavily grazed; *L1–L4* paddocks lightly grazed

The results showed highly significant ($P < 0.05$) aggregation of blowouts around water points with peak densities threefold higher than expected at random, occurring between 90 and 210 m for areas under heavy grazing. However, the aggregation was only weakly significant for the light grazing treatment and occurred only at distances of about 30 m away from the water point. The impact of grazing on soil stability contrasted among sites with different sensitivity to accelerated erosion. In sites with a high density of dune crests close to water points, wind erosion appeared more intense and the density of blowouts increased.

Information and knowledge derived from the aforementioned study can contribute to more sustainable management of these rangelands, and soil conservation. The findings suggest that, for instance, water points should be located in sites not susceptible to accelerated soil erosion Península Valdés. Sensitive areas could be fenced off to exclude livestock. Choices of fencing types could include sand fencing, which reduce wind speed near the ground, thereby causing moving sand to be deposited into a mound on the downwind side of the fence, reducing deposition farther downwind (Gares 1990).

3.5 Soil Erosion Indicators and Vegetation Dynamics: The Application of the State and Transition Model

Sheep grazing affects soil stability directly by *trampling* and indirectly by decreasing vegetation, especially grass cover (Chartier and Rostagno 2006). Changes in vegetation in intensively grazed areas include the decrease of grass and biological (cryptogamic) crust cover and an increase in the cover of unpalatable species. In Península Valdés, the main species increasing in overgrazed areas are the *Ch. avellanedae* and *Mulinum spinosum*. *Ch. avellanedae* has shown to increase in intense grazed areas of fine textured soils (e.g. Calciargids) and *Mulinum spinosum* in soils of coarse texture (Ocariz et al. 2004). Areas dominated by *Ch. avellanedae* transformed original shrub-grass steppes into shrub steppes. Soil quality change in the shrub interspaces limits the recovery of the grass cover, and shrub steppes represent steady states. Degraded soils of the intershrub patches experience decrease of water infiltration as a physical crust develops and the argillic horizon can be exumated (Rostagno and Beeskow 2000).

In Fig. 5 changes in vegetation and soils following intensive grazing and *trampling* are represented in a state and transition model. An effective application of this model is to assess the risks of a given management practice on vegetation and soil erosion dynamics and soil quality (Westoby et al. 1989; Bestelmeyer et al. 2003). This application is based on the assumption that grazing management can prevent soils and plant communities from crossing a threshold by maintaining recovery potential of the ecosystem (Chartier and Rostagno 2006).

The original state (I) is a shrub-grass steppe with vegetation cover between 50 and 60% (a shrub stratum with a cover 20–25% dominated by *Ch. avellanedae*, an

Fig. 5 States and transitions model of the vegetation dynamics under sheep grazing in the central area of Península Valdés (see Sect. 3.5 for further explanation)

herbaceous stratum with a 25–30% cover, dominated by *Nacella tenuis*, *Poa ligularis* and *Piptochaetium napostaense*, and a dwarf shrub stratum with a cover <5% is dominated by *Hoffmanseggia trifoliata* and *Perezia recurvata*) and a high, though variable, litter cover. This plant community dominates in soils with contrasting texture (Xeric Natrargid) with a Btn horizon at 10–15 cm depth. The second state (II) develops following a decrease of the two most *palatable species*, *P. ligularis* and *P. napostaense* cover or the size of the individuals of these species. The state III develops after large bare soil patches form between the shrubs. At this stage the total vegetation cover has decreased to 25–30%. In this state, few palatable grasses grow protected under shrub clumps and *Nassauvia fueguiana*, a dwarf shrub with a shallow root system invade the degraded patches.

Overgrazing or intense sheep grazing is the main cause for the transition 1 or the transition between state I and II (Elissalde and Miravalles 1983). Regardless of the stocking rate, continuous grazing triggers changes in vegetation composition, mainly the reduction of palatable grass cover. Transition 2, unavoidable in areas adjacent to watering points, also occurs in other areas, mainly during long-lasting droughts when wind erosion accelerates. Transition 3 seems to be favored during wet years, mainly wet winters, as vegetation vigor and seed production of grasses recover. The recovery of state III through the transition 4 is highly unlikely. The loss of part of the A horizon and the lack of safe sites for seeds capturing and plant establishment make natural recovery a slow process (Rostagno 1989; Chartier and Rostagno 2006). Thus, the loss in productivity set in motion by accelerated soil erosion is a self-sustaining process. Loss of production on eroded soil further degrades its quality, which, in turn, accelerates soil erosion (Lal 1988). This feedback mechanism maintains or reinforces the degraded plant community and the dominance of unpalatable shrubs (i.e., *Ch. avellanedae*) and limits reversal to the previous conserved plant community where perennial grasses dominated. Recent conceptual advances in community and landscape ecology indicate that positive plant-soil feedbacks are the dominant cause of catastrophic behavior to further desertification (Schlesinger et al. 1990; van de Koppel et al. 1997).

3.6 Fire Occurrence as a Triggering Factor of Soil Degradation: Burned Areas and Wildfire Risk Mapping in Península Valdés

Impacts of wildfires on the environment relay heavily on the intensity, frequency and spatial distribution, which in turn are influenced in complex ways by several natural and *anthropic* factors as stated before (Falk et al. 2007). Knowledge of the spatial and temporal patterns of burned areas at regional scales provides a long-term perspective of fire processes and its effects on ecosystems and vegetation recovery patterns, and it is a key factor to design prevention and post-fire restoration plans and strategies (Murphy et al. 2011).

Assessment of fire patterns depends on accurate information of the timing and extent of fire events. The potential of satellite imagery to that end has been researched (Chuvieco and Kasischke 2007), and a range of global-satellite-derived fire products have been developed overtime. Prior research cautions on the use of these global-scale products for regional and sub-regional applications, and therefore Hardtke et al. (2015) developed a semi-automated wildfire detection algorithm to map wildfire patterns in Península Valdés. Between February 2001 and December 2014, a total of 11 separately burned areas where identified (Fig. 6a). The total area burned was computed as 357 km^2, which represents roughly 10% of the study region. Recorded wildfires size varied between 0.6 and 152.2 km^2 with an average size of 32.45 km^2. The average fire return interval, which represents the average interval between fires in a given area, was 142 years. As expected, wildfire occurrence was strongly seasonal, 9 wildfires occurred in summer, burning 94% of the burned area and 1 in autumn which burned the remaining 6%.

Hardtke et al. (2015) evaluated how weather and climate, along with fuel properties, topography and human activity determine wildfire risk with a probability-based model over Península Valdés. The research fitted a logistic regression which models relationships between a response dichotomous variable and either numerical

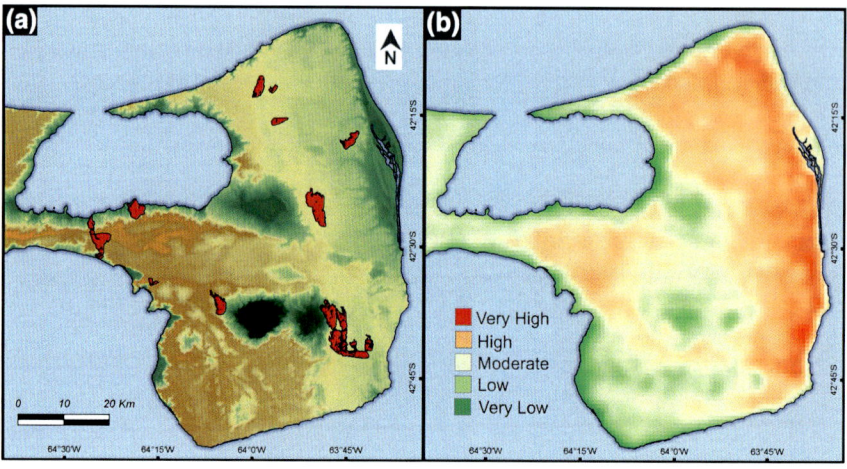

Fig. 6 **a** Burned areas in Península Valdés between February 2001 and December 2014. **b** Wildfire risk in Península Valdés

or categorical explanatory variables without making assumptions about the statistical distribution of the explanatory variables (Hosmer and Lemeshow 1989). MODIS Hot Spot data and pseudo-absence points were used as response variable, with variables described in Table 3 as predictors. Automatic model selection was performed using a stepwise deletion routine, to determine the most parsimonious list of variables. The wildfire risk for each pixel (1 km) was calculated using the parameters of the selected model and the resulting probability was then divided into 5 risk categories; 0–0.20 very low, 0.21–0.40 low, 0.41–0.60 moderate, 0.61–0.80 high, 0.81–1 very high (Fig. 6b). The explanatory variables included in the final model were: the mean net primary production (NPP), summer and spring precipitation, spring temperature, distance to nearest road, elevation, and slope. Contributions of individual predictor variables to the model, show mean NPP to be the greatest contributing variable, followed by spring temperature and summer precipitation, distance to nearest road, spring temperature, slope and altitude. Grouped by variable type, climatic variables were the main contributors, 50% of residual deviance, followed by the vegetation variables with 29% of the residual deviance. Otherwise, topographic variables explained approximately 11% and anthropogenic only 9%. Concerning on the effect of each of the explanatory variables on the wildfire risk, higher mean NPP, spring temperature and spring rainfall are associated with an increasing wildfire risk, while higher summer rainfall, altitude, slope, and distance to road are linked to a decreasing risk. The very high-risk area represented 2.1%, high risk 5.2%, moderate risk 38.8%, low risk 25.4% and the very low risk areas 28.5% of total study area.

Table 3 Datasets included in the wildfire risk analysis

Variable	Type	Source	Resolution	Range
Mean NDVI	Vegetation	Based on MOD13Q1 (2003–2009)	250 m	0–0.91
NDVI range	Vegetation	Based on MOD13Q1 (2003–2009)	250 m	0.29–0.58
Land cover	Vegetation	South America vegetation map (2002)	1 km	25 classes
Seasonal temperature	Meteorological	Spatial interpolation NOAA-GHC (2003–2009)	1 km	4–21 °C
Seasonal precipitation	Meteorological	Based one TRMM NASA (2003–2009)	1 km	10–248 mm
Elevation (DEM)	Topography	CIAT-CSI SRTM	90 m	0–650 m
DEM-slope	Topography	Derivative from DEM	90 m	0°–30°
Distance to nearest road	Anthropogenic	Based on IGN cartography (2005)	250 m	0–240 km
Distance to nearest town	Anthropogenic	Based on IGN cartography (2005)	250 m	0–398 km

4 Perspectives and Future Work

This chapter has described four key causes of soil erosion associated with *overgrazing* in Península Valdés: (1) Poor range management with respect to flock distribution and overstocking, (2) Limited access to information, (3) Top-down and largely ineffective government policy, and (4) Overdependence on grazing systems for sustained livelihoods. Water and wind erosion are the degradation processes that are most strongly evidenced. It was presented a review of soil degradation assessment methods and several soil degradation studies carried out since 1990 in the Península Valdés region. Despite these efforts, there remains a need for soil degradation monitoring networks and decision-support systems aimed at optimization of soil quality in Península Valdés. The pressure on its soils will continue in the future, and a clearly defined regulatory framework is needed. Future work must be done by climate fluctuations and by human impact, and to identify priority areas for action. Important challenges are mentioned hereafter.

It is necessary to engage the scientific community to mount an integrated programme for methods, standards, data collection, and research networks for assessment and monitoring of soil degradation. This means also promoting the involvement of local people in data collection and identifying the technical problems facing them in the field. It is desperately needed to survey and collect the available data on soil degradation and build specialized database utilizing GIS and remote sensing techniques.

From the remote sensing perspective on different scales, several issues need to be addressed. Significant challenges still exist for radar imagery to achieve its full potential for wind-driven land degradation processes. The full range of factors which result in spatial and temporally varying signatures on SAR imagery still have to be quantified, with the influence of wind conditions, and plant cover being the principal uncertainties. Obviously, the combination of optical data with radar data (data fusion) is fundamental for a synergic approach. Data fusion represents a strategic way to increase the potentialities of remote sensing in real-world applications thanks to the promise of improvement in accuracy and reliability of analysis results.

There is an urgent need to develop regional soil erosion assessments that enable estimates of the area that is affected by soil erosion and the expected magnitude in a particular area, both of which are required to formulate sound soil protection policies. A process modeling method is recommended for modeling soil erosion risk in relation to climate and land use changes. Field campaigns are necessary and databases should be made with erosion measurements, field observations, data on underlying factors influencing erosion (climate, topography, soils and land use) and related meta-data (period of record, erosion type, etc.).

Finally, development policies and economic instruments are needed that encourage sustainable land use and assist in the greater use of land resource information for sustainable range management.

Acknowledgements We wish to acknowledge the valuable suggestions made by Dr. Gerardo Bocco Verdinelli (UNAM) and Dr. Juan Esteban Panebianco (INCITAP). Also, Pablo Bouza y Andrés Bilmes provided valuable suggestions that greatly improved the manuscript. Comisión Nacional de Actividades Espaciales from Argentina supplied the satellite images.

Glossary

Erosivity	Potential ability of soil, regolith or other weathered material to be eroded by rain, wind, or surface runoff
Erodibility	Resistance of soils to erosion based on their physical and chemical characteristics such as soil texture, organic matter or structure)
Desert pavements	It is a desert surface covered with closely packed, interlocking angular or rounded rock fragments
Ecological sites	A distinctive kind of land with specific physical characteristics that differs from other kinds of land in its ability to produce a distinctive kind and amount of vegetation
Horton overland flow	Describes the tendency of water to flow horizontally across land surfaces when rainfall has exceeded infiltration capacity and depression storage capacity
Overgrazing	It occurs when plants are exposed to intensive grazing for extended periods of time, or without sufficient recovery periods

Anthropic	Refers to being associated with humans, influenced by humans or taking place during human existence
Pattern	The arrangement and composition of the patches that compose a landscape
Trampling	The mechanical destruction and mortality of ground level vegetation by animals
Piosphere	The zone of influence of grazing on a region's vegetation and soil

References

Abrahams A, Parsons A, Wainwright J (1995) Effects of vegetation change on interrill runoff and erosion, Walnut Gulch, southern Arizona. Geomorphology 13:37–48

Ares J, Beeskow AM, Bertiller M et al (1990) Structural and dynamic characteristics of overgrazed lands of northern Patagonia, Argentina. In: Breymeyer A (ed) Managed grasslands. Elsevier, Amsterdam, pp 149–175

Blanco PD et al (2015) Using the distributed model EROSAR and remotely sensed data for regional assessment of soil water erosion in semi-arid rangelands. 3rd UNCCD Scientific Conference, 9–12 March 2015, Cancun, Mexico

Bagnold RA (1941) The physics of blown sand and desert dunes. Methuen, New York

Barrett E, Hamilton M (1986) Potentialities and problems of satellite remote sensing with special reference to arid and semiarid regions. J Climatic Change 9:167–186

Bastin G, Sparrow A, Pearce G (1993) Grazing gradients in central Australian rangelands: ground verification of remote sensing-based approaches. Rangeland J 15:217–233

Beeskow A, Elissalde N, Rostagno M (1995) Ecosystem changes associated with grazing intensity on the Punta Ninfas rangelands of Patagonia, Argentina. J Range Manage 48:517–522

Bestelmeyer B et al (2003) Development and use of state-and-transition models for rangelands. J Range Manage 56:114–126

Bestelmeyer BT, Ward JP, Havstad KM (2006) Soil-geomorphic heterogeneity governs patchy vegetation dynamics at an arid ecotone. Ecology 87:963–973

Bisigato AJ, Villagra PE, Ares JO (2009) Vegetation heterogeneity in Monte Desert ecosystems: a multi-scale approach linking patterns and processes. J Arid Environ 73:182–191

Blanco P (2010) Ph.D. thesis: soil erosion assessment and spatial modelling in NE of Chubut. Escuela para Graduados Alberto Soriano, Faculty of Agronomy, University of Buenos Aires

Blanco PD et al (2008) Grazing impacts in vegetated dunefields: predictions from spatial pattern analysis. Rangeland Ecol Manage 61:194–203

Bowman D (2009) Fire in the earth system. Science 324:481–484

Cerdà A (2001) Effects of rock fragment cover on soil infiltration, interrill runoff and erosion. Eur J Soil Sci 52:59–68

Chartier M, Rostagno M (2006) Soil erosion thresholds and alternative states in northeastern Patagonian rangelands. Rangeland Ecol Manage 59:616–624

Chartier M, Rostagno M, Roig F (2009) Soil erosion rates in rangelands of northeastern Patagonia: a dendrogeomorphological analysis using exposed shrub roots. Geomorphology 106:344–351

Chuvieco E, Kasischke ES (2007) Remote sensing information for fire management and fire effects assessment. JGR Biogeosciences 112:1–8

Cohen W, Goward S (2004) Landsat's role in ecological applications of remote sensing. Bioscience 54:535–545

Cooke R, Warren A, Goudie A (1993) Desert geomorphology. University College London Press, London
De Stoppelaire G et al (2004) Use of remote sensing techniques to determine the effects of grazing on vegetation cover and dune elevation at Assateague Island National seashore: impact of horses. Environ Manage 34:642–649
del Valle H, Rostagno M, Bouza P (2000) Los médanos del sur de Península Valdés: Su dinámica y los cambios asociados en los suelos y en la vegetación, XVII Congreso Argentino de la Ciencia del Suelo. 11–14 Apr 2000. Asociación Argentina de la Ciencia del Suelo, Mar del Plata, Buenos Aires, Argentina
del Valle H et al (2008) Sand dune activity in north-eastern Patagonia. J Arid Environ 72:411–422
del Valle H et al (2013) Assessment of land degradation using shannon entropy: approach on POLSAR images in patagonian coastal deserts. Geofocus 13–2:84–111
Dregne H (1977) Desertification of arid lands: the human face of desertification. Econ Geogr 53:322–331
Elissalde N, Miravalles H (1983) Evaluación de los campos de pastoreo de Peninsula Valdes. Contribution 70, Central National Patagonico, Puerto Madryn, Chubut, Argentina
Falk D, McKenzie A, Black E (2007) Cross-scale analysis of fire regimes. Ecosystems 10:809–823
Gares P (1990) Eolian processes and dune changes at developed and undeveloped sites, Island Beach, New Jersey. In: Nordstrom K et al (eds) Coastal dunes: form and process. Wiley, Chichester, pp 361–378
Hardtke L, Blanco P, del Valle H (2015) Semi-automated mapping of burned areas in semi-arid ecosystems using MODIS time-series imagery. Int J Appl Earth Obs Geoinf 38:25–35
Herbel C (1979) Utilization of grass and shrublands of the south-western United States. In: Walker BH (ed) Management of semi-arid ecosystems. Elsevier, Amsterdam, pp 161–203
Hesp P (2002) Foredunes and blowouts: initiation, geomorphology and dynamics. Geomorphology 48:245–268
Hesp P, Hyde R (1996) Flow dynamics and geomorphology of a trough blowout. Sedimentology 43:505–525
Holecheck J, Pieper R, Herbel C (2003) Range management: principles and practices. Pearson, USA, 456 pp (5th editions)
Hosmer D, Lemeshow S (1989) Applied logistic regression. Wiley, New York
Johnson D, Watson-Stegner D, Johnson D et al (1987) Proisotropic and proanisotropic processes of pedoturbation. Soil Sci 143:278–292
Koohafkan P, Lantieri D, Nachtergaele F (2003) Land degradation assessment in drylands (LADA): guidelines for a methodological approach. Land Water Develop Div, FAO, Rome
Lal R (1988) Soil erosion by wind and water: problems and proposals. In: Lal R (ed) Soil erosion research methods. The Soil Erosion and Water Conservation Society
Lange R (1969) Grazing impact in relation to livestock watering points. J Range Manag 22:396–400
Livingstone I, Warren A (1996) Aeolian geomorphology: an introduction. Longman, Essex (221 p)
Melton F (1940) A tentative classification of sand dunes: its application to dune history in the southern High Plains. J Geol 48:113–174
Michaelides K et al (2009) Vegetation controls on small-scale runoff and erosion dynamics in a degrading dryland environment. Hydrol Process 23:1617–1630
Mulder V et al (2011) The use of remote sensing in soil and terrain mapping: a review. Geoderma 162:1–19
Murphy B, Williamson G, Bowman D (2011) Fire regimes: moving from a fuzzy concept to geographic entity. New Phytol 192:316–318
Ocariz P, Rostagno M, Degorgue G (2004) Conductoras y pasajeras: El rol del quilembai (Chuquiraga avellanedae) y la flechilla (Stipa tenuis) en la conservación del suelo de un sitio ecológico del noreste de Chubut. II Reunión Binacional de Ecología, Mendoza, Argentina
Parsons A, Abrahams A, Simanton J (1992) Microtopography and soil-surface materials on semi-arid piedmont hillslopes, southern Arizona. J Arid Environ 22:107–115

Paruelo JM et al (1998) The climate of Patagonia general patterns and controls on biotic processes. Ecología Austral 8:85–104

Pickup G, Bastin G, Chewings V (1994) Remote-sensing-based condition assessment for nonequilibrium rangelands under large-scale commercial grazing. Ecol Appl 4:497–517

Poesen J et al (1998) Variation of rock fragment cover and size along semiarid hillslopes: a case-study from southeast Spain. Geomorphology 23:323–335

Prohaska F (1976) The climate of Argentina, Paraguay and Uruguay. In: Schwerdtfeger E (ed) Climate of central and South America. World Survey of Climatology. Elsevier, Amsterdam, pp 57–69

Pye D (1982) Morphological development of coastal dunes in a humid tropical environment Cape Bedford and Cape Flattery, North Queensland. Geografiska Annaler 64(A):212–227

Ravi S et al (2010) Land degradation in drylands: Interactions among hydrologic-aeolian erosion and vegetation dynamics. Geomorphology 116:236–245

Renard K, Simanton J, Osborn H (1974) Applicability of the universal soil loss equation to semiarid rangeland conditions in the Southwest. Hydrology and Water Resources in the South-west, Water Resources Research Center, University of Arizona, Tucson 4:18–31

Rostagno M (1989) Infiltration and sediment production as affected by soil surface conditions in a shrubland of Patagonia, Argentina. J Range Manage 42:382–385

Rostagno C, Beeskow A (2000) Soil erosion, shrub encroachment and ecosystem resilience of two ecological sites of NE Patagonia. XI conference of International Soil Conservation Organization (ISCO) Buenos Aires-22

Rostagno D, Degorgue G (2011) Desert pavements as indicators of soil erosion on aridic soils in north-east Patagonia (Argentina). Geomorphology 134:224–231

Rostagno M, del Valle H, Buschiazzo D (2004) La erosión eólica. In: González M, Bejerman N (eds.). Peligrosidad Geológica en Argentina. ASAGAI. (CD)

Rostagno M, Defosse G, del Valle H (2006) Postfire vegetation dynamics in three rangelands of Northeastern Patagonia, Argentina. Rangeland Ecol Manage 59:163–170

Schlesinger W et al (1990) Biological feedbacks in global desertification. Science 247:1043–1048

Simanton J, Rawitz E, Shirley E (1984) Effects of rock fragments on erosion of semiarid rangelands. In: Nichols J et al (eds) Erosion and productivity of soils containing rock fragments. Soil Science Society of America Special Publication No. 13, Madison, Wisconsin, USA, pp 65–72

Soriano A, Movia CP (1986) Erosión y desertización en la Patagonia. Interciencia 11:77–83

Tueller P, Lorain G (1973) Application of remote sensing techniques for analysis of desert biome validation studies. Utah State University. US/IBP Desert Biome Digital Collection

van de Koppel J, Rietkerk M, Weissing F (1997) Catastrophic vegetation shifts and soil degradation in terrestrial grazing systems. Trends Ecol Evol 12:352–356

Verstappen H (1968) On the origin of longitudinal (seif) dunes. Zeitschrift fur Geomorphologie NF 12:200–220

Villagra P et al (2009) Land use and disturbance effects on the dynamics of natural ecosystems of the Monte Desert: implications for their management. J Arid Environ 73:202–211

Vrieling A (2006) Satellite remote sensing for water erosion assessment: a review. Catena 65(1):2–18

Wainwright J, Parsons A, Abrahams A (1995) A simulation study of the role of rain drop erosion in the formation of desert pavements. Earth Surf Proc Land 20:277–291

Westoby M, Walker B, Noy-Meir I (1989) Opportunistic management for rangelands not a equilibrium. J Range Manag 42:266–274

Wilcox B, Breshears D, Seyfried M (2003) Water balance onrangelands. In: Encyclopedia of water science. Dekker, New York, pp 791–794

Groundwater Resources of Península Valdés

María del Pilar Alvarez and Mario Alberto Hernández

Abstract In this chapter, the groundwater resources of Península Valdés are analyzed in order to recognize the influence of the late Cenozoic stratigraphic record and the geomorphology on the hydrogeology of the region. The geology of the study area defines a hydrogeological sequence that begins with a non-saturated zone of variable thickness and an essentially aquifer-like behaviour. Under the non-saturated zone there is a phreatic aquifer within the Quaternary deposits or in the sandstones of the Puerto Madryn Formation, and below it, one or more semi-confined or confined aquifers in the same formation or in the underlying Gaiman Formation. The aquifer levels located in the Puerto Madryn Formation are main water supply that supports the farming activities in the region. The groundwater hydrodynamic and the associated hydrochemical and isotopic characteristics are closely related to the climate and to the relationship between the geomorphology and lithology of the region. As climate is quite homogeneous all over the peninsula, the different relationships between the geomorphological and lithological unit allows to define four hydrogeological regions: (1) Aeolian landforms, which represents the main recharge zone with the low salinity waters; (2) Terrace levels, as the circulation area with high salinity waters, (3) Endorheic depressions that are the inland discharge sector where the evaporation is the dominant process and hypersaline playa lakes occurs, and (4) Coastal systems which represents both the regional discharge area as well as the local recharge zone with low salinity groundwater reservoirs.

M.d.P. Alvarez (✉)
Instituto Patagónico para el Estudio de los Ecosistemas Continentales (IPEEC), Consejo Nacional de Investigaciones Científicas y Técnicas (CONICET)—CCT Centro Nacional Patagónico (CENPAT), Boulevard Brown 2915, ZC: U9120ACD, Puerto Madryn, Chubut, Argentina
e-mail: alvarez.maria@conicet.gov.ar

M.A. Hernández
Facultad de Ingenieria, Maestría en Evaluación Ambiental de Sistemas Hidrológicos, Universidad Nacional de La Plata, Edificio Central—Calle 1 y 47, La Plata, Argentina
e-mail: mario_h@uolsinectis.com.ar

Keywords Hydrogeology · Hydrochemistry · Hydrodynamic · Hydrogeomorphology

1 Introduction

One of the main natural limitations to the social and economic development of arid and semi-arid regions is their limited water availability. In Península Valdés the lack of permanent watercourses has caused groundwater resources to become the basis of all economic activities. In this area the water demand was historically associated to the extensive sheep farming practices (all throughout the peninsula), and to tourism, which is centralized in Puerto Pirámides.

As is known, geomorphology, geology and climate are first-order determinants of hydrogeological phenomena, not just hydrodynamic, but also hydrochemical ones. In the case of Península Valdés, the rains do not show a strong spatial variability (see Chapter "Miocene Marine Transgressions: Paleoenvironments and Paleobiodiversity"). Thus climate could conditioning the groundwater resources at regional level, but do not answer the groundwater spatial variations that exists in the region. The opposite is the case of the geomorphology and lithology—which have important variations in the regions (see Chapters "Geology of Península Valdés" and "Late Cenozoic Landforms and Landscape Evolution of Península Valdés")—that have being correlated with the regional hydrogeology and hydrochemistry (Alvarez et al. 2010). According to that, in this chapter, the late Cenozoic stratigraphic record and the geomorphological units are analyzed in order to recognize their influence over the groundwater resources.

2 Hydrogeological Units

The first step for the groundwater resources evaluation is the analysis of the hydrogeological units. The geological units depending on their capacity to receive, store and transmit water, are classified into *aquifers*, *aquitard*, *aquicludes* and *aquifuges*. In the case of the aquifers, according to the *hydraulic conductivity*, they can also be subclassified as high, low or moderate permeability aquifers.

In the case of Península Valdés, the stratigraphic units described in Chapter "Late Cenozoic Landforms and Landscape Evolution of Península Valdés" are classified in four hydrolithological groups, (1) High permeability aquifers, (2) Moderate to low permeability aquifers, (3) Aquitards and (4) Aquicludes (Fig. 1).

Before starting with the characterization of each hydrogeological unit it should be state out that, only aquifers in porous media have been described in the study area (Windhausen 1921; Stampone and Cambra 1983; Alvarez et al. 2010).

STRATIGRAPHIC SEQUENCE

CENOZOIC	QUATERNARY	Holocene		5	6	7	8
		Pleistocene	Late–Middle		4		
			Early				
	NEOGENE	Pliocene		3			
		Miocene		2			
				1			

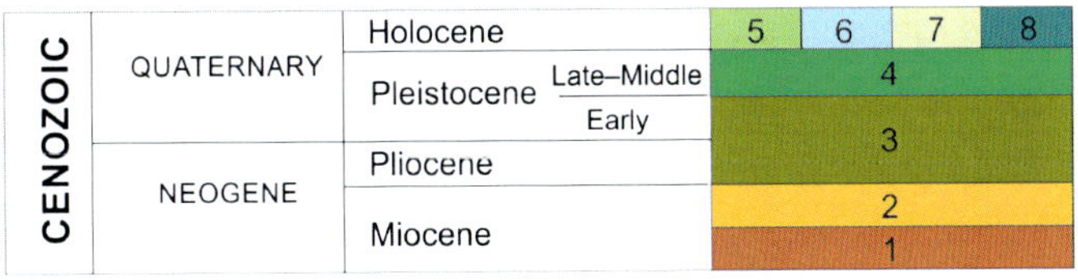

8: Alluvial and colluvial deposits: sand, gravel and silt
7: Aeolian deposits: sand and silt
6: Playa lake, sediments and evaporites: silt, clay and evaporites
5: San Miguel Formation: gravel
4: Caleta Valdés Formation: gravel
3: Rodados Patagónicos *(Patagonian Gravels)*: gravel
2: Puerto Madryn Formation: sandstone, mudstone and coquina
1: Gaiman Formation: siltstone, claystone and sandstone

HYDROLITHOLOGICAL CLASIFICATION

Geologic unit	Hydrolithological units
4 5 7	High permeability aquifers
2 3 8	Moderate to low permeability aquifers
1 2 6	Aquitards/Aquícludes

Fig. 1 Stratigrahic units and their hydrolithological classification

2.1 High Permeability Aquifers

Within this category are included the Quaternary aeolian deposits and the San Miguel Formation.

Aeolian deposits (Active dunefields and Stabilized aeolian field geomorphologic units; see Chapter "Late Cenozoic Landforms and Landscape Evolution of Península Valdés") cover about 700 km^2, representing 40% of the Península Valdés region. This combined with their lithology (well-selected and nonconsolidated sands with a predominantly medium to fine grain size) makes them the most important hydrological unit regarding the possibilities of effective infiltration. In this unit, indicating the behaviour as a preferential recharge zone, interdune wetlands with permanent vegetation were identified (Fig. 2). Unfortunately, the aeolian deposits occupy only the thin upper layer of the stratigraphic sequence, and their reduced and variable thickness is a limiting factor when considering their potential as productive aquifers.

Fig. 2 a Regional view of the wetlands in aeolian landforms (stabilized aeolian field and active dunefields) located at the southern sector of Península Valdés; **b** Wetland in the stabilized Aeolian field. The regional location corresponds to the *yellow box* of (**a**). The water body of the photo has an approximately diameter of 5 m

Among the aeolian quaternary sediments, coastal dunes are also found. Although they are scarce and its dimensions are insignificant within the regional frame (see Chapter "Late Cenozoic Landforms and Landscape Evolution of Península Valdés"), they are locally important as reservoirs of shallow water (2–3 m deep) with a relatively good quality in an environment dominated by brackish water.

The San Miguel Formation deposits constitute a small unit (not reaching to cover 1% of the total area of the study region) of local distribution, restricted to the NE coast of the Península Valdés. It is composed by coarse and very coarse gravel with a small sandy matrix and shells. These features certainly allow to classify it as a high permeability aquifer. The water table depth measured in this unit was always less than 6 m.

2.2 Medium to Low Permeability Aquifers

Under this classification the undifferentiated alluvium and colluvium deposits, the Rodados Patagónicos and the Puerto Madryn Formation are grouped.

The undifferentiated alluvium and colluvium deposits cover a thin sheet of the *bajadas* (in both the Great Endorehic Basin and the Coastal Zone systems; see Chapter "Late Cenozoic Landforms and Landscape Evolution of Península Valdés") occupying an area that represents approximately 12% of the total surface of the Península Valdés region. This unit has heterogeneous particle size distribution, consisting of gravel, sand and silt. Thus, effective porosity variations are expected to be reflected in the *bajadas* hydrogeological unit. Considering that unsaturated levels that would serve to estimate their permeability were not found, it was decided under a conservative criterion, to include the described deposits in this group.

The Rodados Patagónicos deal a relatively large area, reaching up to 35% of the sector but not exceeding 3 m in thickness. The grain size particle is mainly composed by pebbles (<1.5 cm diameter) with a sandy matrix. The fabric varies from *clast-supported* to *matrix-supported* and this grain size anisotropy (vertical and horizontal) also occurs with the cementation generating some uncertainty about the effective porosity and the potential water-bearing that this formation may have. Since it has not been possible to locate a saturated level that allows attributing a certain permeability value, and because it has been verified in many cases a significant percentage of matrix with a high calcitic cementation degree (calcic soil horizons, see Chapter "Soil–Geomorphology Relationships and Pedogenic Processes in Península Valdés"), it was decided to include this unit into the medium/low permeability aquifers.

Regarding the Puerto Madryn Formation, although the outcropping expression is not relevant because it does not reach 4% of the total area, it has an important development in depth and constitutes the main aquifer unit of the area mostly used for farming activities. It is composed by sandstones, mudstones and coquines, where the variations in the fine fraction percentages are reflected in the *hydraulic conductivity* (K) changes. The K coefficient obtained by pumping tests for the productive aquifer levels were 4.9×10^{-1} and 4.08×10^{-2} m/day for the first level and 2.75 m/day for the deepest ones (Alvarez et al. 2012, 2014).

2.3 Aquitards

Within this category are included both fine sediments of the Puerto Madryn Formation, as well as those of the Gaiman Formation. The first lithoestratigraphic unit includes mudstone layers of variable thicknesses that are interbedded with sandstones strata of more than one meter thick. These sections of the profile are interpreted as aquitards. In the case of the Gaiman Formation, which consists mainly of fine material, it is expected that their physical characteristics are much more homogeneous, and except for the limited sandy levels, the hole unit is consider as an aquitard unit. According to well-drilling reports and stratigraphic sections (see Chapter "Miocene Marine Transgressions: Paleoenvironments and Paleobiodiversity"), the aquitards of the Puerto Madryn and Gaiman formations allowed the existence of more than one aquifer level in each lithoestratigraphic unit (Alvarez et al. 2014).

2.4 Aquicludes

The main aquiclude units are composed by the Quaternary playa lake deposits of the Great Endorrehic Basins system and the more clayey levels of the Gaiman Formation. It is important to note that the hydrological behaviour of the Gaiman Formation in Penísnula Valdés was previously described in groundwater well-drilling reports (Windhausen 1921; Alvarez 2010; Alvarez et al. 2014).

3 Hydrodynamics

The regional hydrodynamic characterization of the Península Valdés region was defined by Alvarez et al. (2010) on the basis of the survey of 89 wells. In this study, the main groundwater watersheds as well as the recharge and discharge areas were identified.

Based on the stratigraphic sequence of the study area (Fig. 1), the local hydrogeological system is formed by: (a) a *non-saturated zone* (NSZ) corresponding to the Quaternary deposits and partly to the Miocene sediments; (b) a *phreatic aquifer* that is contained, depending on its spatial position, within these same deposits or in the sandstones of the Puerto Madryn Formation, and which is mainly exploited in the region; (c) one or more *semi-confined* or *confined aquifers*, limited by clayey or silty-clay strata in the same Puerto Madryn Formation or in the underlying Gaiman Formation. The information available on these deeper aquifers is scarce because they are of no qualitative interest for exploitation.

The description of the hydrodynamic cycle begins with the characterisation of the recharge phenomenon, then it focuses on the circulation, ending with the discharge.

3.1 Groundwater Recharge

Despite the fact that recharge is the most important process in the groundwater cycle, it is the one that causes more difficulties when trying to quantify it, especially in arid regions where the necessary data to assess it is usually insufficient (Scanlon et al. 2010; Timms et al. 2012; de Vries and Simmers 2002).

The geographic location of the Península Valdés area rule out the possibility of infiltration from streams and/or outside recharge for the quaternary and Miocene aquifers. Having discarded the possibility of imported water, the only alternative to explain the presence of groundwater in the study area is water of meteoric origin.

A predominantly groundwater radial divergent morphology, observed at the central–southern sector, coinciding with a groundwater watershed (Fig. 3), indicates a preferential recharge area. This corresponds to the aeolian landform unit (Active dunefields and Stabilized aeolian field geomorphologic units).

Considering the average excesses from the groundwater balance (32 mm/a; Alvarez et al. 2013), the groundwater recharge over the sand dunes area (710 km^2) corresponds to 22.7 hm^3/a. This amount represents near 14.5% of the historic average annual precipitation and it is worth mentioning that for other Patagonian regions similar recharge values have been observed (Paruello and Sala 1995). Nevertheless, the annual recharge could notably vary because the main water excesses occur during storm events (Alvarez et al. 2012, 2013; Simmers 1997; Tweed et al. 2011).

Geohydrological map of Península Valdés

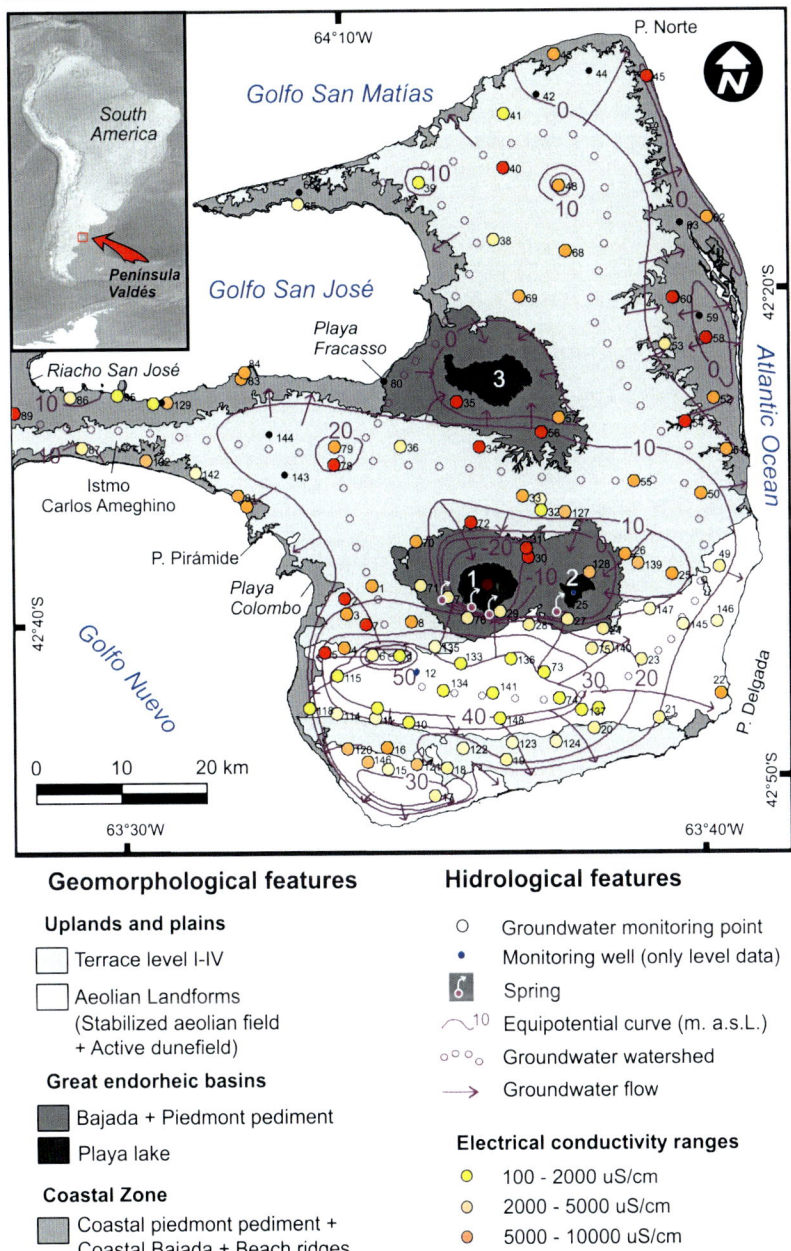

Fig. 3 Groundwater flow map with the associated geomorphological units and the groundwater conductivity values. *1* Salina Grande; *2* Salina Chica; *3* Gran Salitral

3.2 Groundwater Flow

According to Alvarez et al. (2010) two main groundwater watersheds are observed. One runs parallel to the coast of the peninsula and separates the regional discharge towards the sea from the local discharge towards the endorheic depressions (Fig. 3). The other watershed divides the area into a northern and a southern sector, at the latitude of the Istmo Carlos Ameghino (Fig. 3).

In the southern sector water flows from the aeolian landform towards the Golfo Nuevo and the Atlantic Ocean, and towards the Salina Grande and Salina Chica. In the northern sector of the Península Valdés water flows from the terrace levels towards the Gran Salitral, the Golfo San José and the Atlantic Ocean. It is important to highlight that terrace levels (i.e. geomorphic units that includes the Rodados Patagónicos) are one of the geomorphological units that better represents circulation in the hydrodynamic cycle.

3.3 Groundwater Discharge

The regional discharge, interpreted from the equipotential map and evidenced by wet spots and vegetation changes founded over the coastal cliffs, runs towards the Golfo San José, Golfo Nuevo, and the Atlantic Ocean (Fig. 4). At the coastal zone some marshes are found like Playa Colombo, Caleta Valdes, Riacho San José and Playa Fracasso (Fig. 3). In these marshes the groundwater flow have a marine component, but also a continental component. The last is represented by the regional discharge from the peninsula and/or the local discharge from the coastal dunes and ridges (Alvarez et al. 2014; Fig. 3). The discharges have also a significant local components represented by the internal drainage system of the Salina Grande, Salina Chica and Gran Salitral great enhoreic basins (Fig. 3). The wetlands and springs, such as the one originally used to supply the town of Puerto Pirámides, evidence this local discharge and are mainly located at the southern margin of Salina Grande and Salina Chica (Stampone and Cambra 1983; Alvarez et al. 2008, 2013; Fig. 3). The equipotential line that encloses both playa lakes is very noticeable on the map, and its central sector reaches −40 m a.s.l. (Fig. 3).

4 Hydrgeochemistry

4.1 Groundwater Salinity and Major Ionic Composition

As regards, the hydrochemistry of groundwater, the electric conductivity (EC) values—directly related to salt contents—reflects two clearly identifiable zones: a freshwater zone (low salinity) and another one with brackish to saline characteristics.

Fig. 4 **a** Wetlands at the southern margin of the Salina Grande; **b** Wet sectors in the coastal cliffs of the Golfo Nuevo near to Puerto Piramides locality. *White arrows* mark groundwater discharge areas

The low salinity zone—develops in the southern sector—matches with aeolian landform units and the radial divergent morphology of the water table, defined as the main recharge area. The electric conductivity values between 100 and 2000 µS/cm are located in the area with equipotential curves between 50 and 40 m a.s.l. and towards the limit of the aeolian landform units the values reaches 5000 µS/cm (Fig. 3). The high salinity zone, with electric conductivity values above 5000 µS/cm, occurs in the northern and western sectors, coinciding with the circulation and discharge areas (Alvarez et al. 2008; Fig. 3). At the Terrace levels most of the values are in the range 5000–10,000 µS/cm, and surface and subsurface waters in playa lakes have values above 30,000 µS/cm (Fig. 3).

Regarding the ionic composition of the water, the predominant major ion water type is sodium chloride, with sodium bicarbonate type water occurring only in the low salinity zone (Alvarez et al. 2010). In order to clarify the illustration of the different geomorphological units in the *Piper plot*, a selection of representative water samples were done. The ionic classification of the springs samples were included as well (Fig. 5).

The aeolian landform units presents sodium chloride–bicarbonate and sodium bicarbonate–chloride water types, with only one sample showing the presence of the calcium/sodium chloride–bicarbonate facies.

In the case of the Great Endorheic Basin system, the groundwater chemistry is mainly a sodium chloride water type, with localized occurrences of sodium chloride–sulphate water facies. The prevalence of chlorides marks the evolution of water and corroborates the discharge properties of the unit.

Despite the fact that geomorphologically the spring samples belong to the Great Endorheic basin system in the piper plot they are classified as sodium chloride–sulphate water type, with a larger proportion of bicarbonate and a smaller proportion of total salts than the water from the geomorphological unit containing them.

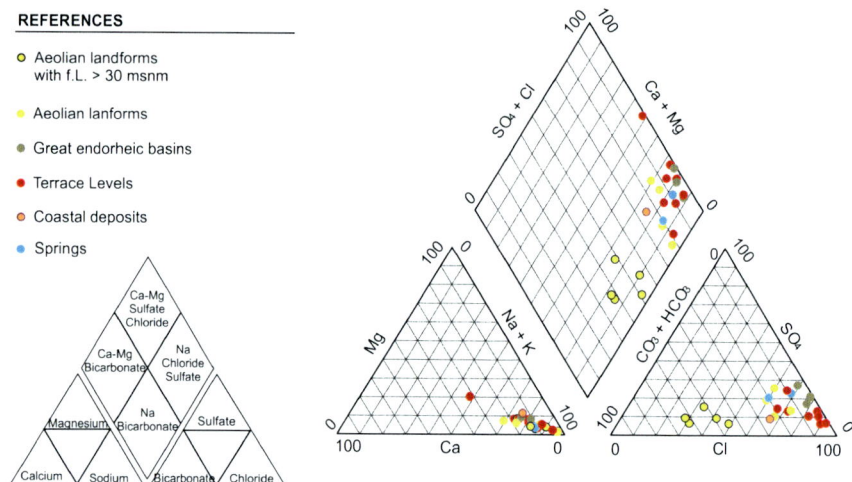

Fig. 5 Piper plot showing the major ionic composition of water samples on each geomorphological unit

The ionic composition of the groundwater from the Puerto Madryn Formation below the Terrace Levels unit does not seem to correspond to the circulation phenomenon essentially. In a normal evolution of the groundwater composition, the water type varies from bicarbonate at the recharge zone to sulphate in the circulation area and finally to chloride type at the discharge zone. In the case of Península Valdés, the circulation zone is characterized by sodium chloride–sulphate water type, quite similar to those waters observed in endorheic depressions. This is possibly due to the fact that there are smaller basins within the plain, as well as the low flow velocity of the groundwater.

In the Coastal Zone system the deposits of the geomorphic units composed of sandy material (i.e. coastal dunes) varies from sodium chloride–bicarbonate water type to sodium chloride–sulphate water type. On the other hand the geomorphic units constituted by gravel material (i.e. beach ridges) have sodium chloride–sulphate water type (Fig. 5).

4.2 Groundwater Isotopic Composition

The isotope hydrology was used as a complementary tool to analyze the source of the groundwater contained in the different landforms. Selected points of the different geomorphological units as well as the subsurface water samples of the Salina Grande and Salina Chica were analyzed (Fig. 6). In this analysis, seawater, individual rains, and the 1999–2000 weighted average precipitation of the Puerto Madryn collected station were included (Fig. 6).

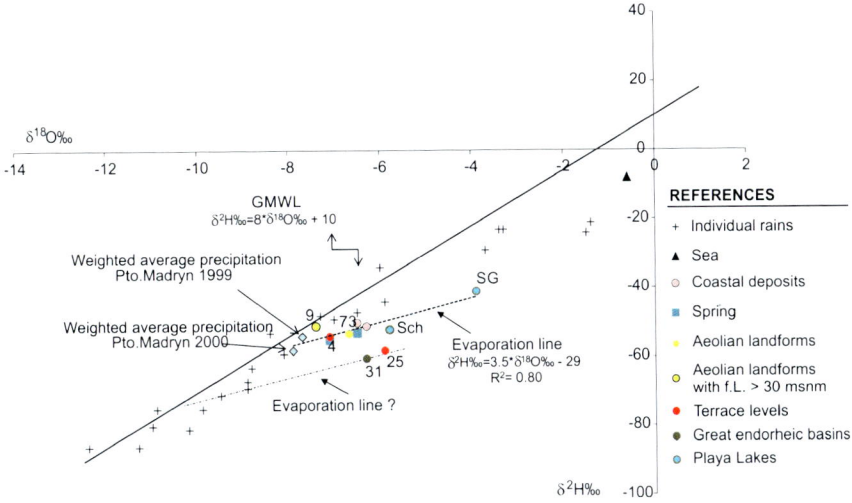

Fig. 6 δ^2H versus $\delta^{18}O$ diagram with representative water samples of the region. Samples include geomorphological units, springs, seawater and rainwater from Puerto Madryn station (belonging to the National Network of Argentina and Global Network for Isotope in Precipitation; IAEA/WMO 2002; Dapeña and Panarello 2008). Salina Grande: SG and Salina Chica: Sch

The isotopic composition of all, most all the analyzed groundwater samples shows an alignment from a near-average rainwater origin towards the Salina Grande (SG) composition (Fig. 6). This alignment would indicate evaporation processes occurred during infiltration through the unsaturated zone, or in surface previously to its infiltration. The extreme evaporation takes place in the playa lakes as it is evidenced by the isotopic composition of the Salina Grande (Fig. 6).

Regarding the aeolian landforms (with phreatic levels at 30 m a.s.l.), their position on the δ^2H versus $\delta^{18}O$ diagram, near to the global meteoric water line, indicates an origin of direct recharge from rainwater without evaporation or mix with other waters.

Another situation that could occur is the evaporation process from rainwater of different composition as could be interpreted for the samples number 25 and 31 (Fig. 6). The possibility of water mixing with deepest groundwater levels with different isotopic composition is not ruled out but cannot be verified.

5 Geohidrological Regions

The integration of geomorphological, lithological and hydrogeological studies of the Península Valdés allows to define four geohydrological regions (Fig. 7): Two in the Uplands and plains system (aeolian landforms and terrace levels) one in the Great endorheic basin system and one in the Coastal zone system (Fig. 7;

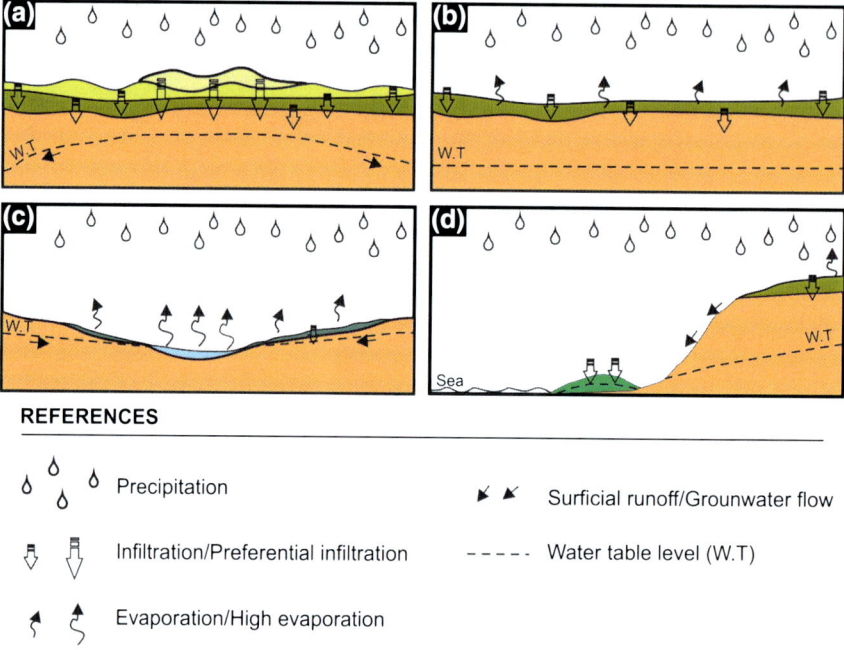

Fig. 7 Geohydrological regions. **a** Aeolian landforms; **b** Terrace Levels; **c** Great Endorheic Basin; **d** Coastal Zone. The references of the geological units are according to that showed in Fig. 1

see Chapter "Late Cenozoic Landforms and Landscape Evolution of Península Valdés"). Based on the groundwater hydrodynamics (recharge, flow and discharge), the hydrochemical composition and the isotopic characteristics of the groundwater of each geomorphologic system, a conceptual model is defined in order to explain the functioning of the hydrogeological system.

5.1 Aeolian Landforms

As it was said before, aeolian landforms (i.e. stabilized aeolian field + Active dunefield of the Uplands and plains system) constitutes the main recharge area (Figs. 3 and 7a). A hypothetical cross section of this region starts with a high primary permeability unit, constituted by a stabilized aeolian field and active dunefileds deposits that overlies the Rodados Patagónicos and/or the Puerto Madryn Formation. The recharge takes place in this area due to the already mentioned soil moisture excess, but it is favoured by the fast infiltration process. A special mechanism that is a determining factor of the recharge process in arid regions (Hernández et al. 2002; Hernández 2005).

The fast infiltration is an effective recharge mechanism for arid environments as the evapotranspiration is minimized. It is favoured by the superficial grain size sediments (as the present aeolian sands) and the low retention capacity of the soils like the one present in this unit (Alvarez et al. 2013). The lack of vegetation over the active aeolian landforms gives an extra chance to the fast infiltration mechanism. Likewise the vegetation adaptation to arid climate leads to the plant to survive with less water favouring the soil moisture excesses, and thus the leaching.

According to the groundwater hydrochemistry, the aeolian landforms region is typified by low salinity waters where the bicarbonate anion is more abundant. This ionic type, added to the hydrodynamic characteristics of the area, indicates that the infiltration of rainwater is the hydrological process that prevails over this region (Fig. 5).

5.2 Terrace Levels

Terrace Levels—all of them in the Uplands and plains system and composed by the deposits of the Rodados Patagonicos unit—represents the most extensive circulation areas, including local recharge–discharge zones. The condition of low recharge and poor groundwater circulation (as could be interpreted from the hydrodynamic map, Fig. 3) gives to this unit a general poor groundwater quality. Groundwater of this unit is mainly salty with local freshwater sites (1200–9500 µS/cm) and mainly a sodium chloride–sulphate water type. Additionally, an increase of the saline content could be due to an evaporation effect (interpreted from the enrichment in the isotopic composition; Fig. 6). The rainwater that occasionally infiltrate over this unit could suffer an evaporation process previously to reach the aquifer (Fig. 7b). This is a common process in arid climate regions, where the superficial sediments have low permeability (Ahmed et al. 2013). Another mechanism that increases the groundwater saline content is the evaporite precipitation by complete evaporation of rainwater over less permeable sediments and then dissolution by rainwater and leaching towards the aquifer (Fitzpatrick et al. 2000; Tweed et al. 2011). As the soils that characterized the terrace levels are the Natrargids (soils with alluvial clay accumulation and high exchangeable sodium percentage; see Chapter "Soil–Geomorphology Relationships and Pedogenic Processes in Península Valdés"), the presence of salt precipitation in the unsaturated zone is very common and the described processes is highly probable. The flat topography with little to medium endorheic basins, combined with the hydrolithological configuration (low permeability sediments overlaying the Miocene sediments), favours the mentioned processes as it makes the rainwater remains on the surface and evaporate before its infiltration (Fig. 7b).

5.3 Great Endorheic Basin System

The great endorheic basin system (Fig. 7c) represents the most relevant inland discharge areas (i.e. Salina Grande, Salina Chica and Gran Salitral salt pans) showing a typical radial phreatic morphology with a tendency towards a convergent cylindrical pattern and high hydric gradients. Discharge is also evidenced by the presence of stratigraphic springs at the hillslope of the great endorheic basins (Fig. 4a), where groundwater emerges and runs towards the playa lakes (Alvarez et al. 2008). Playa lakes are areas where all the groundwater from the discharge evaporates creating salt deposits (Brodtkorb 1999) and saline to hypersaline water. This evaporation process that takes place over the playa lakes is corroborated by the enrichment in the isotopic composition of both δ^2H and $\delta^{18}O$ (Fig. 6).

Furthermore, on the hillslopes of the endorehic depressions, salt dissolution–precipitation occurs, since the presence of Natrargids (see Chapter "Soil-Geomorphology Relationships and Pedogenic Processes in Península Valdés") facilitates this process, as in terrace levels.

In the case of the springs that flow on the southern margin of the endorheic depressions (Salina Grande and Salina Chica) they are less evolved and their conductivities are of the order of 3000–4000 µS/cm, due to their proximity to the main recharge located in the sand dune area.

5.4 Coastal Zone Region

The sand and gravel deposits at the Coastal zone conforms a local recharge area, whose morphology is that of a divergent radial flow and it is independent from the regional hydrogeomorphological behaviour (Fig. 7d). Its waters are chloride–bicarbonate to chloride types having the sand dunes lower salinity content than the coastal ridges. The recharge mechanism that operates is the same described for the aeolian landforms, but the extra saline content could be due to the proximity to the sea and the marine aerosols that increase the saline content of the rain water (Appelo and Postma 2005; Salama et al. 1999). The presence of these local recharge areas are due to the hidrolithological configuration, characterized by superficial high permeable deposits overlaying less permeable sediments. This allows to the fast infiltration of rainwater and then accumulation over the less permeable unit conforming a shallow aquifer. This unit, in which the groundwater has a relatively good quality and where the water table is at a shallow position, it is of particular interest to understand the hunter-gatherer populations that occupied the Península Valdés more than 5000 years ago (see Chapter "Archaeology of the Península Valdés: Spatial and Temporal Variability in the Human Use of the Landscape and Geological Resources").

6 Perspective and Future Works

The groundwater resources of Península Valdés are closely related to the climate and to the relationship between the geomorphology and lithology of the region. As climate is quite homogeneous all over the peninsula, this control mainly conditioned the recharge phenomena at a regional scale. The spatial geohidrological variations are more closely related with the variations in the geomorphology and in the relationships between the lithological units.

Ongoing research over the hydrogeology of the Peninsula Valdes is needed to understand with more detail hydrodynamics and hydrochemistry in the region. Mathematical modelling of the recharge phenomena and of the hydrochemistry processes that modifies groundwater quality would probably tackle this topics. It would be also interesting to orient future works into the study of the aquifer porosity variations, and the delimitation and study of deeper aquifer levels in order to find new water reservoirs for this waterless region.

Acknowledgements The authors would like to thank to Dr. Eleonora Carol, Dr. Cristina Dapeña and Dr. Daniel Martinez for their helpful and constructive comments that allowed to improve this chapter. They would also like to thank the Editors for their generous comments and support during the review process.

Glossary

Aquicludes	Rocks or unconsolidated material of low hydraulic conductivity which yield inappreciable quantities of water to drains, wells, springs and seep
Aquifer	Permeable water-bearing rocks or unconsolidated material capable of yielding exploitable quantities of water
Aquifuges	Rocks or unconsolidated material that cannot absorb or transmit water
Aquitard	Rocks or unconsolidated material that transmits water at a very slow rate compared with an aquifer
Clast-supported	Conglomerate or sand in which most individual clasts are in contact with each other. The matrix (finer material between the gravel clasts, such as sand, silt or mud) fills the spaces between the clasts
Confined aquifers	Aquifer overlain and underlain by an aquifuge, aquiclude or acuiclude
Hydraulic conductivity	Is a measure of the soil ock or unconsolidated material's ability to transmit water when submitted to a hydraulic gradient.

Matrix-supported	Conglomerate or sand where the individual clasts are held together by the matrix (i.e. they are not necessarily in contact with each other). The rock may appear disorganized without any clear internal structure
Non-saturated zone or unsaturated zone	Portion of the lithosphere in which the interstices are filled partly with air and partly with water
Phreatic aquifer	An aquifer that is physically in contact with the atmosphere through the groundwater surface
Piper plot	Ternary diagram useful to visualize the chemistry of a rock, soil, or water sample. The apex of the diagram include major cations and anions founded in nature
Saturated zone	Zone of the water-bearing material in which all voids, large and small, are filled with water
Semi-confined	Aquifer overlain and/or underlain by a relatively thin semi-pervious layer, through which flow into or out of the aquifer can take place

References

Ahmed MA et al (2013) Factors controlling mechanisms of groundwater salinization and hydrogeochemical processes in the quaternary aquifer of the Eastern Nile Delta, Egypt. Environ Earth Sci 68:369–394

Alvarez MP et al (2008) Geohidrología de humedales cercanos a la costa con cota bajo nivel del mar. Península Valdés, Argentina In: Rev Latinoam Hidrogeologia 6(1):35–42

Alvarez MP (2010) Investigación geohidrológica en un sector de Península Valdés, provincia de Chubut. PhD Thesis. Universidad Nacional de La Plata. http://sedici.unlp.edu.ar/handle/10915/5321

Alvarez MP et al (2010) Linking geomorphology and hydrodynamics: a case study from Península Valdés, Patagonia, Argentina. Hydrogeol J 18:473–486

Alvarez MP et al (2012) Groundwater flow model, recharge estimation and sustainability in an arid region of Patagonia, Argentina. Environ Earth Sci 66(7):2097–2108. ISSN 1866-6280

Alvarez MP et al (2013) Estimación de la recarga en zonas áridas según distintos métodos. Área medanosa de Península Valdés. In: Gonzalez N et al (eds) En Agua Subterránea recurso estratégico vol 1, pp 15–22. ISBN978-987-1985-03-6

Alvarez MP et al (2014) Hidrogeología del Sur de Península Valdés, Chubut, Argentina. Revista Latino-Americana de Hidrogeología. ALSHUD-Univ. Fed. Paraná 9(1):9–24

Appelo C, Postma D (2005) Geochemistry, groundwater and pollution, 2nd edn. Balkema Publishers, p 479

Brodtkorb A (1999) Salinas Grande y Chica de la Península de Valdés, Chubut. In: Zappettini EO (ed) Recursos Minerales de la República Argentina (SEGEMAR)

Dapeña C, Panarello H (2008) Isótopos en precipitación en Argentina. Aplicaciones en estudios Hidrológicos e Hidrogeológicos. 9° Congreso Latinoamericano de Hidrología Subterránea. ALHSUD CD T-100. 8 p. Quito, Ecuador

de Vries JJ, Simmers I (2002) Groundwater recharge: an overview of processes and challenges. Hydrogeol J 10:5–17

Fitzpatrick RW et al (2000) What are saline soils and what happens when they are drained? J Aust Assoc Nat Resour Manag 6:26–30

Hernández MA (2005) Mecanismos de recarga de acuíferos en regiones áridas (síntesis) Proceedings of the II Seminario Hispano Latinoamericano sobre Temas Actuales de la Hidrología Subterránea. Río Cuarto, Córdoba, Argentina pp 249–254

Hernández MA et al (2002) Mecanismos de recarga de acuíferos en regiones áridas. Cuenca del Río Seco, Provincia de Santa Cruz, Argentina. Proceedings of the XXXII IAH Congress and VI Congreso ALHSUD, CD-ROM. Mar del Plata, Buenos Aires, Argentina

IAEA/WMO (2002) Global network for isotopes in precipitation. The GNIP Database. http://isohis.iaea.org

Paruello JM, Sala O (1995) Water losses in the Patagonian steppe: a modelling approach. Ecology 76(2):510–520

Salama RB et al (1999) Contributions of groundwater conditions to soil and water salinization. Hydrogeol J 7:46–64

Scanlon BR et al (2010) Groundwater recharge in natural dune systems and agricultural ecosystems in the Thar Desert Region, Rajasthan, India. Hydrogeol J 18(4):959–972

Simmers I (1997) Groundwater recharge principles, problems and developments. In: Recharge of phreatic aquifers in (Semi-) arid areas, vol 19. IAH, A.A. Balkema, Rotterdam, Brookfield, pp 1–18

Stampone J, Cambra H (1983) Estudio hidrogeológico del área Sur de Península Valdés. XI Congreso Nacional del agua. Córdoba, pp 139–171

Timms et al (2012) Implications of deep drainage through saline clay for groundwater recharge and sustainable cropping in a semi-arid catchment, Australia. Hydrol Earth Syst Sci 16:1203–1219

Tweed S et al (2011) Arid zone groundwater recharge and salinisation processes; an example from the Lake Eyre Basin, Australia. J Hydrol 408:257–275

Windhausen A (1921) Informe sobre un viaje de reconocimiento geológico en la parte noreste del Territorio de Chubut, con referencia especial a la provisión de agua de Puerto Madryn. Con un estudio petrográfico de algunas rocas por R. Beder Bulletin 24B. Dirección General de Minas, Buenos Aires, Argentina

Archaeology of the Península Valdés: Spatial and Temporal Variability in the Human Use of the Landscape and Geological Resources

Julieta Gómez Otero, Verónica Schuster and Anahí Banegas

Abstract The archaeological record of Península Valdés shows this area was intensively used by native hunter-gatherers since at least 5000 years BP to the nineteenth century. These populations located their settlements in sandfields across the littoral zone, primarily on coastal bajadas and low marine terraces near fixed shoals of molluscs. Archaeofaunal studies and stable isotope analyses (^{13}C and ^{15}N) of human bone samples indicate that they had a terrestrial-marine diet including guanaco meat, land plants, mollusks, fishes and pinnipeds. The old inhabitants of the peninsula profited local rocks and clay minerals to manufacture their artefacts. To lithic technology, they used small pebbles of silica and basalt, big pebbles of riolites and granites, consolidated sandstones and fossil cetacean bones. The basic toolkit comprised knives, end-scrapers, side-scrapers, drills, burins, notches, fishing weights and a variety of projectile points. The big pebbles were used as *manos*, hammer stones and anvils, while sandstones and fossil bones were primarily employed in milling activities. With respect to pottery technology, the abundance and good quality of local clay sources allowed its important development in the final late Holocene. Most vessels present oval or spherical shapes, straight sides and concave bases. All these features are suitable for domestic activities, e.g. preparing, storing and/or cooking food and liquids. These hunter-gatherers did not live isolated from other populations: the presence of foreign rocks and of three ceramic vessels, which might have been manufactured in northwest Patagonia or perhaps central Chile, indicates that they took part in an extensive trading net since at least the late Holocene.

J.G. Otero (✉) · V. Schuster · A. Banegas
Instituto de Diversidad y Evolución Austral (IDEAus), Consejo Nacional de Investigaciones Científicas y Técnicas (CONICET) - CCT Centro Nacional Patagónico (CENPAT), Boulevard Almirante Brown 2915, ZC: U9120ACD Puerto Madryn, Chubut, Argentina
e-mail: julietagomezotero@yahoo.com.ar

V. Schuster
e-mail: veroschus@hotmail.com

A. Banegas
e-mail: banegas.anahi@yahoo.com.ar

Keywords Archaeology · Península Valdés · Human use · Landscape · Geological resources

1 Introduction

Radiocarbon record of eastern Patagonia (between the Colorado River and the Strait of Magellan) shows that human beings reached this region *circa* 12,000–11,000 years BP (Borrero and McEwan 1997; Politis et al. 2008). The history of archaeological research shows that until the mid-80s decade the eldest ages had been obtained from sites located in the Patagonian Andes and the central foreland area, while the datations in the Atlantic coast were all younger than 3000 years old (Orquera 1987). This fact might be related to the scarce knowledge about the archaeology of the littoral area as all research work in Patagonia, with rare exceptions, had generally focused on inland ecosystems far from the coast (synthesis in Orquera and Gómez Otero 2007). Consequently, the prevailing conception was that the Atlantic coast was occupied lately by inland hunters that would only reach the coast randomly or seasonally. It was hard to accept, however, that hunter-gatherers in Patagonia would have failed to take advantage, on a regular basis, of an environment that was abundant and diverse in terms of food supplies, which made it favourable for survival as against others (Yesner 1980; Perlman 1980; Bailey and Parkington 1988).

With respect to Península Valdés, the first systematic research began in 1993 and continues to this day framed in a project whose main purpose is to know the temporal and spatial variability in the human–environment relationship in the Northeast area of the province of Chubut, including the marine coast and the lower valley of Chubut river (Gómez Otero 2006; Gómez Otero et al. 1999). Before then, the information available was limited to a few discoveries of isolated artefacts: the study of a multiple burial site and an unpublished summary by Menghin and Bormida (circa 1995) of the results of unsystematic archaeological prospections (Gómez Otero et al. 1999). In light of the above, research was now focused, on the one hand, on proving that the ocean and its resources were much more important for hunter-gatherers in the study area than expected; and, on the other, on confirming that the littoral area had been occupied before the late Holocene and that there had been groups that lived all year round, or most of it, in the coastal strip.

In this chapter, we summarize the results of 20 years of archaeological studies in the Península Valdés region, with an emphasis on the human use of landscape (considered yet another resource), rocks, clay minerals and soil components.

2 The Península Valdés and Its Advantages for Human Life

The foreland region next to the Patagonian Atlantic coast runs across 2500 km from the Colorado River in the north, to the Strait of Magellan, in the south, 40° S and 52° S (Fig. 1). Generally speaking, it is formed by a series of tertiary-aged plateau—distributed approximately 200 km to the west of the coast—that are disrupted by only seven rivers (the Chubut river among them), that represent the only permanent and significant sources of fresh water. This region is characterized by a relative environmental homogeneity based on low topography, arid climate and xeric vegetation.

In this context, Península Valdés has several advantages for human life. First, it is formed by several types of coasts: the gulf coasts of Golfo San José, Golfo San Matías and Golfo Nuevo; the barrier spits and islands of the Caleta Valdés (Valdés

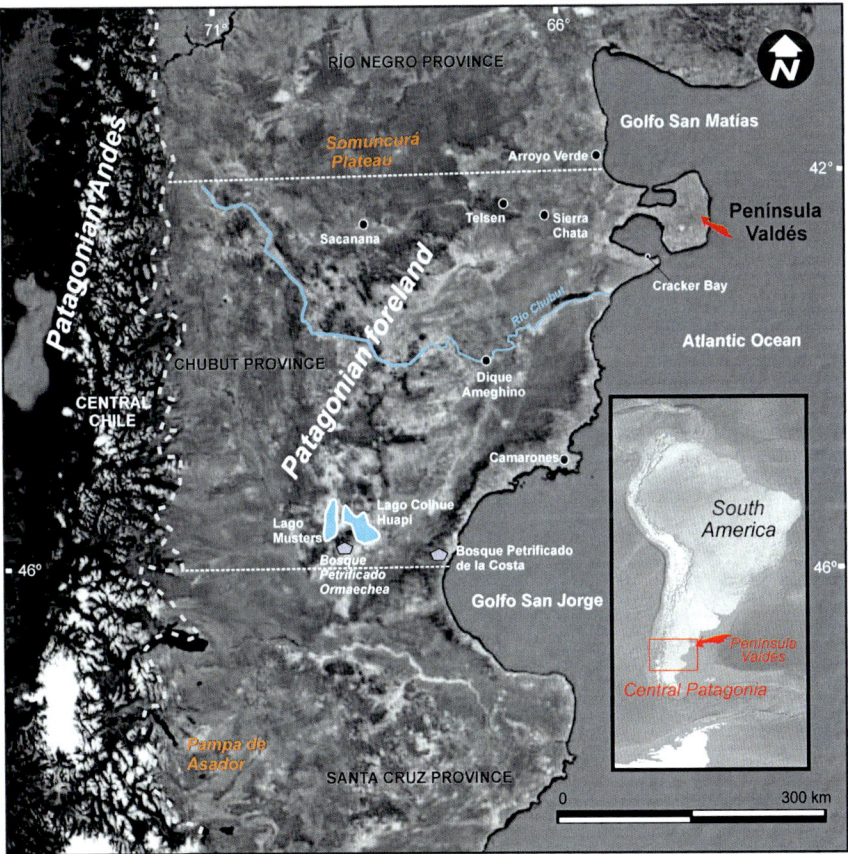

Fig. 1 The Península Valdés region in a regional context

inlet), and the open sea coast between Punta Delgada and Morro Nuevo (Fig. 2). On the other hand, the peninsula may be regarded virtually as an island of semi-circular shape, with a maximum radius of 45 km and no internal geographical

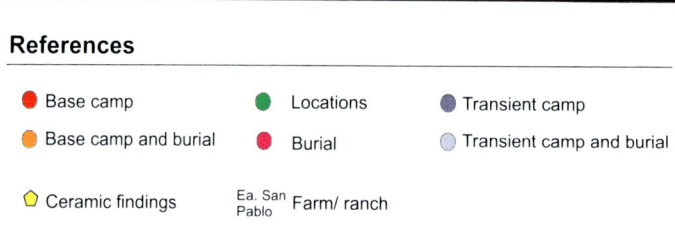

Fig. 2 Site spatial distribution and ceramic findings in the Península Valdés

barriers. This allows for quick access to any point and fast connection between coasts (Gómez Otero et al. 1999) (Fig. 2). Topography does not rise above more than 110 m a.s.l. and access to the ocean is possible and easy in areas with wide littoral *bajadas* or downward slopes (see Chapter "Late Cenozoic Landforms and Landscape Evolution of Península Valdés"), low topography or through *cañadones* or gullies whose heads are several kilometres inland (Súnico 1996). In addition, the neogene deposits of the Gaiman and Puerto Madryn Formations (see Chapter "Geology of Península Valdés") show cavities and small overhangs that may have been used in the past as shelter for both animals and humans and that were in fact used as burial sites (Outes 1915). Moreover, the sandy environments that discontinuously run over the coastal perimeter covering a large portion of the southern area of the peninsula (Súnico 1996; see Chapters "Late Cenozoic Landforms and Landscape Evolution of Península Valdés" and "Groundwater Resources of Península Valdés") also provide shelter as well as an adequate surface for settlement. What distinguishes it from other arid Atlantic littoral sectors is the permanent fresh water supply through several springs available in the Salina Grande and Salina Chica, both located in the central area of the peninsula (Fig. 2; see Chapter "Groundwater Resources of Península Valdés"). This water supply is supplemented by other reservoirs such as sandbanks and temporary small lakes that retain water after the short yet stormy rains present in the area (see Chapter "Groundwater Resources of Península Valdés").

Bathymetry near the coast presents a smooth relief; during ordinary low tides, large intertidal areas remain exposed, some with wave-cut platforms (informally called restingas) that enable the development of shoals of molluscs and other invertebrates. Moreover, there is a 6-hour difference between tides from the Golfo San José and the Golfo Nuevo, and this increases the possibilities for daily extraction of intertidal resources in each gulf. In addition, the Península Valdés is influenced by the tidal fronts (i.e. Península Valdés tidal front; Acha et al. 2004) a mechanism that determines the existence of high coastal productivity, especially in spring and summer.

The Península Valdés also features rocks and minerals suitable for different technologies (Fig. 3). There are raw materials of good quality for flaking, such as basalts and silicas, available in the gravels of the *Rodados Patagónicos deposits* and in the Pleistocene–Holocene gravel beach ridges (see Chapters "Geology of Península Valdés" and "Late Cenozoic Landforms and Landscape Evolution of Península Valdés"). Sandstones flat boulders from the Puerto Madryn Formation and cetacean fossilized bones are also present, which may be used as devices for grinding. Finally, clay minerals suitable for pottery technology are available in soil argillic horizons (Bouza et al. 2007; Schuster 2014; see Chapters "Geology of Península Valdés" and "Soil-Geomorphology Relationships and Pedogenic Processes in Península Valdés").

With regard to weather conditions, the low altitude and, especially, insularity provide the area with undoubtedly oceanic properties (see Chapter "The Climate of Península Valdés Within a Regional Frame"). The mean annual precipitation values in the Península Valdés area are within the 100–300 mm/a, with some years of very

Fig. 3 Lithic and mineral resources. **a** Flat boulders of the Puerto Madryn Formation; **b** small and medium-sized gravels from beach ridges; **c** Pleistocene big gravels from the Caleta Valdés area; **d** clay source available in temporary small ponds

low precipitation (about 50 mm) and other with exceptionally high values (>500 mm) (see Chapter "Miocene Marine Transgressions: Paleoenvironments and Paleobiodiversity"). By way of comparison, in other coastal region of central Patagonian, mean annual values range from 100 to 180 mm.

As for vegetation, there is an association between shrub steppe and grass steppe (see Chapter "Vegetation of Península Valdés: Priority Sites for Conservation"), which means there is an abundance of firewood and good pastures for *guanaco*, the largest herbivore in the area and the main prey of hunter-gatherers in Patagonia. In areas where aridization and over-exploitation have had a lesser impact, woody species are abundant, with good caloric potential (up to 700 °C; see Bouza et al. 2007). Primary woody species are *molle* (*Schinus* sp.), the two kinds of *algarrobo* (*Prosopis alpataco* and *P. denudens*), the *jume* (*Suaeda divaricata*), the *quilembay* (*Chuquiraga avellanedae*), the *mata sebo* (*Monttea aphylla*) and the *piquillín* (*Condalia microphylla*). *Quilembay* was also used to send smoke signals (Jardín Botánico de la Patagonia Extraandina 2002). The most important edible vegetation includes *macachín* tubers (*Arjona tuberosa*), *piquillin* berries, *algarrobo* pods, as well as fruits and leaves from cactaceae of the *Opuntia* and *Maihueniopsis* genus. In addition, several species were and still are used for their medicinal or therapeutic properties, including *molle*, *barilla* (*Larrea divaricata*), *quilembay*, *matasebo*, *botón de oro* (*Grindellia chiloensis*) and *carqueja* (*Bacharis crispa*) (Jardín Botánico de la Patagonia Argentina 2002).

As far as zoogeography is concerned, the Península Valdés region is one of the largest zoological parks in the world. Primary terrestrial vertebrates are *guanaco*, rheidae (*Pterocnemia pennata* or *choique*), dasipodidae or *armadillos* (*Zaedyus* sp. and *Chaetophractus* sp.), lagomorphs (*Dolichotis australis* or *maras*), and, to a lesser extent, canidae and felidae (see Chapter "Animal Diversity, Distribution and Conservation"). Marine fauna stands out for the diversity and abundance of molluscs, fish, seabirds and mammals. The variety of molluscs is higher than in the rest of the Patagonian Atlantic coast because two malacological provinces converge: the Argentine Province, which runs from Río de Janeiro to 42° parallel S, and the Magellan Province, which runs from such parallel to the province of Tierra del Fuego and the south of Chile (Escofet et al. 1978). This means that the richest gulfs in terms of malacofauna are the Golfo San Matías and the Golfo San José (Fig. 2). As for fish, there are species from rocky shores and from gravelly and sandy beaches (Cousseau and Perrotta 2000). In the case of seabirds and marine mammals valuable for human consumption, there are cormorants (*Phalacrocorax* spp.), Magellanic penguins (*Spheniscus magellanicus*), South American sea lions (*Otaria flavescens*), sea elephants (*Mirounga leonina*) and several cetaceans, primarily the southern right whale (*Eubalaena australis*). There are no current records of South American fur seals (*Arctocephalus australis*). Cormorants, penguins and pinnipeds are grouped in fixed and irregularly scattered colonies (Carrara 1952). Sea lions and cormorants remain in the area all year round; sea elephants and penguins—albeit migratory—are stable during their time of residence, i.e. spring–summer. Colonies that are closest to each other are in the Golfo San Matías. From the seventeenth century to mid-twentieth century, they were exploited by foreign and local *loberos* or sea lion hunters (Crespo and Pedraza 1991).

3 Creation of Predictive Models for Settlement and Diet

In the absence of archaeological background, at an early stage of the research, a preliminary model was designed applicable to the last 5000 years which proposed that the Península Valdés was occupied all year round by highly mobile hunter-gatherers whose diet was based on *guanacos* and sea lions (Gómez Otero et al. 1999). These hunters possibly moved, on a regular basis, alongside the coastal perimeter with short incursions into the centre of the peninsula for water supply from the Salina Grande and Salina Chica. On the basis of a high supply and of lithic raw material from gravels, record of expedient artefacts (sensu Binford 1979) was expected, especially throughout the coastal area, where there is more availability of gravels (Gómez Otero et al. 1999).

Later, as research moved forwards and by way of heuristic tools, some models were designed derived from the "*Optimal Foraging Theory*" (Bettinger 1991; Winterhalder and Smith 1992, inter alia). In this research, the Diet Breadth Model was applied, as well as models that analyse the settlement system in connection with the costs of searching and handling the food: the *Patch Choice Model* and the *Central Place Model* (Bettinger 1991: pp 83–90). The *Diet Breath Model* assumes that a prey will be chosen if the net energy it provides is higher or at least equal to the energy spent during foraging (Winterhalder and Smith 1992; Bettinger 1991). As for the choice of the best place to settle in, the *Central Place Model* claims that the hierarchization of diet elements—including fresh water—will vary based on the distance between resources and the place where they are searched, secured and brought from. For its part, the *Patch Choice Model* proposes that the highest ranked patch or parcel type will produce the best return per unit of foraging time (the sum of all search and handling time in the patch). But, when patches of that kind are widely spaced, time spent in travels between them may cause overall rate of energetic return to be suboptimal.

By applying these theoretical tools, another predictive model was designed applicable to the entire northern coast of the Chubut province (Fig. 1). This model proposed that the human use of landscape was subject to three variables: (a) the coastal productivity, (b) the relation between the resource supply offered by the marine environment and its adjacent terrestrial environment and (c) the fresh water availability (Gómez Otero 2006).

In terms of coastal productivity, not all types of coasts that make up the Península Valdés region feature the same oceanographic and topographic characteristics (see Chapter "Late Cenozoic Landforms and Landscape Evolution of Península Valdés"). Moreover, the influence of marine fronts is also uneven. If closeness to the Peninsula Valdés tidal front is considered, as well as the circulation of marine waters in the distribution of nutrients, the most productive coast is that located between Punta Norte and Punta Delgada and the least productive is that of the Golfo Nuevo, because the anticyclonic turns of its current prevent the intrusion of nutrient-rich water from such front (see Chapter "The Climate of Península

Valdés Within a Regional Frame"). The coasts of San Matías and San José Gulfs are somewhere in the middle (Gómez Otero 2006).

In addition to coastal productivity, the relative value and nutritious performance of adjacent terrestrial communities were considered which, in turn, vary on the basis of latitude and topography (Yesner 1980; Jones 1991). According to the respective energy performance as well as the easiness to search and secure preys, it was proposed that the optimal, annual average diet was possibly sustained on a relatively even combination of two preys—pinnipeds and *guanacos*—supplemented with mollusc consumption (Gómez Otero 2006). This average diet may have been seasonally reinforced with other types of food by reason of annual variations in resource availability and the nutritional needs of hunter-gatherers. Thus, *choiques*, small vertebrates (fish, birds, mammals) from the marine and terrestrial ecosystems, eggs and plants may have been incorporated to the basic diet throughout the year.

If *guanacos* and South American sea lions were the two most valuable preys and molluscs served as a daily dietary supplement, the highest ranking parcels would have been those offering the three types of resources. Since *guanacos* and molluscs have a wide and relatively even distribution throughout the coastal strip, parcel selection should have been determined by *Otaria* settlement distribution that is linear and discontinuous. In this regard, the coast of the Golfo San Matías features the highest number of sea lion colonies (Carrara 1952; Crespo and Pedraza 1991). However, fresh water availability, a critical resource in the Atlantic coast, should also be considered. Therefore, there are three possible options: one, a central settlement in the parcels that provided primary basic diet resources: sea lions, *guanacos* and molluscs; two, settling near permanent fresh water sources in the Salina Grande and Salina Chica; and three, choosing a location equidistant from both types of resources.

In sum, based on the three basic restrictions and the application of Optimal Foraging Theory models, the following hierarchy of landscape-use intensity was proposed, in a decreasing order, taking into account the different coastal sectors: Golfo San Matías, Golfo San José, Valdés inlet, Golfo Nuevo and the open sea coast between Punta Delgada and Morro Nuevo (Gómez Otero 2006).

4 Findings

4.1 The Use of the Landscape

The chronological record shows that hunter-gatherers used the Peninsula at least since the middle Holocene: one site in Punta Pardelas and another one in Punta Cormoranes (Fig. 2) indicated a conventional radiocarbon age of 5580 ^{14}C BP and 4340 ^{14}C BP, respectively (Gómez Otero 2006). An additional three dating studies performed in sites located in adjacent coast areas support such evidence: 7400 years ^{14}C BP at the mouth of the Arroyo Verde (41° 54′ S, Fig. 1, and 5500 ^{14}C BP and

5390 ^{14}C BP at Cracker Bay (42° 55′ S, Fig. 1; Gómez Otero et al. 2013a). This period matches the maximum marine transgression of the middle Holocene, that is, the sea level was above the current level (Weiler 1998; see Chapters "Climatic, Tectonic, Eustatic and Volcanic Controls on the Stratigraphic Record of Península Valdés" and "Late Cenozoic Landforms and Landscape Evolution of Península Valdés"). The number of sites considerably increases in the late Holocene (after 3200 BP) and abruptly decreases after the *Equestrian Period* (seventeenth–eighteenth centuries) when horse adoption promoted that mobility circuits changed and moved further from the coast (Gómez Otero 2007). At this time, in 1779, the Spanish Crown established a military settlement in the Península Valdés, occupying the Golfo San José and the Salina Grande area. This had a high impact on the social and economic system of hunter-gatherers (Bianchi Villelli and Buscaglia 2015).

Site distribution shows that most of them are concentrated in the coastal zone, especially in the coastal *bajadas* near fixed banks of molluscs. Sites were also recorded in the *bajadas* of the Salina Grande and Salina Chica and on the edge of cliffs, although these were not frequent (Fig. 2). As for topographic conditions, the highest archaeological density was observed in dunefields (52%), mainly in the coastal perimeter, followed in degree of importance by Holocene beach ridges (16%), coastal pediments (13%), the *salinas*' margins (7%) and around the banks of temporary pans (4%).

A diversity of sites were registered which included different functions and duration (sensu Binford 1980) such as base and transient camps, "locations" or special purpose occupations and human burials (Fig. 4). The preservation and visibility of the archaeological record is often irregular because of the highly dynamic environment that has unfolded since the Pleistocene–Holocene transition. First, the earliest coastal sites might have been submerged or destroyed due to early middle Holocene sea level variations. Second, as most sites are currently on the surface of a sandy environment, they are impacted by aeolian and water erosion as well as by other post-depositional processes, which include modern human activities. In these sites, all remains are widespread and mingled, forming palimpsests. The presence of thin and isolated, stratified *concheros* or *shellmiddens* in dunes and bajadas was also established. However, no high or big shell middens have yet been found.

With regard to the current height above sea level, sites located between 6 and 10 m a.s.l. prevail, which means parcels close to marine resources were chosen. Sites recorded between 11 and 20 m a.s.l. were also numerous. The relation between the sea level height and the functionality of occupations was explored, verifying that all sites located above 20 m a.s.l. were used for restricted activities (lithic quarrying, mollusc consumption or burials) while those located below 20 m a.s.l. were base camps and transient camps. This would be related to the type and time duration of the activities performed. But, in some cases the location of sites might be also correlated to sea level changes: the latest sites are situated at lower levels, while those from the middle Holocene are above 10 m, in line with a higher sea level than the current one at that time (Weiler 1998, see Chapter "Late Cenozoic Landforms and Landscape Evolution of Península Valdés"). As to the distance from

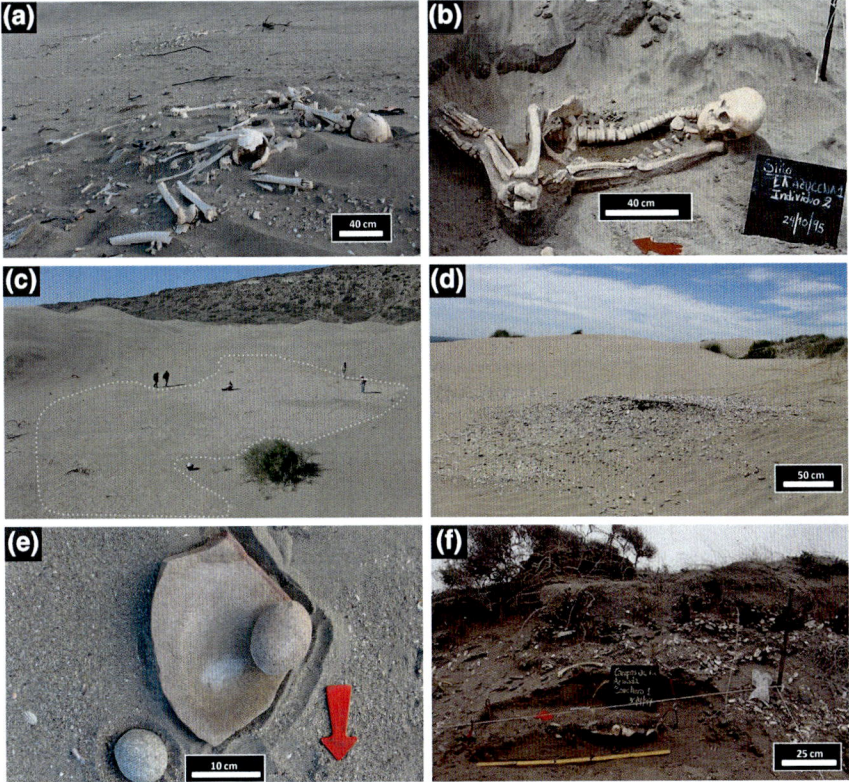

Fig. 4 Site diversity of the Península Valdés region. **a, b** Burial sites; **c, d** shellmiddens; **e** isolate findings; **f** stratified hearth

the high water line, sites were concentrated in a strip located within less than one hundred meters from the ocean. An exception was found in the Caleta Valdés (with a peak between 350 and 500 m far away from the coast) which might be the consequence of the decrease in the sea level after the mid-Holocene maximum transgression (see Chapter "Late Cenozoic Landforms and Landscape Evolution of Península Valdés"), which would have increased the distance between the original site location and the ocean.

As to fresh water, as it was mentioned above, the archaeological record shows few sites (most of them transient camps), in the area of the Salina Grande and Salina Chica and isolated materials around temporary pans associated to small to medium closed basins (Fig. 2; see Chapter "Late Cenozoic Landforms and Landscape Evolution of Península Valdés"). This supports evidence that the human populations that occupied the Peninsula chose to settle in parcels that ensured the acquisition of fixed and predictable marine resources. From those parcels, they travelled to the *salinas* in search of water and salt. Since 20 km run from the different coasts to the *salinas*, the groups that fetched water must have stayed at least one night outside the base camp; during that time, they probably hunted and consumed the terrestrial resources located in the surrounding area. Evidences of this are the scattered projectile points, *bolas* and knives found in different sites. Ceramic vessels were also found, whose presence suggests these vessels were used to transport or store water in late Holocene times. Before that, leather water skin, choique eggs, bladders and/or stomachs were probably used, although no evidence of these manufactures has been still registered in this area.

Nevertheless, basic water needs in the coastal area might have been satisfied by the use of rainwater found in sandbanks (see Chapter "Groundwater Resources of Península Valdés") In this regard, D'Orbigny (1999: pp 289–290) documented that in the Northern Patagonian coast water could be removed from sandbanks and that this water had a much better quality than that from drillings. Therefore, coastal dunefields might have offered not only adequate surfaces for settlement but also minimum fresh water supply.

Site concentration was observed in parcels near fixed shoals of molluscs from which there was easy access to both evenly distributed and scattered resources such as *guanacos*, small terrestrial vertebrates, fish and plants, and resources located in patches, such as pinnipeds, seabirds and stranded cetaceans. This might have ensured a minimum daily supply of minerals, vitamins, proteins and carbohydrates. In sum, the best option might have been to settle base camps in strategic locations that made it possible to forage without high costs in terms of searching and securing basic resources for an optimal diet.

As to the comparison between coastal sectors, results supported partially the preliminary model, because the highest archaeological density and richness were found in the coast of the Golfo San Matías, followed in a decreasing order by the Golfo San José and the Golfo Nuevo. But, against the preliminary expectations, the Caleta Valdés area and the open sea between Punta Delgada and Morro Nuevo, which are precisely the one that receives greater influence from the *Península Valdés tidal front*, showed the lesser archaeological density. Therefore, landscape occupation intensity might have been primarily determined by the relation between supply of terrestrial and marine resources and, secondly, by permanent fresh water supply and coastal productivity.

4.2 The Diet

The diet was explored through archaeofaunal and taphonomic studies and ^{13}C and ^{15}N stable isotope analyses in human collagen and apatite (Gómez Otero et al. 2000; Gómez Otero 2007). Archaeofaunistic and taphonomic studies were carried out following the methods of Behrensmeyer (1978), Mengoni Goñalons (1988), and Lyman (1994); isotope analyses were performed according to Van der Merwe (1992), Ambrose (1993), and Koch et al. (1997).

The results showed that since the middle Holocene up to the introduction of horses (late seventeenth century to early eighteenth century), food consumption by hunter-gatherers who occupied the Península Valdés comprised a wide array of terrestrial and marine resources (Gómez Otero 2006). Nevertheless, the isotopic values obtained from five human individuals discovered in the Peninsula indicated that marine preys with high trophic levels—pinnipeds—had greater relevance in food consumption than in other sectors of the Northern Chubut Province coast (Gómez Otero 2007). As it was mentioned above, during the equestrian period, diet became essentially terrestrial, due to the changes in the hunter-gatherers mobility after horse adoption.

Based on the comparison between the optimal diet preliminary model and archaeological records, both discrepancies and coincidences were observed. First, the two resources ranking on top of the food hierarchy—*guanacos* and sea lions— lacked representation in all archaeofaunal sites and samples. Second, there remains of *guanaco* in 57% of the samples and of pinnipeds in 27% (Gómez Otero 2006). This is an indication that the *guanaco* had more significance in food consumption than pinnipeds. Both types of prey are represented by all parts of the skeleton, which indicates they were hunted in the vicinity and were taken in whole pieces to the camps. Several indicators of seasonal zooarchaeological features, such as the estimation of both prey individuals' age and the observation of thin sections of adult pinniped canine teeth, show that *guanaco* hunting took place all year round and pinniped hunting occurred during the summer season until early autumn. Pinniped bones were primarily concentrated at sites on the coasts of the Golfo San Matías, Punta Delgada (open sea) and Golfo Nuevo (Punta Pirámide and Bajada Colombo, Fig. 2). It is worth noting that sea lion colonies are found along these coasts nowadays, which suggests that such colonies already existed and were exploited when native peoples occupied the area.

As regards molluscs—the resource ranking second in the optimal diet model— remains were found in all archaeofaunal sites and samples. Moreover, 58% of samples solely comprised molluscs (Gómez Otero 2006). Prevailing taxa are ribbed mussels (*Aulacomya atra*) and limpets of the *Patella* genus, followed by small gastropods like *Buccinanops* spp. and *Trophon* spp. Mollusc valves were also used in the making of *beads*, spoons and *containers*. From these findings it follows that molluscs were a regularly exploited resource both as food and as raw material.

As far as *choique* is concerned—ranking third in the proposed hierarchy—and contrary to expectations, only one bone specimen was found in a site of the Golfo San Matías. This finding concurs with the scarce archaeological records of remains of this prey across Patagonia. However, there are plenty of sites with *choique* eggshells—some of them burnt—that show the local hunter-gatherers ate these large-sized eggs, which are available by the middle of spring and might have represented an abundant, nutritious, energetic and easy-to-obtain food source.

As put forwards in the preliminary model, minor vertebrates such as fish, birds and small mammals were scarce in archaeofaunal sets. Among the samples containing remains of fish, taxonomic diversity was observed but with a very few individuals per species. The largest variety was found in the coast of the Golfo San Matías: skates, bluefish (*Pomatomus saltatrix*), mero (*Acanthistius patachonicus*), Argentine seabass (*Acanthistius brasilianus*), Argentine hake (*Merluccious hubbsi*),

Patagonian blenny, (*Eleginops maclovinus*), Argentinian sandperch (*Pseudopercis semifasciata*) and Brazilian sandperch (*Pinguipes brasilianus*).

With respect to marine birds, Magellanic penguins and cormorants are predominant and, like fish, they are represented by a very low number of individuals. As for terrestrial birds, no taxon was identified.

The archaeofaunal record of minor terrestrial mammals is also scarce. The more abundant ones are *armadillos* (*Zaedyus* sp. and *Chaetophractus* sp.) and small rodents (Cricetidae and Ctenomydae). *Armadillos* are mainly represented by the plates in the shell, which might have been used as a container. As for rodents, the absence of marks of human processing suggests that, in most cases, their presence is due to natural causes. It was also determined the presence of skunk bones (*Conepatus* sp.) in one site of the Golfo San Matías. An interesting fact is the finding of cetaceans remains, recorded on surface sites in certain coastal areas of the Golfo San José, Golfo San Matías and Golfo Nuevo (Fig. 2), where whales and dolphins currently tend to strand. This allows us to infer that there was an opportunistic use of these animals with extremely high caloric content.

Finally, and contrary to the hypothesis of the preliminary optimal diet model, $\delta^{13}C$ and $\delta^{15}N$ isotope analyses of human bone remains show that plants had much more significance in the diet of these hunter-gatherers than originally expected (Gómez Otero 2007). In all individuals examined, type C_3 or CAM photosynthetic pathways plants were present. In Península Valdés C_3 plants are the majority, while the CAM plants are represented by cactus of *Maihueniopsis* and *Opuntia* genus. Likewise, fatty acids gas chromatography studies and ^{13}C stable isotope analyses of organic waste in pottery potsherds indicate the presence of plants in all of them (Gómez Otero et al. 2015).

With regard to the technological record, in some cases it endorses the results of archaeofaunal and isotope studies and in other it contradicts them (see Lithic and Pottery Technologies, infra). Coincidences include the higher degree of energy used to make artefacts for hunting and processing *guanaco* prey, such as projectile tips, *bolas*, knives and scrapers. No tool was found which might have been related to pinniped hunting. Another consistency would be the presence of milling and pottery artefacts in late Holocene sites, which contributes evidence to the isotope record of C_3 and *CAM plant* consumption. These tools may have enabled a more efficient and intensive exploitation of plants, especially calorie-rich carob tree (*algarrobo*) pods, which can be transformed into flour. Nevertheless, milling stones may have also been used to process meat (for example, to extract *charqui*) and fish products. Furthermore, the existence of fishing-related technology—lithic weights and a wooden hook—is inconsistent with the low record of fish remains in the area (Gómez Otero et al. 2015). It should be considered, however, that this lack of fish bones might be due to post-depositional factors, as indicated by actualistic studies in the area (Gómez Otero et al. 2015). The low number of fish and the high ichthyological diversity observed in several samples suggest that weights were mostly used in lines or fishing rods rather than in nets. However, caution should also be taken in the case of the technological record: on the one hand, no special technology is necessary to gather molluscs; on the other, it was possible to hunt pinnipeds in the Atlantic coast on land with a simple cudgel, or their pups captured after causing a stampede in breeding colonies; finally, there may have been technology in organic materials which was not preserved.

5 The Use of Rocks and Minerals

5.1 The Lithic Technology

To assess the ways in which hunter-gatherers that settled in the Península Valdés made use of rocks, the regional structure of lithic resources (Ericson 1984) was investigated first. This made it possible to discuss supply strategies based on mobility and the use of space, technological choices and also the preferences or restrictions involved in rock selection (Nelson 1991). The methodology consisted in counting and characterizing the *nodules* lithology collected by one person for a pre-defined period of time (Franco and Borrero 1999). Then, a lithological identification of the hand sample was performed, as well as an analysis of thin sections of natural and archaeological samples. Finally, for each sample, the percentage of rock of good to excellent quality for flaking purposes was estimated.

According to the geological information available (see Chapter "Geology of Península Valdés") the local supply of hard rocks consists mainly of Pliocene–Holocene gravels. The rock samples allowed for verification that the nodules potentially used for flaking were located in the San Miguel Formation (Haller 1981), in the Caleta Valdés Formation (Rovereto 1921), and in *the Rodados Patagónicos* (Fidalgo and Riggi 1970). This prevents relating the origin of the banks to a specific geological formation. Except for the San Miguel Formation, which arose after the maximum Holocene transgression, all other formations in the coast area already existed before the beginning of human occupation in the area about at least 7400 years BP (Gómez Otero 2006).

The systematic sampling of gravels helped to determine the presence of ten lithological varieties, where basalts prevail (56%), followed by siliceous rocks (18%) and riolites (15%). As for the properties of each raw material for flaked stone technology, basalts are of good to excellent quality, siliceous rocks are of good quality and rhyolites are regular quality rocks. Beach ridges and berms associated to present-day beaches (see Chapter "Late Cenozoic Landforms and Landscape Evolution of Península Valdés") are the landforms that feature the greatest percentage of raw materials with qualities ranging from good to very good: 94 and 72%, respectively (Banegas et al. 2015). In contrast, the quality of spit barrier nodules is regular for *flaking* but acceptable for *percussion*.

As in the rest of Patagonia, the lithic record is the most abundant of the entire archaeological record and features the greatest spatial and temporal distribution in the Peninsula: 97.8% of the 94 archaeological sites documented include stone artefacts (Gómez Otero 2006; Gómez Otero et al. 2013b). However, the artefact density among samples turned out to be variable and would apparently be related to the duration and form of spatial occupation, the site functionality and/or the occurrence of postdepositional processes as well (Goye et al. 2015). In this respect, taphonomic observations in mobile dunes showed that small lithic artefacts (<50 mm) tend to be displaced or buried in the sand substrate within very short periods of time (even minutes) depending on the speed and intensity of the wind. As to inferred activities, by way of example, the inexistence or little presence of lithic artefacts in *shell middens* might be related to specific and short-lived activities such as in situ gathering and consumption of molluscs (Goye et al. 2015). On the other hand, the existence of locations with medium to high artefact density and richness might be related to craftwork, tool manufacture or repair and/or the processing of fauna and vegetal resources (Gómez Otero 2006).

To the present day, the technological record of 17 archaeological sites have been studied (n = 2526 lithic artefacts) and analysed through morphological and technical variables following the Aschero method (1975–1983) and the Aschero and Hocsman method (2004) (Gómez Otero et al. 1999; Banegas and Goye 2015; Banegas et al. 2015). For lithological determinations, different specialized analyses were conducted: thin section petrography, geochemical analyses of obsidians (Gómez Otero and Stern 2005) and fossil wood observation through Scanning Electron Microscope (SEM) and Dispersive X-ray Analysis (EDAX) (Banegas et al. 2015).

The results showed the intensive use of local raw materials for different purposes (Gómez Otero 2006; Gómez Otero et al. 1999; Banegas and Goye 2015; Banegas et al. 2015). Good quality medium basalt gravels (31%) and siliceous rocks (35%) were selected for *flaking*; large porphyritic and granitic rocks were used as hammer stones, anvils and milling hand stones (i.e. *manos*); small and flat pebbles from different rocks were used as stone weights for fishing; consolidated sandstones of Puerto Madryn Formation were used as blank for *metates* (milling flat stones) and leather scrapers; and cetacean fossil bones were used as *manos*. Evidence was found that large gravels from Caleta Valdés were moved to other places in the Península Valdés region (Fig. 2).

Two lithic reduction techniques were evenly used: direct percussion and *bipolar flaking*, mainly applied to thinning the smallest basalt nodules (Banegas and Goye 2015). In line with this, there is an abundance of *bipolar cores*, followed by *multidirectional extraction cores*, and third by cores with one or two extractions. Prevailing flaked products include *flakes*, although every site presented blades in very variable proportions (between 3 and 33%) (Gómez Otero et al. 1999; Gómez Otero and Stern 2005; Banegas and Goye 2015). Given the abundance and wide distribution of lithologies adequate for different types of artefacts, it is inferred that the supply of these local rocks might have been direct and might have implied low collecting costs. This might have resulted in the predominance of expedient artefacts (Binford 1979).

In addition to local rocks, allochthonous raw materials were used, but not frequently (Gómez Otero 2006; Gómez Otero et al. 1999; Banegas et al. 2015). Six varieties of obsidian stand out. Three derive from known sources: Telsen-Sierra Chata I (T/SCI), Sacanana I (SI), and Pampa del Asador I (PDAI)—the former two are located 180 and 330 km west, respectively, from the beginning of the Istmo Carlos Ameghino, and the latter is about 800 km south, in the central-northern portion of the Province of Santa Cruz (Gómez Otero and Stern 2005; Fig. 1). Other excellent quality rocks for flaking which also come from other regions are chalcedonies and xylopals or fossil wood rocks. The former are found as veins associated with the Marifil Formation (Jurassic) (Malvicini and Llambías 1974) which outcrops at the border of the Province of Río Negro, 100 km west from the Istmo Carlos Ameghino (Haller et al. 2001; Massaferro and Haller 2000; see Chapter "Geology of Península Valdés"). Moreover, the petrified woods closest to the Península Valdés are located 150 km south west from it, in the area of Dique Ameghino (Fig. 1). Farther locations should also be considered, such as the "*Bosque Petrificado de la Costa*" (Petrified Coast Woods), located in Golfo San Jorge, 450 km south, and the "*Bosque Petrificado Ormaechea*" (Ormaechea Petrified Woods), near the Lago Musters and Lago Colhue Huapi (Fig. 1; Banegas et al. 2014). Furthermore, hematites, olivine and vesicular basalts, granites, slates and porphyritic rocks were also found. Porphyritic rocks are available in the volcanic outcrops of the Marifil Formation less than 100 km north, west and south of the Península Valdés region, while the rest of the rocks may be obtained from the west and north, in the Somuncurá plateau, within radiuses ranging from 100 to 300 km (Fig. 1). As expected, the implementation of curated technological strategies (Binford 1979) for obsidians and chalcedonies was also determined (Gómez Otero and Stern 2005; Banegas and Goye 2015).

A high degree of artefact diversity was observed, which is indicative of the development of extraction and maintenance activities related to hunting and flaking, and also to meat, leather and vegetable processing. The flaked stone tools found include knives, end-scrapers, side-scrapers, denticulates, drills, burins, gouves, notches, fishing weights, and a variety of projectile points (Fig. 5). Ground stone artefacts are represented by *bolas* with and without equatorial groove, star-shaped maces, mortars, and milling hand stones. Moreover, large pebbles were used as hammer stones, hand stones and anvils (Fig. 6). It is worth noting that heavy and very large tools were left at various settlements as site furniture. In several base camps, the simultaneous presence of milling and ceramic tools was observed, which indicates plants and other foods processing and consumption in situ (Gómez Otero 2006; Gómez Otero et al. 2015).

By the final late Holocene (after 2000 BP), a gradual increase in the selection of non-local rocks was evidenced, probably related to the intensification of social and economic interaction which consisted in the exchange of objects, raw materials, designs and information (Gómez Otero 2006). A study on the morphometric variability of projectile points revealed similarities among designs recorded both in northern and southern Patagonia (Gómez Otero et al. 2011).

Fig. 5 The flaked stone tools founded in Península Valdés. **a** End-scrapers; **b** projectile points (scale 1 for **a** and **b**); **c** retouched knives; **d** hammer stone; **e** anvil; **f** stone weights for fishing; and **g** cores (scale 2 for **c–g**)

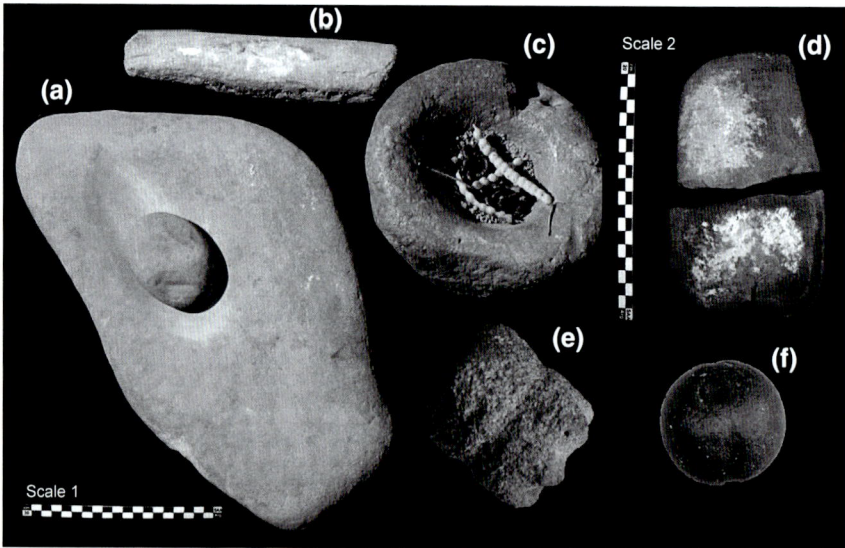

Fig. 6 Grinding and chipping tools founded in Península Valdés. **a** *Metate* or milling flat stone; **b** *mano* made of a cetacean fossil bone; **c** milling mortar; **d** star-shaped maces; and **e** *bola*

5.2 The Pottery Technology

The pottery record in Península Valdés has been one of the most researched across Argentine Patagonia since 1998: to date, over 1289 potsherds have been analysed from 36 archaeological sites (Gómez Otero et al. 1998; Schuster and Gómez Otero 2009; Schuster 2014), most of which would have served as base camps for hunter-gatherer groups) (Gómez Otero 2006) (Fig. 7). Radiocarbon dating shows that pottery technology probably developed in the Peninsula at least 900 years ^{14}C BP (Gómez Otero et al. 2013b), that is, during the final late Holocene.

Several analytical techniques were applied in pottery study: macroscopic and microscopic analyses (petrography), X-ray diffraction, 3-D digital reconstructions and X-ray images (see infra). Furthermore, organic remains adhered to or absorbed by the pores of ceramic pieces were examined through different physical and chemical analyses: isotope analyses and gas chromatography. With regard to the selection of raw materials for pottery production, microscopic analyses and X-ray diffraction showed much resemblance in the mineralogical associations of pastes and soils of the Peninsula (Bouza et al. 2007). In decreasing order of abundance, the following were identified: illite, interstratified illite-smectites, quartz, plagioclase, potassium feldspar, acid vulcanite, opaque minerals (Bouza et al. 2007; Schuster 2009, 2014, 2015) (Fig. 7). Calcitic-clay aggregates and the fine material of the ceramic matrix probably derived from soils with soil horizons rich in clay (Btk), so the calcareous material and sand grains of the pastes would constitute natural anti-plastics (Bouza et al. 2007). These features indicate that pottery technology had a local development, which is to be expected given the abundance, wide distribution and accessibility to several sources of clay materials (Fig. 3), in addition to the availability—albeit more restricted—of other indispensable resources such as firewood and fresh water.

As to the manufacture of pieces, macroscopic studies determined that the most frequent primary technique to raise the body of the pieces was ring building (Schuster 2014) while X-ray analyses indicated that in some cases pinching was used (Schuster and Banegas 2010; Schuster et al. 2013).

As far as morphology is concerned, in addition to a single, entire vessel, the form of twelve vessels was partially reconstructed (Schuster 2014). A wide prevalence of ovaloid or spheroid pieces was observed, with simple shapes, straight sides and concave bases (Fig. 7). Most of these features favoured heat transmission (Rice 1987; Skibo 1992; Orton et al. 1997); therefore, they were possibly fit for domestic tasks such as cooking food. With less frequency, compound-shaped vessels were discovered that feature sides with external reinforcement, necks and/or handles (Schuster 2014). As to mouth opening, while closed or restricted mouths prevail, there are also open-mouthed or unrestricted-mouthed vessels whose diameter does not generally exceed 16 cm. This trend probably indicates openings that, on the one hand, enabled easy access to their content while food was being mixed or stirred and that, on the other hand, made it difficult for content to rapidly evaporate,

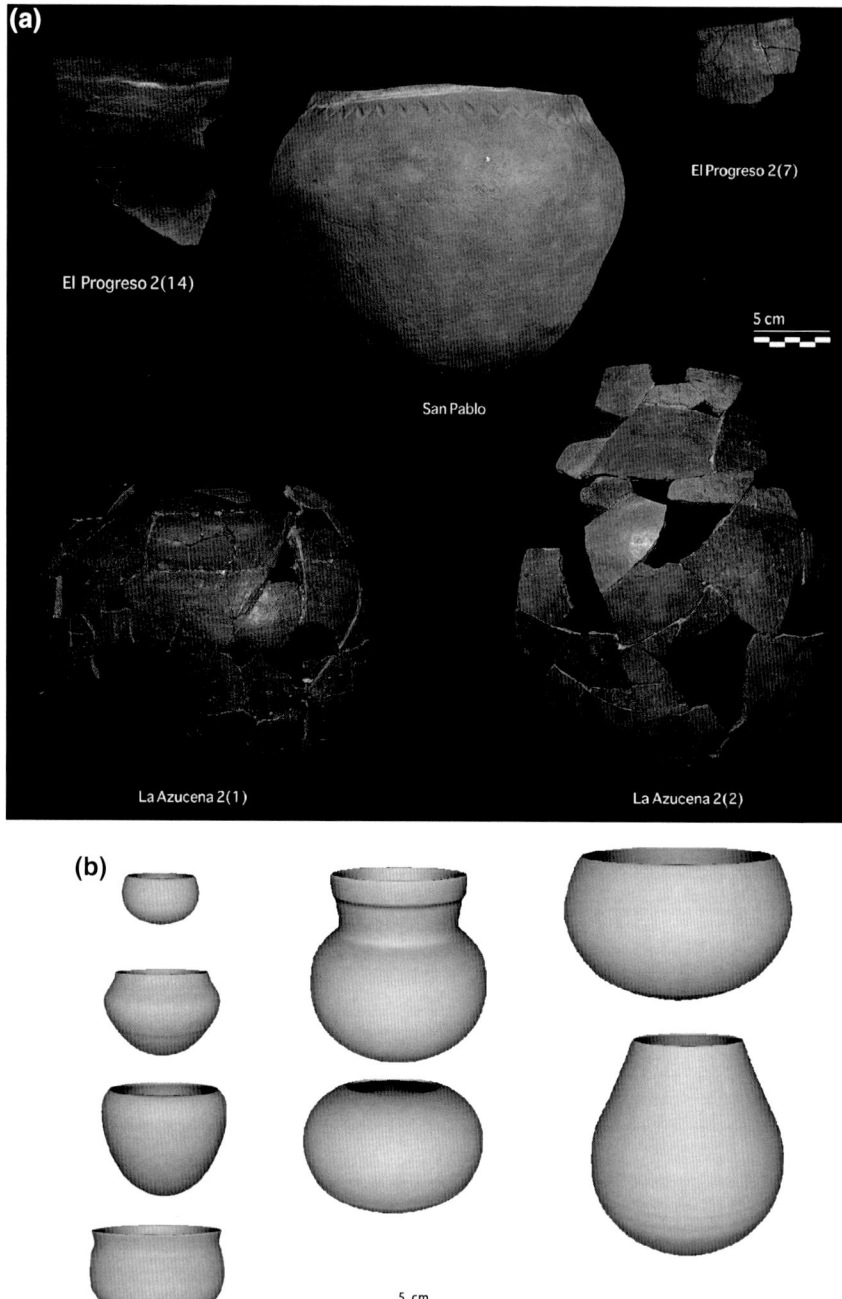

Fig. 7 Pottery founded in Península Valdés. **a** Ceramic archaeological vessels reconstructed; **b** 3-D reconstructed models of potsherd

especially liquids. This would have shortened cooking time and reduced firewood consumption while performing such tasks.

Another study involved 3-D digital reconstructions that secured not only three-dimensional models but also the opportunity to calculate volume or capacity, as well as the weight of reconstructed pieces (Fig. 7; Schuster and Quinto Sánchez 2016). This is the first time ever this technique is applied in the study of pottery of hunter-gatherers from Patagonia. Based on the data obtained, it was determined that the ceramic set is represented by ceramic pieces whose capacity (or gross volume) ranges from 0.21 to 10.87, with a mean of 3.2 L. The relationship between size and capacity or volume makes it possible to put forwards certain considerations related to the "movement" or "mobility" of pieces while performing several tasks or giving them different uses in the past. Empty container weight estimation (or paste emptiness) also makes it possible to put forwards and explore the transportability of pottery among mobile hunter-gatherer societies such as those considered here. Based on the proposal by Sopena Vicién (2006), most of the ceramic set under study in the Península Valdés consists of highly "dynamic" pieces because their weight and volume are below 0.50 kg and 2.57 L, respectively (Schuster and Quinto Sánchez 2016).

With respect to decoration, in the set being analysed, decorated vessels or potsherds are scarce, although several execution techniques were identified: incisions, light *engobe*, and red and black painting (Schuster 2014; Fig. 7). While decorated pottery can be used for several domestic purposes, its use is expected in activities that do not involve cooking, such as serving or consuming, since these pieces require more work (Rye 1981). However, paint coating or *engobe* on the sides serves an important functional purpose as it increases resistance to breaks and/or fractures and, also, waterproofs pieces avoiding the loss of liquids by evaporation through pores (e.g. Rice 1987; Rye 1981).

Petrographic studies also show that some vessels could have been made outside the Península Valdés. Three pieces feature an abundant proportion of mica (muscovite and biotite) in their paste and they stand out from the rest because of the finish on the surface (e.g. intense polishing, red paint, incised decoration and/or scraping) and certain morphological features (e.g. reinforced lips). Furthermore, there are other pieces whose paste composition does not feature elements that are probably allochthonous such as mica but, nevertheless, show potentially extra-regional morphological features, such as handles or neck vessels (Schuster 2014, 2015). These vessels may have entered the Península Valdés region after being made and decorated in another production location, although, for now, it is not possible to rule out their local production based on replication or imitation of foreign techniques and designs (Schuster 2015). In this regard, as Gómez Otero (2006) puts forwards, for the last 1000 years—when pottery was probably adopted in the region—interethnic contact and the existence of extensive exchange networks outside the Patagonian region intensified. Therefore, if in the study area several objects were found that possibly come from the Cuyo area, the Northwestern area of Argentina and central Chile (Gómez Otero 2006) some ceramic vessels with different morphology and decoration from the rest of the set—as well as non-local aggregates in the paste—are very likely to have reached Península Valdés through some kind of mechanism of exchange or circulation of goods, products and ideas. This fits with the lithic technological studies, which demonstrated the higher record of foreign rocks for late Holocene times.

On the basis of isotopic and chromatographic studies (Gómez Otero et al. 2014) on both local and probably allochthonous pieces, the presence of processed and/or preserved food remains was identified. The results obtained allowed for estimation of the following trend: there is presence of vegetables in all cases, followed by terrestrial proteins and fat and, to a lesser extent, marine proteins and fat. Some of this marine fat would correspond to fish. With regard to a possible link between morphological and technological features of ceramic vessels and the type of food identified through physical and chemical analyses, no association could be demonstrated (Schuster 2014; Gómez Otero et al. 2015). This suggests that no given form was selected to process different types of food.

Based on the general morphology of ceramic vessels, the data retrieved from isotopic and chromatographic studies, and from the ethnographic compilation of different cooking techniques, most of these types of food was possibly processed through boiling. Thus, stews or soups were probably cooked, made with vegetables only, or with meat from terrestrial mammals (e.g. *guanaco*) or sea fish, although fat from fatty pinnipeds or fish may have been removed or stored (Gómez Otero et al. 2014).

6 Synthesis and Conclusion

The archaeological record of the Península Valdés shows that this area was intensively used by native hunter-gatherers since at least the mid-Holocene to the nineteenth century. These populations located their settlements in sandfields across the littoral zone, primarily on coastal *bajadas* and low marine terraces near fixed shoals of molluscs. In the coast environment, they could obtain a variety of vital resources such as marine and terrestrial food, good quality rock nodules for flaking and other uses, clay minerals for pottery technology, and also a minimal fresh water reserve in the dunes. Mobility patterns included short trips to diverse resource supply sources: fresh water and salt in the Salina Grande and Salina Chica and big pebbles in the Caleta Valdés area.

It follows from the archaeofaunal, isotope and technological information that the *guanaco*, a variety of molluscs, and plants may have been the most systematically consumed resources in the study area within the time frame under analysis. Pinniped consumption, albeit frequent, was probably neither systematic nor intense, except in specific cases. The basic diet—namely, *guanacos*, molluscs and plants— apparently contributed all the necessary protein, vitamin, mineral and carbohydrate requirements for adequate nutrition, but not lipids. Lipids may have been provided by other resources: pinnipeds, *armadillos*, some fatty fish, cetaceans and rhea eggs. Moreover, marine birds may have contributed linoleic acid. Some fat may have been stored as a strategy to withstand seasonal shortages. This might have been favoured by salt availability in the two *salinas* of the Península Valdés. In sum, when the native peoples occupied the area, the abundance and diversity of food may have allowed them to live by a varied, complete diet in terms of basic nutrients. As a result, ancient populations probably did not face a risk of death by starvation or food stress situations, as evidenced by health and nutritional studies (Gómez Otero and Novellino 2011).

As regards lithic technology, these hunter-gatherers used mainly rocks that were locally available through pebbles, consolidated sandstones and fossil cetacean bones. For flaking technology, they selected basalt and silica nodules and applied two reduction techniques: direct and bipolar percussion. The basic toolkit comprised knives, end-scrapers, side-scrapers, denticulates, drills, burins, gouves, notches, fishing weights, and a variety of projectile points. To make side-scrappers and projectile points, they selected best quality rocks, some of them allochthonous, like chalcedonies and xylopals. The big pebbles of Caleta Valdés were used as *manos*, hammer stones and anvils, while sandstones and fossil bones were primarily employed in milling activities. With regard to pottery technology, the abundance and good quality of local clay sources allowed its important development in the final late Holocene. The native populations manufactured vessels with generalised or multifunctional purposes, since most of their technical and morphological variables were probably fit to serve a wide range of domestic activities, e.g. preparing, storing and/or cooking food and liquids (Schuster 2014). Based on the foregoing, it was probably unnecessary to make many ceramic pieces because each could be used for several, different purposes. In addition, specialised analyses made it possible to identify food remains in all samples analysed which would support the hypothesis that pottery technology in the study area was probably related to domestic activities (Schuster 2014; Schuster and Gómez Otero 2009). Specifically, the discovery of vegetables in all samples examined supports the hypothesis that, in the area, pottery was adopted as a technological strategy to leverage their consumption more intensively (Gómez Otero 2006). This technology, which would have been functional to the lifestyle of hunter-gatherers, probably favoured an improvement in the sanitary and preservation conditions of food, as well as a broader variety of options to prepare and consume several animal and vegetable resources available in the area (Gómez Otero 2006).

These hunter-gatherers did not live isolated from other populations: the presence of foreign rocks (obsidians, slates, hematite, xilopals, chalcedonies, vesicular basalts), and of three ceramic vessels, which might have been manufactured in northwest Patagonia or perhaps central Chile, indicate that they took part in an extensive trading net since at least the late Holocene. Local native inhabitants could exchange shells, dry molluscs, dry fish, fat from marine mammalians and cetacean fossil bones. This trade would have comprised not only raw materials and goods, but methods, designs, ritual practices and information, among others. To sum up, the favourable environmental conditions of Península Valdés were beneficial to human life since at least the middle Hocolene and population growth for thousands of years, until 1779 when the Spanish Crown occupied the San José Gulf and the Salina Grande area. And then, another history began.

7 Perspectives and Future Work

Although the knowledge of the Península Valdés archaeology has improved remarkably in the last 20 years, further research needs to be done. The agenda involves the extension of the different studies that have just been made and also new analyses. Further work needs to explore the spatial and temporal cultural diversity, as well as the range of the social relationships and trade nets these populations sustained inside and outside the Patagonia region. In addition, new DNA studies in human skeletal remains, ^{18}O isotope analysis in fresh water sources, mass spectrometry to determine the origin of organic residues in ceramics, identification of phytoliths and fatty acids recovered in grinding tools and actualistic studies are necessary.

New survey studies and analytical techniques will allow us to improve the understanding of the way of life and cultural changes of the hunter-gatherer populations that occupied the Northern coast of Patagonia since at least the middle Holocene.

Acknowledgements We wish to thank to all those that have been of much help to our research: Bobby Taylor, Delfina Palleres, Juan Bautista Belardi, Alejandro Súnico, Nilda Weiler, Vasco Aguerrebere, Gabriela Massaferro, Mirsha Quinto Sánchez, Jorge Briguglio and the rural inhabitants of Península Valdés. We also are grateful to Pablo Bouza and Andrés Bilmes for their invitation to participate in this book, to the English translator Eugenia Monedero and to the two reviewers, Gustavo Martínez and Cristina Bellelli, for their helpful suggestions. Financial support was provided by diverse grants from the National Research Council of Argentina (CONICET): PIP 0321, 02786, 6470, and 11401000100210 and the Grant 5453/95 from the National Geographic Society.

Glossary

Bead	A small piece of glass, stone, shell or similar material that is threaded with others to make a necklace or rosary or sewn on to fabric
Bipolar core	Core smashed by bipolar reduction or percussion, likely by placing it on a stone anvil or floor and chipping it with a hammer stone

Bola	Weighted ball of stone which are either grooved or pierced for fastening to rawhide thongs and used for hunting. Usually consists in two or more lobular or pear-shaped stones attached each other with long thongs
Conchero ("shellmidden" in English)	A mound of domestic waste consisting mainly of shells, common at prehistoric sites. They can be of different size and height
Container	An object for holding or transporting something
Core	A piece of flint from which flakes or blades has been removed by direct or bipolar percussion
Curated technological strategies	Strategies applied in order to preserve tools life. Tools are carried from place to place, maintained, reused and sometimes recycled into other tool forms. These strategies are expected when good quality raw materials are scarce
Engobe	Clay diluted in water and sometimes mixed with minerals pigments. It is used in pottery decoration
Expedient artefacts	Tools made and used in response to an immediate need and not maintained for future use. This kind of artefacts is expected when lithic raw material is abundant
Flake	A piece of hard stone chipped off for use as a tool by prehistoric humans
Flaking	Action of extracting flakes from a nodule. Bipolar flaking: action of extracting flakes
Multidirectional extraction core	Core chipped in all its faces
Nodule	Piece of stone used as raw material for lithic technology
Potsherd	A broken piece of ceramic material, especially one found on an archaeological site.

References

Acha EM, Mianzán HW, Guerrero RA, Favero M, Bava J (2004) Marine fronts at the continental shelves of Austral South America. Physical and ecological processes. J Mar Syst 44:83–105

Ambrose SH (1993) Isotopic analysis of paleodiets: methodological and interpretive considerations. In: Sandford MK (ed) Investigations of ancient human tissue: chemical analysis in anthropology. Gordon and Breach Science Publishers, Pensylvania, pp 59–129

Aschero C (1975–1983) Ensayo para una clasificación morfológica de artefactos líticos aplicada a estudios tipológicos comparativos. Informe al CONICET, Buenos Aires. MS

Aschero CA, Hocsman S (2004) Revisando cuestiones tipológicas en torno a la clasificación de artefactos bifaciales. In: Ramos M, Acosta A, Loponte D (eds) Temas de Arqueología. Análisis Lítico. Universidad Nacional de Luján, Luján, pp 7–25

Bailey G, Parkington J (1988) The archaeology of prehistoric coastlines: an introduction. The archaeology of prehistoric coastlines. Cambridge University Press, New Directions in Archaeology, New York, pp 1–10

Banegas A, Goye MS (2015) Spatial and temporal variability in the use of lithic raw materials for flaked stone technology in the northeast of Chubut Province (North Patagonia) during Late Holocene. In: Ambrústolo P, Zubimendi MA (eds) Archaeology of coastal hunter-gatherer occupations in the Southern Cone (Quaternary International 373: 55–62)

Banegas A, Pujana R, Gómez Otero J (2014) Caracterización tecnológica de xilópalos de la costa centro-septentrional de Patagonia: tendencias temporales y potenciales fuentes de aprovisionamiento. In Libro de Resúmenes del las IX Jornadas de Arqueología de la Patagonia, 14. CIEP, Coyhaique

Banegas A, Goye S, Gómez Otero J (2015) Caracterización regional de recursos líticos en el nordeste de la provincia del Chubut (Argentina). In: Alberti J, Fernandez V (eds) Materias primas líticas en Patagonia. Localización, circulación y métodos de estudio de las fuentes de rocas de la Patagonia argentino-chilena. Intersecciones en Antropología, Dossier 2:39–50

Behrensmeyer AK (1978) Taphonomic and ecologic information from bones weathering. Paleobiology 4:150–162

Bettinger RL (1991) Hunter-gatherers. Archaeological and evolutionary theory. Plenum Press, New York

Bianchi Villelli M, Buscaglia S (2015) De gestas, de salvajes y de mártires. El relato maestro sobre el Fuerte San José repensado desde la arqueología histórica (Península Valdés, pcia. de Chubut, siglo XVIII). Revista del Museo de Antropología 8(1):187–200

Binford L (1979) Organization and formation processes: looking at curated technologies. J Archaeol Res 35:255–270

Binford L (1980) Willow smoke and dog's tails: hunter-gatherer settlement systems and archaeological site formation. Am Antiq 45(1):4–20

Borrero LA, McEwan C (1997) The peopling of Patagonia. The first human occupation. In: McEwan C, Borrero LA, Prieto A (eds) Patagonia. Natural history, prehistory and ethnography at the uttermost end of the earth. The British Museum Press, London, pp 32–45

Bouza P, Gómez Otero J, Taylor R, Schuster V, Melatini MS (2007) Tecnología de cerámicas arqueológicas en el nordeste de la provincia del Chubut. In: Actas del XVI Congreso Nacional de Arqueología Argentina, vol. III. Universidad Nacional de Jujuy, Jujuy, pp 447–452

Carrara IS (1952) Lobos marinos, pingüinos y guaneras del litoral marítimo e islas adyacentes de la República Argentina. Ministerio de Educación, Facultad de Ciencias Veterinarias (Publicación especial), UNLP, La Plata

Cousseau MB, Perrotta RG (2000) Peces marinos de Argentina. Biología, distribución, pesca. INIDEP, Mar del Plata

Crespo EA, Pedraza SN (1991) Estado actual y tendencia de la población de lobos marinos de un pelo (*Otaria flavescens*) en el litoral norpatagónico. Ecología austral 1:87–95

D'Orbigny A (1999) Viaje por América meridional II. Emecé Editores, Buenos Aires

Ericson J (1984) Towards the analysis of lithic reduction systems. In: Ericson JE, Purdy B (eds) Prehistoric quarries and lithic production. Cambridge University Press, Cambridge, pp 11–22

Escofet AM, Orensanz JM, Olivier SR, Scarabino V (1978) Biocenología bentónica del golfo San Matías (Río Negro, Argentina): metodología, experiencias y resultados del estudio ecológico de un gran espacio geográfico en América Latina. Anales del Centro Científico del Mar y Limnológico 5(1):59–82, Universidad Nacional Autónoma de México

Fidalgo F, Riggi JC (1970) Consideraciones geomórficas y sedimentológicas sobre los Rodados Patagónicos. Asociación Geológica Argentina 25:430–443

Franco N, Borrero LA (1999) Metodología de análisis de la estructura regional de recursos líticos. In: Aschero CA, Korstanje MA, Vuoto PM (eds) En los tres reinos: Prácticas de recolección en el cono Sur de América. Magna Publicaciones, San Miguel de Tucumán, pp 27–37

Gómez Otero J (2006) Recursos, dieta y movilidad en la costa centro-septentrional de Patagonia durante el Holoceno medio y tardío. Unpublished Doctoral Thesis. Facultad de Filosofía y Letras, Universidad de Buenos Aires, Buenos Aires

Gómez Otero J (2007) Isótopos estables, dieta y uso del espacio en la costa atlántica centro septentrional y el valle inferior del río Chubut (Patagonia argentina). In: Morello F, Martinic M, Prieto A, Bahamondes G (eds) Arqueología de Fuego-Patagonia. Levantando piedras, desenterrando huesos... y develando arcanos. Universidad de Magallanes, Punta Arenas, pp 151–161

Gómez Otero J, Novellino P (2011) Diet, nutritional status and oral health in hunter-gatherers from the central-northern coast of Patagonia and the Chubut river valley, Argentina. J Osteoarchaeol 21:643–659. Disponible online Doi:10.1002/oa.1171

Gómez Otero J, Stern C (2005) Circulación, intercambio y uso de obsidianas en la costa de la provincia del Chubut (Patagonia, Argentina) durante el Holoceno tardío. Intersecciones en Antropología 6:93–108

Gómez Otero J, Bouza P, Taylor R (1998) Primeros estudios sobre tecnología cerámica arqueológica en Península Valdés, costa centro-norte de Patagonia. In: Resúmenes de ponencias de las IV Jornadas de Arqueología de la Patagonia. Universidad Nacional de la Patagonia Austral, Río Gallegos, pp 7–8

Gómez Otero J, Belardi J, Súnico A, Taylor R (1999) Arqueología de cazadores –recolectores en península Valdés (costa central de Patagonia): primeros resultados. In: Soplando en el viento. Universidad Nacional del Comahue, Neuquén, pp 393–417

Gómez Otero J, Belardi JB, Tykot R, Grammer S (2000) Dieta y poblaciones humanas en la costa norte del Chubut (Patagonia argentina). In: Desde el País de los Gigantes. Perspectivas arqueológicas en Patagonia. Universidad Nacional de la Patagonia Austral, Río Gallegos, pp 109–122

Gómez Otero J, Banegas A, Goye MS, Franco NV (2011) Variabilidad morfológica de puntas de proyectil en la costa centro-septentrional de Patagonia argentina: primeros estudios y primeras preguntas. In: Las fuentes en la construcción de una Historia Patagónica. Secretaría de Cultura de la Provincia del Chubut, Rawson, pp 110–118

Gómez Otero J, Weiler N, Banegas A, Moreno E (2013a) Ocupaciones del Holoceno medio en Bahía Cracker, costa atlántica de Patagonia central. In: Zangrando AF, Barberena R, Gil A, Neme G, Giardina M, Luna L, Otaola C, Paulides S, Salgán L, Tivoli A (eds) Tendencias teórico-metodológicas y casos de estudio en la arqueología de la Patagonia. Museo de Historia Natural de San Rafael, San Rafael, pp 77–186

Gómez Otero J, Banegas A, Goye S, Palleres D, Reyes M, Schuster V, Svoboda A (2013b) Nuevas investigaciones arqueológicas en la estancia San Pablo (costa del golfo Nuevo, Península Valdés). In: Bárcena JR, Martín SE(eds) Libro de Resúmenes del XVIII Congreso Nacional de Arqueología Argentina. Universidad Nacional de la Rioja, La Rioja, p 523

Gómez Otero J, Constenla D, Schuster V (2014) Isótopos estables de carbono y nitrógeno y cromatografía gaseosa en cerámica arqueológica del nordeste de la provincia del Chubut (Patagonia Argentina). Arqueología 20(2):263–284

Gómez Otero J, Schuster V, Svoboda A (2015) Fish and plants: the "hidden" resources in the archaeological record of the north-central Patagonian coast (Argentina). In: Ambrústolo P, Zubimendi MA (eds) Archaeology of coastal hunter-gatherer occupations in the Southern Cone (Quaternary International 373:72–81)

Goye S, Banegas A, Gómez Otero J (2015) Abundancia y diversidad lítica en concheros de la costa Norte de la Provincia del Chubut, Patagonia Argentina. In: Pifferetti A, Dosztal I (eds) Arqueometría Argentina. Metodologías científicas aplicadas al estudio de los bienes culturales: Datación, caracterización, prospección y conservación. Aspha Ediciones, Ciudad Autónoma de Buenos Aires, pp 181–194

Haller M (1981) Descripción Geológica de la Hoja 43 h - Puerto Madryn. Servicio Geológico Nacional, Boletín 184:41

Haller M, Monti A, Meister C (2001) Hoja Geológica 4363-I- Península Valdés. Servicio Geológico Minero Argentino (SEGEMAR), Boletin 266

Jardín Botánico de la Patagonia Argentina (2002) Usos tradicionales de las plantas en la meseta patagónica. Centro Nacional Patagónico-CONICET-ICGB, Puerto Madryn

Jones TL (1991) Marine-resource value and the priority of coastal settlement: a California perspective. Am Antiq 56(3):419–443

Koch PL, Tuross N, Fogel MI (1997) The effects of sample treatment and diagenesis on the isotopic integrity of carbonate in biogenic hydroxylapatite. J Archaeol Sci 24:417–429

Lyman RL (1994) Vertebrate taphonomy. Cambridge University Press, Cambridge

Malvicini L, Llambías E (1974) Geología y génesis del depósito de manganeso Arroyo Verde, provincia del Chubut, República Argentina. In Actas del 5° Congreso Geológico Argentino T 2:185–202 (Buenos Aires)

Massaferro GI, Haller MJ (2000) Texturas de las vetas epitermales del Macizo Norpatagónico. In Actas del 5° Congreso de Mineralogía y Metalogenia, La Plata, pp 312–319

Menghin OF, Bormida M (circa 1995). Arqueología de la costa patagónica. Unpublished manuscript

Mengoni Goñalons G (1988) Análisis de materiales faunísticos de sitios arqueológicos. Xama 1:71–120 (Mendoza)

Nelson M (1991) The study of technological organization. Archaeol Method Theory 3:57–100

Orquera LA (1987) Advances in the archaeology of the Pampa and Patagonia. J World Prehistory I 4:333–413

Orquera LA, Gómez J Otero. 2007. Los cazadores-recolectores de las costas de Pampa, Patagonia y Tierra del Fuego. Relaciones de la Sociedad Argentina de Antropología XXXII:45–63

Orton C, Tyers P, Vince A (1997) [1993] La Cerámica en Arqueología. Editorial Crítica, Barcelona

Outes FF (1915) La gruta sepulcral del cerrito de Las Calaveras. Con un examen anátomo-patológico por Angel H. Roffo. Anales del Museo Nacional de Historia Natural de Buenos Aires XXVII:365–400

Perlman SM (1980) An optimun diet model, coastal variability and hunter-gatherer behavior. In: Schiffer MB (ed) Advances in archaeological method and theory, vol 3. Academic Press, New York, pp 257–310

Politis G, Prates L, Pérez SI (2008) El poblamiento de América. Arqueología y bioantropología de los primeros americanos. Eudeba, Colección Ciencia Joven, Buenos Aires, p 35

Rice PM (1987) Pottery analysis. A sourcebook. University of Chicago Press, Chicago

Rovereto G (1921) Studi di geomorfología Argentina: La Peninsula Valdés. Societá. Geológica Italiana, Bolletino 40:1–47

Rye O (1981) Pottery technology. Principles and reconstruction. Manuals on Archaeology 4. Taraxacum, Washington

Schuster V (2009) Petrografía de la cerámica arqueológica del nordeste del Chubut (Patagonia Argentina). Primeros resultados. In: La Arqueometría en Argentina y Latinoamérica. Facultad de Filosofía y Humanidades, Córdoba, pp 103–108

Schuster V (2014) La organización tecnológica de la cerámica de cazadores-recolectores. Costa norte de la Provincia del Chubut (Patagonia Argentina). Relaciones de la Sociedad Argentina de Antropología XXXIX(1):203–231

Schuster V (2015) Cerámica arqueológica de la costa, valle y meseta de la provincia del Chubut (Patagonia Argentina): estudio comparativo preliminar de la composición de las pastas a través de la petrografía. Intersecciones en Antropología 16(2):353–366

Schuster V, Banegas A (2010) Rayos X en la cerámica arqueológica de Patagonia: primeras experiencias para la costa y meseta central del Chubut. In: Barcena JR, Chiavazza H (eds) Arqueología Argentina en el Bicentenario de la Revolución de Mayo, vol V. Universidad Nacional de Cuyo and INCIHUSA-CONICET, Mendoza, pp 1987–1992

Schuster V, Gómez Otero J (2009) Aportes al conocimiento de la tecnología cerámica en grupos cazadores-recolectores de la costa centro-septentrional de Patagonia. In: Azar PF, Cúneo EM, Rodríguez SN (eds) Tras la senda de los ancestros: Arqueología de Patagonia. Editorial EDUCO Cd-Room, San Carlos de Bariloche

Schuster V, Quinto Sánchez M (2016) Reconstrucciones virtuales 3D del repertorio cerámico de cazadores-recolectores de la costa nordeste de la provincia del Chubut (Patagonia Argentina). Unpublished report, Beca Posdoctoral, Consejo Nacional de Investigaciones Científicas y Técnicas (CONICET)

Schuster V, Banegas A, Taylor R (2013) Revelando imágenes...Rayos X en cerámicas arqueológicas y piezas experimentales. In: Zangrando AF, Barberena R, Gil A, Giardina M, Luna L, Otaola C, Paulides S, Salgán L, Tivoli A (eds) Tendencias teórico metodológicas y casos de estudio en la arqueología de Patagonia. Museo de Historia Natural de San Rafael, San Rafael, pp 233–242

Skibo JM (1992) Pottery function. A use-alteration perspective. Plenum Press, New York

Sopena Vicién MC (2006) La investigación arqueológica a partir del dibujo informatizado de cerámica. Salduie. Estudios de Prehistoria y Arqueología 6:13–27

Súnico CA (1996) Geología del Cuaternario y Ciencia del Suelo: relaciones geomórficas y estratigráficas con suelos y paleosuelos. Unpublished Doctoral Thesis, Facultad de Ciencias Exactas y Naturales, Universidad de Buenos Aires

Van der Merwe NJ (1992) Light stable isotopes and the reconstruction of prehistoric diets. Proc Br Acad 77:247–264

Weiler NE (1998) Mid-holocene littorals deposits at southwest of the golfo San José, Península Valdés, Argentine Republic. International Coastal Symposium 26:33–38. Palm Beach

Winterhalder B, Smith E (1992) Evolutionary ecology and the social sciences. In: Smith EA, Winterhalder B (eds) Evolutionary ecology and human behavior. Aldine de Gruyter, New York, pp 3–24

Yesner D (1980) Maritime hunter-gatherers: ecology and prehistory. Curr Anthropol 21(6): 727–750

Animal Diversity, Distribution and Conservation

Ricardo Baldi, Germán Cheli, Daniel E. Udrizar Sauthier, Alejandro Gatto, Gustavo E. Pazos and Luciano Javier Avila

Abstract In this chapter, we summarize the ecological information available on the species of arthropods, reptiles, terrestrial birds and mammals known to occur at Península Valdés, within the context of the Monte and Patagonia eco-regions. Two hundred species of insects and spiders, 12 species of reptiles, 139 species of birds and 23 species of native mammals inhabit today the island-like peninsula. We describe the community structure, distribution and abundance of the different taxa according to current knowledge. As the Península Valdés region is a World Natural Heritage Site and a provincial protected area of high importance for the conservation of biodiversity and the regional economy, we found relevant to summarize knowledge on the effects of human activities on different components of biodiversity. Habitat degradation, grazing by domestic sheep and poaching are major

---

R. Baldi (✉) · G. Cheli · D.E. Udrizar Sauthier · G.E. Pazos · L.J. Avila
Instituto Patagónico para el Estudio de los Ecosistemas Continentales (IPEEC), Consejo Nacional de Investigaciones Científicas y Técnicas (CONICET)—CCT Centro Nacional Patagónico (CENPAT), Boulevard Brown 2915, ZC: U9120ACD, Puerto Madryn, Chubut, Argentina
e-mail: rbaldi@cenpat-conicet.gob.ar

G. Cheli
e-mail: cheli@cenpat-conicet.gob.ar

D.E. Udrizar Sauthier
e-mail: dsauthier18@gmail.com

G.E. Pazos
e-mail: gpazos@cenpat-conicet.gob.ar

L.J. Avila
e-mail: avilacnp@gmail.com

A. Gatto
Centro para el Estudio de Sistemas Marinos (CESIMAR), Centro Nacional Patagónico—CONICET, Boulevard Brown 2915, 9120 Puerto Madryn, Chubut, Argentina
e-mail: alegatto@cenpat-conicet.gob.ar

G.E. Pazos
Facultad de Ciencias Naturales, Universidad Nacional de la Patagonia San Juan Bosco, Boulevard Brown 3051, 9120 Puerto Madryn, Chubut, Argentina

© Springer International Publishing AG 2017
P. Bouza and A. Bilmes (eds.), *Late Cenozoic of Península Valdés, Patagonia, Argentina*, Springer Earth System Sciences,
DOI 10.1007/978-3-319-48508-9_11

threats to wildlife, although Península Valdés still harbours some of the most abundant populations of wild species in Patagonia. It is a priority to implement the management plan available for Península Valdés, taking into account the interactions between biodiversity, the different human activities and the physical environment. At the same time, it is necessary to identify and implement actions to conserve wild species and habitats, and also to develop programmes for the coexistence of responsible human activities and healthy wildlife populations.

Keywords Biodiversity · Human activities · Conservation · Península Valdés · Patagonia

1 Introduction

Península Valdés is a provincial protected area "with managed resources" according to the criteria of the International Union for the Conservation of Nature (IUCN). It was declared a World Natural Heritage Site by UNESCO in 1999 and is intended to protect the landscape, natural and cultural patrimony of the area, as well as to facilitate research, promote sustainable activities compatible with conservation and maintain representative samples of terrestrial coastal and marine ecosystems (Plan de Manejo del Área Protegida Sistema Península Valdés 1999).

Península Valdés is located in the southeastern part of the Monte eco-region (Burkart et al. 1999) which comprises shrublands, grasslands and scattered, mainly temporary inland-wetlands (see Chapter "Vegetation of Península Valdés: Priority Sites for Conservation"). In terms of vegetation physiognomy and floristic composition, the Península is considered an ecotone (León et al. 1998), containing elements from both the Monte and Patagonian phytogeographic provinces (Soriano 1956; Roig et al. 2009). Accordingly, some of the major taxa of terrestrial animals found in Península Valdés are represented by species associated either to the Monte or Patagonia biota.

As an arid land, Península Valdés is home to a diverse and abundant group of arthropods—including insects and spiders—which play an important role across the trophic web. They represent a mosaic of species from both Patagonia and Monte biogeographical regions, although the arthropod assemblage as a whole is most representative of the latter.

The herpetofauna is also representative of the Monte, with some widespread species, that are common in the Patagonian steppe. Whereas, most of the birds described in the area occupy the southern part of the extensive Monte ornithological eco-region. Mammalian diversity in the Península Valdés region is today the result of a dynamic interchange of species from the two eco-regions which has taken place during the last few hundred years, being dominant Monte species.

In this chapter, we summarize the available information on the arthropods, reptiles, birds and mammals that inhabit Península Valdés, while providing an insight on the main issues resulting from the interactions between human activities and wildlife which ultimately affect ecosystem functioning and conservation.

2 Patterns of Animal Diversity and Distribution

2.1 Arthropods

The terrestrial arthropods—insects and arachnids—are the most diverse and abundant animals in arid lands worldwide (Polis 1991; Ayal 2007). They act either as herbivores, predators or decomposers and thus play multiple roles across the trophic web (Flores 1998; Ayal 2007). In the Península Valdés region, around 200 arthropods species or *morphospecies* have been recorded. The use of morphospecies is the most common method implemented to improve the cost efficiency in the assessment of invertebrate biodiversity. A *morphospecies* consists of a non-formal taxonomical identity arising from invertebrate specimen sorting, using external morphological features by personnel with minimal training in formal taxonomy (non-specialist personnel), but validated by the input from specialists on critical phases of the process (Oliver and Beattie 1993, 1996, 1997; Pik et al. 1999). Several authors have shown that estimates of species richness and species turnover produced using morphospecies can be very similar to formal species estimates for a number of arthropod taxa (Oliver and Beattie 1993, 1996, 1997; Pik et al. 1999). The arthropods morphospecies comprise 18 orders and approximately 60 families of insects and arachnids (Appendix 1). Ants (Hymenoptera, Formicidae) are the most abundant arthropod taxa of Península Valdés, a group which is known by their remarkable breeding success associated with their social behaviour. Ants are followed in terms of abundance by a co-dominance of spiders (Aranea) and beetles (Coleoptera), then by *grasshoppers* and crickets (Orthoptera), *springtails* (Collembola) and sun spiders or *solifuges* (Solifugae). This pattern is coincident with the findings across other arid lands of Argentina like Chaco and Mendoza provinces, and other continents such as in South Africa and Australia (Gardner et al. 1995; Bromham et al. 1999; Molina et al. 1999; Seymour and Dean 1999; Lagos 2004). *Co-dominance patterns* are similar at lower taxonomic levels, as 60% of the hexapod abundance is accounted for by only six taxa: Sminthuridae, Tenebrionidae, Acrididae, Phloeothripidae, Carabidae y Mummusidae (Cheli et al. 2010, 2013).

Arthropod taxa *endemic* of the Península Valdés have been described recently, such as the beetle-like true bug *Anomaloptera patagonica* (Hemiptera, Oxycarenidae) (Fig. 1a) (Dellapé and Cheli 2007) and a new genus and species *Valdesiana curiosa* (Hemiptera, Miridae) (Fig. 1b) (Carpintero et al. 2008). As a result of the same investigations, a new species of scorpion, *Urophonius martinezi* (Scorpiones, Bothriuridae) (Fig. 1f) (Ojanguren-Affilastro and Cheli 2009) was described for the Península Valdés region and found outside the protected area afterwards. More recently, Flores et al. (2011) have described a new species endemic to Península Valdés, *Calymmophorus peninsularis* (Fig. 1e), and two subspecies of *Praocis* (*Hemipraocis*) *sellata* Berg 1889 [*P. (H.) sellata granulipennis* (Fig. 1c) and *P. (H.) sellata peninsularis* (Fig. 1d)], the first is distributed throughout northern Patagonia while the second is endemic to the peninsula. In fact, the knowledge of the arthropods of Península Valdés has increased significantly as a result of recent work. Studies

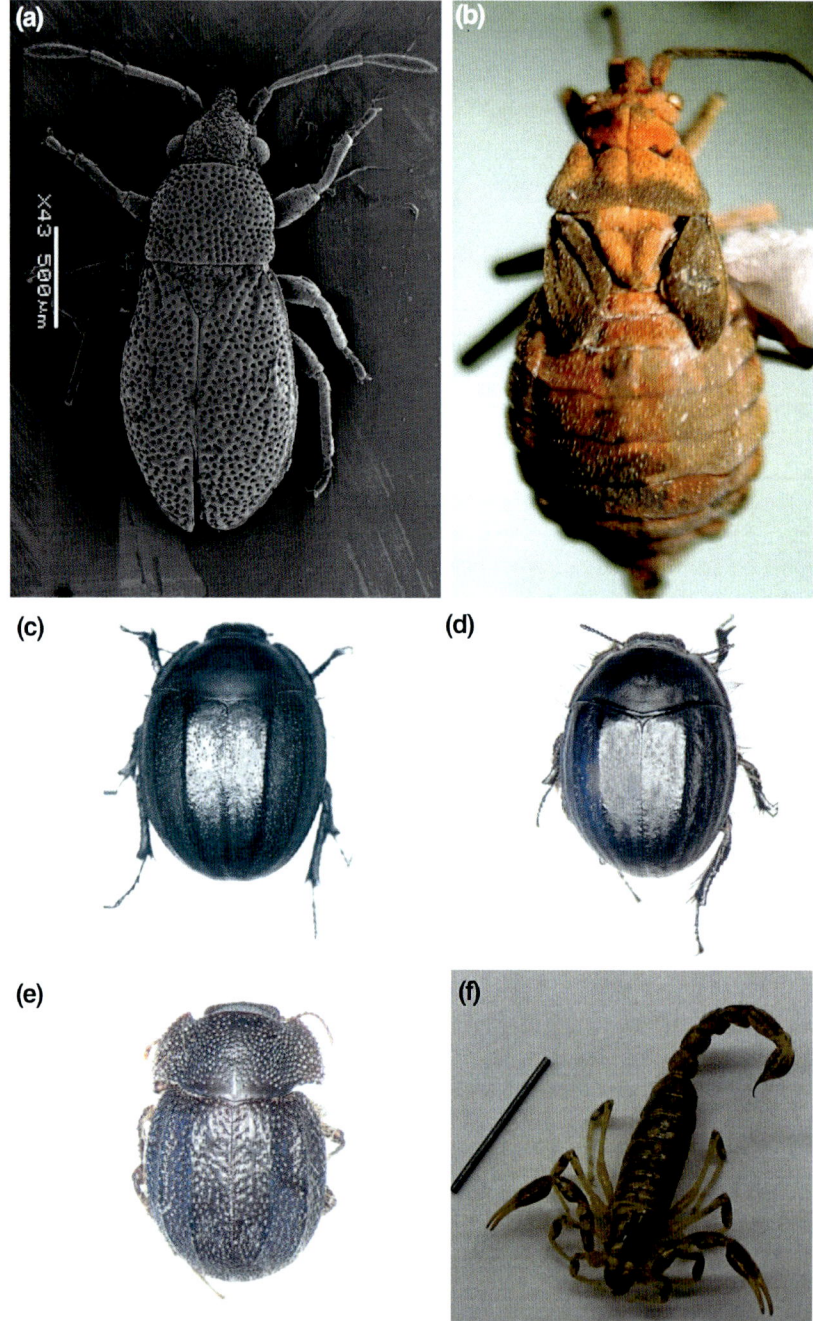

◀ **Fig. 1** Arthropod taxonomic novelties recorded in Península Valdés. **a** *Anomaloptera patagonica* (Hemiptera, Oxycarenidae; previously published in Dellapé and Cheli (2007), Zootaxa 1528: Fig. 1; copyright Magnolia Press, reproduced with permission); **b** The remarkable new monospecific genus of true bug *Valdesiana curiosa* (Hemiptera, Miridae; previously published in Carpintero et al. 2008, Zootaxa 1672: Fig. 1; copyright Magnolia Press, reproduced with permission); **c** *Praocis (Hemipraocis) sellata granulipennis* (Coleoptera, Tenebrionidae), paratype (previously published in Flores et al. 2011, Zootaxa 2965: Fig. 5; copyright Magnolia Press, reproduced with permission); **d** *Praocis (Hemipraocis) sellata peninsularis*, holotype (previously published in Flores et al. 2011, Zootaxa 2965: Fig. 3; copyright Magnolia Press, reproduced with permission); **e** *Calymmophorus peninsularis* (Coleoptera, Tenebrionidae), paratype (previously published in Flores et al. (2011), Zootaxa 2965: Fig. 6; copyright Magnolia Press, reproduced with permission); **f** *Urophonius martinezi* (Bothriuridae), the only scorpion species of the peninsula found to be active during the winter

conducted by Dellapé and Cheli (2007), Carpintero et al. (2008), Cheli (2009), Ojanguren and Cheli (2009), Cheli et al. (2010), Carrara et al. (2011) and Flores et al. (2011) have contributed to the first species lists of terrestrial arthropods (Appendix 1) and to the understanding of their biological diversity.

Similar to the patterns of co-dominance, diversity of terrestrial arthropods described for Península Valdés is similar to other arid lands of Argentina, such as the arid Chaco (Gardner et al. 1995; Molina et al. 1999) and Central Monte (Lagos 2004). Nevertheless, species and family richness are lower in Península Valdés than in the other areas, while the dominance of some taxa is higher—hence evenness is lower—at both taxonomic levels. Lower richness of species and families together with a higher dominance could be the result of the harsh climatic conditions prevailing in Patagonia during the last 10,000 years, which would impose limits to the survival of some of the northern species, and at the same time would favour the ability of other species like the beetle *Blapstinus punctulatus* (Coleoptera, Tenebrionidae) to use more efficiently the limited resources (Cheli et al. 2010). By applying theoretical models, Cheli et al. (2010) described a system in which an unsaturated habitat would have favoured the occupation by species with intermediate levels of niche preferences, while the intensity of migration between communities should have been important (Magurran 2004).

Studies on the theoretical distribution of species abundances suggest that few factors are dominating the ecological interactions among the arthropods in the Península Valdés region (Cheli et al. 2010). In the arid Patagonia, as in other arid lands, the main factors modelling the structure of the arthropod communities are related to the vegetation. The influence of vegetation is well documented for both Península Valdés (Cheli 2009; Martínez 2013) and other sites (Farji-Brener et al. 2002; Folgarait and Sala 2002; Mazía et al. 2006; Martínez Román 2014). Across deserts and semi-deserts, the diversity, dominance and abundance of epigeal arthropods (i.e. arthropods whose main activities are conducted above the surface of the ground, similar sense than "epigean", "epigeic", "epigeous" or "ground-dwelling") are strongly correlated with plant cover (Crawford 1988; Seymour and Dean 1999). Usually, vegetation structure is determinant of the arthropod assemblage, as it provides shelter, food, microsites for oviposition and other resources (Dennis et al. 1998; Seymour and Dean 1999; Mazía et al. 2006).

Spatial heterogeneity strongly influences the distribution of the arthropods within Península Valdés, showing different assemblages in different habitats. For example, the shrublands of the central Península Valdés region represent the most xeric environment in the area (see Chapter "Vegetation of Península Valdés: Priority Sites for Conservation"), adapted to lower rainfall. Consequently, the soils are characterized by the presence of argillic horizons of hard consistence in dry conditions (see Chapter "Soil–Geomorphology Relationships and Pedogenic Processes in Península Valdés") and support a low primary productivity, dominated by a low and homogeneous shrubland. This vegetation community is associated with a particular arthropod assemblage where the beetle *Nyctelia darwini* (Tenebronidae) is a species found solely in the central shrublands of Península Valdés. In contrast, the highest diversity of arthropods is found in coastal zones and the southern part of the area, where the precipitation is higher, and the sand-dominated soils sustain higher plant productivity (Carrara et al. 2011; Martínez 2013).

2.2 Reptiles

The knowledge on reptile diversity in Patagonia has increased considerably in recent decades after numerous studies carried out across the Monte and Patagonian steppe regions (see Minoli et al. 2015 for a general review), although Península Valdés is still not well known. After the pioneering publication by Scolaro (1976) on lizard species, Daciuk and Miranda (1980) added information about snakes and amphisbaenians. However, there is still a lack of well-designed studies to survey the herpetofauna and obtain accurate information about species diversity and distribution. In addition, the knowledge of reptiles was blurred by confusing nomenclatural problems or the lack of appropriate systematic studies for some groups, and in the last years due to changing taxonomy. The studies by Scolaro (1976), Daciuk and Miranda (1980) and Cruz et al. (1999) were affected by these changes although they are useful for a general introduction on the herpetofauna of the Península Valdés region. Recently, Minoli et al. (2015) made comments about the present status of some species in Chubut province and rectified some mistakes related to previous taxonomical classification, although the work is still at an early stage.

A total of 12 species of reptiles belonging to six families were registered in the Península Valdés region so far, including an *amphisbaenian*, five lizards and six snakes (Appendix 2). They belong to the families Amphisbaenidae (1 sp.), Liolaemidae (or clade Liolaemini, 3 spp.), Leiosauridae (1 sp.), Phyllodactylidae (1 sp.), Dipsadidae (4) and Viperidae (1). To our knowledge, most of the reptiles of Península Valdés are characteristic of the Monte biogeographic region, probably with the exception of *Homonota darwinii* and *Leiosaurus belli* (Fig. 2c) which seem to be very common across the Patagonian steppe.

Amphisbaenians are secretive, burrowing animals that usually live in bare soils. They seem to be very common in more temperate areas but surprisingly two species

Fig. 2 Most common species of reptiles found in Península Valdés: **a** *Phylodrias trilineata*; **b** *Liolaemus melanops* (male); **c** *Liolaemus darwinii* (male); **d** *Liolaemus gracilis*; **e** *Phylodrias patagoniensis*; **f** *Leiosaurus bellii*. Pictures (**b**) and (**f**) by Jared A. Grummer, all others by Grupo de Herpetología Patagónica

reach Patagonia, *Anops kingii* and *Amphisbaena plumbea*. The first species, *A. kingii* is the only amphisbaenian ever registered in Península Valdés, as Daciuk and Miranda (1980) recorded this small, fossorial, earthworm-like reptile at the Istmo Carlos Ameghino. *A. kingii* is usually found in bare soil sites, and frequently along the sandy coastal areas. The other amphisbaenean species, *A. plumbea*, is slightly larger and was registered as far as Punta Tombo—over 200 km south from Península Valdés—by Cruz et al. (1999) and in northern Chubut by Avila et al. (2007) but never around the Península Valdés, although this could be a matter of lack of collected specimens rather than the absence of suitable habitats.

Other Squamata of Península Valdés belong to three families: Phyllodactylidae (*H. darwinii*), Leiosauridae (*L. belli*) and Liolaemidae (*Liolaemus melanops*, *L. darwinii* and *L. gracilis*, Fig. 2b, c, d). All species are common in sandy habitats along the coastal areas (i.e. Coastal Zone system; see Chapter "Late Cenozoic Landforms and Landscape Evolution of Península Valdés"), but less common in the central areas of the peninsula (i.e. Upland and plains system and Grreat endhoreic basin system; see Chapter "Late Cenozoic Landforms and Landscape Evolution of Península Valdés"). However, this could result from an insufficient sampling effort, the lack of suitable sites or else habitat modification after overgrazing by the domestic sheep. *Homonota darwinii* is a small, crepuscular, oviparous and insectivorous species that is usually found below natural or artificial objects, but probably lives in burrows, in cavities around roots of shrubs or grass bunches, as well as in rock crevices on coastal cliffs. Nominal species have a wide geographic distribution across Patagonia, spanning from Santa Cruz to Mendoza provinces, but recent studies evidence highly fragmented populations, as populations from the Atlantic coast north of the Río Chubut—which reaches the sea 100 km south from Península Valdés—showed clear genetic differences with the southernmost Patagonian Steppe populations (Morando, unpublished data). The only representative of the Leiosauridae family is a large, stout, oviparous, insectivorous and diurnal but very inconspicuous species of Leiosaurus, *L. belli* (Fig. 2f). It is common in shrub-dominated areas where numbers can be very high (Udrizar Sauthier et al. 2007) but difficult to detect due to its behaviour and well-adapted body coloration to shrubby habitats. As described for *H. darwinii*, *L. belli* has a wide geographic distribution across the Austral Monte and the Patagonian Steppe. Liolaemids are the most important group of lizards of southern South America in terms of species richness, but species diversity and population densities seem to be higher in the central plateaus of Patagonia than in the Atlantic coast. Despite the citations of several other species for Península Valdés, only three are confirmed: *L. melanops*, *L. darwinii* and *L. gracilis* (Fig. 2b, c, d). The first two species belong to the Eulaemus subclade (or subgenera), while *L. gracilis* is part of the *Liolaemus* sensu stricto clade or subgenera (or *chiliensis* group), all species endemic to the Austral Monte biogeographical region. As well as the majority of the species of this group they are diurnal, insectivorous, and oviparous species, the first two showing a remarkable sexual dichromatism almost unnoticeable in *L. gracilis*. Usually, the five species of lizards (Squamata) can be found together in the same location when conditions seems to be suitable, mainly at sites where dunes or sandy soils are combined with the presence of well preserved grasses and shrubs, but is more common to observe only one or sometimes two species in syntopy (i.e. the joint occurrence of two species in the same habitat o geographic place at the same time).

As discussed by Minoli et al. (2015), previous records of species such as *L. boulengeri*, *L. kingii*, *L. xanthoviridis* or *L. goestchi* at Península Valdés are erroneous. Likewise, species found at very close localities in the continental area were not found in the peninsula, such as *Cnemidophorus longicauda* which is probably the most austral teiid lizard in the world (Yokes et al. 2006). Although the southernmost record is relatively recent (Frutos et al. 2005), the species was observed several years ago in coastal dunes of Golfo Nuevo close to Puerto Madryn (Avila, unpublished data). Thus, we can expect to find the species in similar habitats along Península Valdés; however, populations of this species seem to be very small and difficult to find. *Liolaemus boulengeri* was cited by Scolaro (1976) for Salina Grande but never recorded afterwards. *Liolaemus boulengeri* was observed along the coast of Puerto Madryn over 20 years ago (Avila, unpublished data) but was extirpated by human development, although today it is present in some locations next to the city (Cruz et al. 1999). There is no evidence of the presence of the species in the Península Valdés region.

Snakes in Península Valdés belong to two families, Dipsadidae (*Clelia rustica*, *Pseudotomodon trigonatus*, *Phylodrias trilineata* (Fig. 2a), *P. patagoniensis*) (Fig. 2e) and Viperidae (*Bothrops ammodytoides*). The last three species are commonly found in Península Valdés. Previous work indicates that the first two reach only the isthmus (Daciuk and Miranda 1980). However, this could be due to the lack of sampling instead of a real absence of *C. rustica* and *Pseudotomodon* from Península Valdés. *Phylodrias* species are diurnal, oviparous and active hunters, they feed on small mammals, birds and lizards, preys that are very common around the area. *Bothrops ammodytoides* is the southernmost Viperidae snake of the world and seems to be very common in this region, but southern populations of this species are barely known. In general, snake populations occur at very low densities in Patagonian habitats, records of *C. rustica* and *P. trigonatus* are scarce and limited to the isthmus region, as stated above. Other species were registered along the years in continental localities very close to Península Valdés but never found inside the area, as *Phalotris bilineatus*, *Oxyrhophus rhombifer* and *Erythrolamprus saggitifer saggitifer* (Cei 1986; Scolaro and Cei 1979; Avila et al. 2001; Scolaro 2006; Carrera and Avila 2008a, b; Avila 2009).

2.3 Birds

Although the Monte and Patagonian Steppe eco-regions are arid ecosystems which differ in terms of physiognomy and dominant plant species (see Chapter "Vegetation of Península Valdés: Priority Sites for Conservation"), they do not impose geographic barriers for birds. Thus, several bird species common to both regions can be found at Península Valdés (Haene 2004). Terrestrial birds of these environments are coloured similar to the landscape which is dominated by patterns of browns and greys; they rarely form aggregations but occur generally dispersed, showing evasive behaviours. In contrast, diverse and coloured waterbird species—some of them forming large flocks—can be found associated with the inland-wetlands across the landscape.

There are no published studies assessing the occurrence of land bird species throughout Península Valdés, and most of the available information is limited either to non-exhaustive checklists, studies restricted to some particular groups (e.g. Pruscini et al. 2014) or species (e.g. Daciuk 1978; Baldi et al. 2015), restricted to limited areas (e.g. Daciuk 1977, 1979; Codesido et al. 2005; Krapovickas et al., unpublished data), or species observed occasionally or potentially present in the area due to their expected distribution ranges (Narosky and Izurieta 2004; Kovacs et al. 2005; Harris 2008; De la Peña 2013). We provide a plausible and updated list of terrestrial and inland-wetland birds for the Península Valdés region by integrating published and unpublished sources (see Appendix 3). It is important to note that the species of seabirds, shorebirds and water birds that usually occur in salt marshes, intertidal zones and adjacent marine habitats were not included in this list. There are 139 species of birds registered regularly in Península Valdés, belonging to 36 families and 18 orders [following the systematic by Remsen et al. (2015)]. The most represented families are Tyrannidae, Furnariidae and Thraupidae for terrestrial birds and Anatidae and Scolopacidae for inland-wetlands birds. The shrublands and grasslands of Península Valdés offer a suitable habitat for ratite birds like the lesser rhea (*Rhea pennata*) (Fig. 3e) and tinamous (Fig. 3g), insectivorous birds like Earthcreepers (Fig. 3c, d), Canasteros and Monjitas; granivorous birds like finches and the Rufous-collared sparrow (*Zonotrichia capensis*) (Fig. 3b); and several species of diurnal and nocturnal raptors. Salinas (Playa lake; see Chapter "Late Cenozoic Landforms and Landscape Evolution of Península Valdés") are regularly deluged attracting water birds like flamingos and phalaropes, while the adjacent wetlands or 'mallines' associated with the drain systems offer habitat for geese and ducks. After rainfall, several temporary ponds and small lagoons (i.e. small to medium closed basins; see Chapter "Late Cenozoic Landforms and Landscape Evolution of Península Valdés") become flooded and it is possible to find several anatids, grebes and coots. Some persistent ponds and permanent water points provide habitat for birds as aquatic plants are well established and it is likely to observe rails, such as the Austral Rail (*Rallus antarticus*) which was rediscovered in Patagonia and reported for Península Valdés (Mazar Barnett et al. 1998; Pugnali et al. 2004).

Four endemic bird species are commonly found in Península Valdés: the Band-tailed Earthcreeper (*Ochetorhynchus phoenicurus*) (Fig. 3d), the Patagonian Canastero (*Pseudasthenes patagonica*), the White-throated Cacholote (*Pseudoseisura gutturalis*) (Fig. 3a) and the Rusty-backed Monjita (*Xolmis rubetra*), which make up 25% of the bird species endemic of Argentina (see López-Lanús et al. 2008). Approximately 70% (100 species) of the species recorded at Península Valdés are considered resident, while three show partial, seasonal movements, mainly northwards during the autumn (see Appendix 3). The other 36 species are

Fig. 3 Representative land birds of Península Valdés. **a** White-throated Cacholote (*Pseudoseisura gutturalis*); **b** Rufous-collared Sparrow (*Zonotrichia capensis*); **c** Scale-throated Earthcreeper (*Upucerthia dumetaria*); **d** Band-tailed Earthcreeper (*Ochetorhynchus phoenicurus*); **e** Lesser Rhea (*Rhea pennata pennata*); **f** Long-tailed Meadowlark (*Sturnella loyca*); **g** Elegant Crested-Tinamou (*Eudromia elegans*). Photographs by Darío Podestá

migratory: ten Neoartic, four Neotropical and 22 Austral migrants (see Appendix 3). At least two species found in the Península Valdés region fall within the range of categories of conservation concern according to the *IUCN criteria* (López-Lanús et al. 2008), as the Austral rail and the Yellow cardinal (*Gubernatrix cristata*) are classified as endangered (BirdLife International 2012). Although classified as of least concern, the Ruddy-headed goose (*Chloephaga rubidiceps*), the Ashy-headed Goose (*Chloephaga poliocephala*), the Rufous-chested dotterel (*Charadrius modestus*), the Upland goose (*Chloephaga picta*), the Elegant crested-tinamou (*Eudromia elegans*) (Fig. 3g), the Hudson's black-tyrant (*Knipolegus hudsoni*), the White-throated cacholote and the Rusty-backed monjita were reported to be decreasing (BirdLife International 2012). There are two introduced species forming wild populations associated to human settlements, like the Rock pigeon (*Columba livia*) and the House sparrow (*Passer domesticus*). Other species, like thrushes, are currently expanding their distribution southwards due mainly to habitat modification by humans (Pérez et al. 2006; Veiga et al. 2010).

Information on the abundance of terrestrial and inland-wetland birds of Península Valdés is scarce, and limited to conspicuous species like the lesser rhea. Recently, Baldi et al. (2015) found that the abundance of this species in the Península Valdés region was low, averaging 0.28 adults/km^2. Lesser rheas tended to be more common in open grasslands than in shrublands, and also where sheep ranching does not take place (Baldi et al. 2015). For example, ongoing work conducted at a sheep-free site —a private reserve (San Pablo de Valdés)—estimates densities between 2.2 and 3.6 rheas/km^2 (Fernández and Geremías Toscano, unpublished data). The Elegant crested-tinamou appears to be much more abundant than the lesser rhea although this is to be expected as the body size of the tinamou is much smaller than that of the rhea. Densities of passerines found in southwest of the Península Valdés were estimated in 10–160 individuals/km^2 (Krapovickas et al. unpublished data), lower to the estimates available for the central part of the Monte eco-region (131–1091 indivuduals/km^2, Lopez de Casenave 2001). The most abundant passerines in Península Valdés included the Rufous-collared sparrow, the Patagonian mockingbird (*Mimus patagonicus*), the Plain-mantled Tit-spinetail (*Leptasthenura aegithaloides*), the Short-billed pipit (*Anthus furcatus*), the Common miner (*Geositta cunicularia*), the Long-tailed Meadowlark (*Sturnella loyca*) (Fig. 3f) and the austral migrant Common Diuca-finch (*Diuca diuca*) (Pruscini et al. 2014; Krapovickas et al. unpublished data). There is no information on the abundance neither of raptors nor inland-wetlands birds in Península Valdés. However, some of the species commonly observed include the Variable hawk (*Geranoaetus polyosoma*), the American kestrel (*Falco sparverius*), the Chimango caracara (*Milvago chimango*), the Burrowing owl (*Athene cunicularia*), the Chilean flamingo (*Phoenicopterus chilensis*) and the Crested duck (*Lophonetta specularioides*).

Independently of the season, the mixed steppes of shrublands and grasslands showed the highest species richness and abundance (Krapovickas et al. unpublished data). In contrast, the herbaceous steppes appeared associated with the lowest species richness and abundance, possibly due to the limited availability of perching and nesting substrates (Pruscini et al. 2014; Krapovickas et al. unpublished data).

Seasonal and inter-annual fluctuations in the structure and abundance of birds assemblages found in the southwestern part of Península Valdés are remarkable (Krapovickas et al. unpublished data). Spring is the most diverse season, as it includes the presence of several migrant species. Although restricted to one spring season studied, Pruscini et al. (2014) found two different assemblages of passerines which could be characterized, one for the herbaceous steppe and another for the shrub steppes, while two additional assemblages combining species from the former two were described for mixed steppes. Nevertheless, it is important to notice that the natural history of the birds of Península Valdés is poorly known, and it is a priority to increase the scientific knowledge on these species' ecology, their adaptations to a rigorous environment, and to assess the threats they are facing to plan appropriate conservation actions.

2.4 Mammals

Twenty three species of native terrestrial mammals have been recorded in Península Valdés, as well as five species of wild introduced mammals (Appendix 3). The native mammals comprise 6 orders and 12 families. Rodents are the most diverse group with 10 species, there are seven species of carnivores, two species of bats, two species of armadillos, one species of marsupial and one species of artiodactyl (Nabte 2010). According to recent studies, the mammalian assemblage of the Península Valdés region has changed significantly during the last 1000 years (Udrizar Sauthier 2009; Udrizar Sauthier and Nabte 2012; Abba et al. 2014; Prevosti et al. 2015; Formoso et al. 2015). Moreover, substantial changes in species composition involved different groups without close phylogenetic relationship among them like the marsupials, carnivores and armadillos.

Almost 1600 years ago, the southernmost living marsupial, the Patagonian opossum (*Lestodelphys halli*), inhabited Península Valdés since its osteological remains were frequent in Holocene mammal assemblages (Udrizar Sauthier 2009; Formoso et al. 2015). Although *L. halli* is still common across the Patagonian steppe, it is not part of the extant mammalian fauna of Península Valdés (Massoia et al. 1988; Pardiñas et al. 2001; Nabte et al. 2008; Udrizar Sauthier and Pardiñas 2006; Trejo and Lambertucci 2007). Among the carnivores, the Patagonian weasel (*Lyncodon patagonicus*) is frequently associated with Patagonian steppes (Schiaffini et al. 2013) but it has not been recorded at Península Valdés recently, despite intensive surveys and interviews to local residents. Instead, the Patagonian weasel was abundant in Península Valdés a few hundreds to a few thousands years ago (Udrizar Sauthier and Nabte 2012). The fox *Dusicyon avus* became extinct in South America during the last few hundred years (Prevosti et al. 2015) and it is the canid most frequently found in the Holocene assemblages of the Península Valdés region (Carrera and Udrizar Sauthier 2011), sharing the area with other two foxes, the South American grey fox (*Lycalopex gymnocercus*) (Fig. 4c) and the culpeo (*Lycalopex culpaeus*) are both present in Península Valdés today (Nabte 2010).

Fig. 4 Charismatic terrestrial mammals of Península Valdés. **a** Tuco-tuco (*Ctenomys* sp.); **b** Mara (*Dolichotis patagonum*); **c** South american grey fox (*Lycalopex gymnocercus*); **d** Big hairly armadillo (*Chaetophractus villosus*); **e** Guanaco (*Lama guanicoe*); **f** Southern mountain cavy (*Microcavia australis*); **g** Palid fat-tailed opossum (*Thylamys pallidior*); **h** Molina's hog-nosed skunk (*Conepatus chinga*). Photographs by Darío Podestá

The culpeo fox would have occupied Península Valdés a few hundred years ago, as suggested by its absence or occasional presence in the carnivore assemblages of the past, and in accordance with the hypotheses of a recent expansion of the species towards eastern Patagonia (Crespo and de Carlo 1963; Novaro 1997). The other recent incomer in the Península Valdés region is the Big hairy armadillo (*Chaetophractus villosus*) (Fig. 4d), perhaps sometime during the last 300 years as suggested by historic data and the study of remains from the Holocene (Crespo 1974; Abba et al. 2014), sharing the area with the other armadillo species, the pichi (*Zaedyus pichiy*), which is apparently declining in numbers (Nabte 2010).

Although there is no evidence of changes in the composition in the assemblages of rodents, some species often related to the Monte Province like the Eastern Patagonian laucha (*Eligmodontia typus*), the Drylands vesper mouse (*Calomys musculinus*), the dolores grass mouse (*Akodon dolores*) and the grey leaf-eared mouse (*Graomys griseoflavus*) have increased their abundance in present times, while others like the Bunny rat (*Reithrodon auritus*) and the Tuco-tucos (*Ctenomys* spp.) (Fig. 4a) have decreased compared to their abundance in the assemblages of the past, 1000 year ago (Udrizar Sauthier 2009, unpublished data).

The distribution and abundance of mammals inside Península Valdés is not homogeneous and result from several factors such as vegetation structure, geomorphology and human activities among others interacting elements. For example the Pallid fat-tailed opossum *Thylamys pallidior* (Fig. 4g) occupies most of the area, although it is clearly more frequent in the western and northern parts of the peninsula where the Monte vegetation (see Chapter "Vegetation of Península Valdés: Priority Sites for Conservation") prevails (Formoso et al. 2011). While the Big hairy armadillo is more common in the north, the Pichi is more abundant in the southern portion of Penínusla Valdés (Abba et al. 2010), where grasslands are abundant (see Chapter "Vegetation of Península Valdés: Priority Sites for Conservation"). Instead, the distribution of the two species of bats *Tadarida brasiliensis* and *Histiotus montanus* is affected mainly by the availability of shelter sites which are mainly provided by human infrastructure like the shearing sheds, where these animals are commonly found (Udrizar Sauthier et al. 2013). While some carnivore species like the Grey fox, the Hog-nosed skunk (*Conepatus chinga*) (Fig. 4h) and the Lesser grison are widely distributed within Península Valdés (Carrera et al. 2012); the felids show a different pattern. For example, the Puma (*Puma concolor*) which occurs occasionally in Península Valdés, is restricted to sites where the topography is uneven, with canyons and ravines leading to the beaches surrounding the peninsula (D'Agostino et al. 2015) and avoids the more flat, central steppes. Instead, the smaller Geoffroy's and Pampas cats (*Leopardus geoffroyi* and *L. colocolo*) are distributed in the northern and southern sectors of the peninsula, respectively (Nabte 2010). Regarding the Culpeo fox, records are scarce and limited to the southern grasslands of Península Valdés (D'Agostino et al. 2015).

The small mammal assemblage is similar to that of the Monte eco-region (Udrizar Sauthier and Pardiñas 2014). Although species distribution and habitat selection patterns are not well known, ongoing studies by Udrizar Sauthier (unpublished data) are revealing that the Eastern Patagonia laucha (*Eligmodontia typus*) is the most abundant and commonly distributed across the whole area. While the Bunny rat and Tuco-tucos are more abundant in the grass-dominated sandy soils of southern Península Valdés, the Grey leaf-eared mouse prefers the shrub-dominated communities of the west and north part of the peninsula. The Dolores grass mouse (*Akodon dolores*) is common in coastal halophytic shrublands of Península Valdés (Nabte et al. 2009), while other mouse species such as the Drylands vesper mouse (*Calomys musculinus*) and the Intelligent field mouse (*Akodon iniscatus*), which have been associated to the Monte eco-region (Pardiñas 2009; De Tommaso et al. 2014), are found in most habitat types across the peninsula (Udrizar Sauthier unpublished data). The small cavies *Galea leucoblephara* and *Microcavia australis* (Fig. 4f) are not very abundant and they are mostly found in shrub-dominated areas (Udrizar Sauthier et al. 2015), while the largest rodents, maras *Dolichotis patagonum* (Fig. 4b) are abundant across Península Valdés and prefer open grass-dominated sites to set their breeding dens (Baldi 2007). The only native ungulate is the guanaco (*Lama guanicoe*) (Fig. 4e) which is abundant and can be found in most of the area, although variation in numbers is mostly related to the presence of sheep and human activities (Baldi et al. 2001).

3 Human Activities and Their Effects on Wildlife Populations

The patterns of animal diversity and the ecological processes taking place today in Península Valdés cannot be described without considering the human factor. Although Península Valdés was inhabited by humans more than 5000 years ago (see Chapter "Archaeology of the Península Valdés: Spatial and Temporal Variability in the Human Use of the Landscape and Geological Resources"), a major transformation started in the 1880s when the first European settlers arrived to the area bringing the domestic sheep (Coronato 2010). Since then, sheep ranching has been the main economic activity carried out in the terrestrial ecosystems of the Península Valdés region. As for most of the arid Patagonia, the division of the land fenced into properties and paddocks (the minimum units of land divided by fences), and the extensive grazing by thousands of sheep on the native vegetation resulted in an entirely new environment, both in terms of habitat modification and the relationships between humans and wildlife. The native herbivores were perceived as potential competitors and the carnivores as predators of the sheep; therefore they were chased or hunted. The significant increase in the biomass of herbivores affected the vegetation, the soil and the major ecological processes as in other arid lands in the world (Milchunas and Lauenroth 1993; Reynolds and Stafford Smith 2002). However, during the 1960s Península Valdés became a world attraction for tourists and called the attention on its conservation values, which led to the creation of the first coastal reserves and subsequently to the declaration of UNESCO World Natural Heritage Site for the entire area. At the same time, the rapid growth of human populations in Puerto Madryn and Puerto Pirámides resulted in an increased pressure for the recreational use of the area. Likewise, the conservation of its wildlife, the habitat and the ecological processes involved needs to be undertaken within the framework of a complex socio-ecological system, which is still unique for its wilderness and natural value.

In the Monte ecoregion, overgrazing by livestock is the main human-related factor causing ecosystem disturbance, followed by wildfires and logging (Pol et al. 2006). Although the effects of sheep grazing on the vegetation of Península Valdés have been scarcely addressed in the ecological literature, some studies allowed the identification of specific effects at different scales. Elissalde and Miravalles (1983) surveyed all the vegetation units (see Chapter "Vegetation of Península Valdés: Priority Sites for Conservation") and reported a reduction in the abundance of herbivore-preferred plant species at sites with higher sheep-stocking rates. On the basis of a four-category scale of vegetation condition, they concluded that more than 80% of the surveyed areas were either in poor or regular condition. These patterns at community scale are coincident with recent findings by Burgi et al. (2012), who found that ground cover and species richness of palatable grass species across four vegetation communities were higher in sites after 5 years of sheep exclusion than in neighbouring rangelands, where sheep grazing continued.

At landscape scale, two studies revealed patterns of environmental degradation related to the distance to artificial watering points [i.e. *piosphere* sensu Andrew (1988)]. Blanco et al. (2008) found that the abundance and spatial aggregation of blowouts occurring in grass steppes of the southern Península Valdés were higher near artificial water points, where grazing pressure is higher (see Chapter "Soil Degradation in Peninsula Valdes: Causes, Factors, Processes, and Assessment Methods"). Moreover, in northern shrub-grass steppes of the area, Cheli et al. (unpublished data) found increasing cover of perennial grasses, a higher vertical complexity of vegetation patches and microtopography with increasing distance from the water point. On the contrary, the cover of bare soil and soil compaction were higher in areas up to 800 m away from permanent water points.

The principal threats to native wildlife in the Península Valdés region are habitat degradation due to overgrazing by livestock, competition for resources with and disease transmission from livestock and poaching. Sheep ranching, including the direct effects of sheep on the habitat and the human activities associated with ranch and livestock management, is an important factor explaining patterns of animal abundance.

Habitat loss due to changes in vegetation communities and soil erosion, following sheep overgrazing or other human activities, are major threats for the arthropods of Península Valdés. Cheli et al. (2010) found that the communities of epigeal arthropods in the Península Valdés regions are dominated by predators. Comparative studies across areas with and without livestock grazing in Argentina have shown that the dominant arthropods are detritivorous and predators, respectively (Gardner et al. 1995; Molina et al. 1999; Lagos 2004). Thus, sheep grazing is likely to have had an important effect on the arthropod assemblage in Península Valdés. In fact, the general response in relation to grazing pressure appears to be an increase in the abundance of some taxa such as Coleoptera and Formicidae in areas with higher sheep abundance, while community structure and diversity varied among the different groups of arthropods, probably due to the effects of habitat modification on the life strategies of different species (Cheli 2009).

In semiarid and arid ecosystems, the vegetation structure and composition, i.e. the architecture of shrubs, grass and herbaceous plants, offer a great variety of habitats and niches to sustain reptile diversity. However, intensive human land use practices like widespread sheep ranching and, at specific sites, use of all terrain vehicles and other recreational activities have altered the vegetation composition and structure in several places of Península Valdés. Extensive grazing leads to a reduction of the perennial and herbaceous vegetation cover, which results in an increased availability of bare soil, allowing for rapid wind and water erosion with the subsequent loss of habitat that can maintain populations of reptiles, arthropods, birds and small mammals. There are no assessments on the response of the reptile community of this region, but by inferring results obtained for other sites we can assume that wild populations are smaller and probably threatened in some areas.

As it was described above, lesser rheas are large, conspicuous birds for which recent estimates of abundance have been reported as low for Península Valdés and other sites in Patagonia (Baldi et al. 2015). Ongoing studies modelling the spatial variation in their abundance within Península Valdés are showing that human-related factors such as the distance to ranch buildings are negatively related to the abundance of lesser rheas (Antún and Baldi, unpublished data). The avoidance of human dwellings by lesser rheas could be associated with activities such as hunting and egg collection for consumption. Regarding smaller birds like the passerines, selective shrub logging for firewood could be affecting the abundance and quality of potential nest substrates. The annual rate of clearance of native vegetation in northeastern Patagonia has been estimated at 3.7%, a rate ten times higher than the average 0.4% rate of loss for global tropical rainforest (Balmford et al. 2003; Pezzola et al. 2004).

Studies on maras conducted in Península Valdés have shown that they are monogamous and *breed communally*, an unusual combination among mammals (Taber and MacDonald 1992a). Clusters of breeding warrens were associated with relatively large clearings, usually surrounding shepherds outstations (Taber and MacDonald 1992b). Although isolated warrens occurred also in shrubby areas, they were frequently located in small clearings in the bush (Taber and MacDonald 1992b). Also, mara pups born in warrens located in open habitats had higher survival than pups born in warrens located in closed habitats (Baldi 2007). More recently, it was shown that the location of breeding warrens was associated both with the presence of clearings and elements of infrastructure, but also close to shrub-dominated vegetation patches (Alonso Roldán 2012). Communal breeding at open sites would allow improved vigilance against predators and the early response both of mara pups to enter to the breeding warrens and the adults to run and hide in the bush. Thus, the availability of human dwellings and infrastructure could favour the local presence of maras in Península Valdés, although the costs associated with human activities are not yet known. Even when there is some evidence that maras are hunted by locals (Taber and MacDonald 1992b), they are not perceived as a pest as other wild species like guanacos and carnivores.

In Península Valdés, the guanaco was the solely ruminant ungulate since the Pleistocene until the introduction of the domestic sheep. Although sheep of the merino breed weigh around 40–60 kg compared to 80–120 kg for adult guanacos, body sizes of both are well within the range of what Jarman (1974) defined as species of *intermediate selectivity* in terms of their foraging strategies. Both guanacos and sheep include important proportions of mono and dicotyledoneous plants —roughly, grasses and shrubs—in their diets. In a comparative study across different sites, some of them located in shrublands and grasslands of Península Valdés, Baldi et al. (2004) concluded that the potential for competition between guanacos and sheep was high, since the diet of both herbivores overlaps markedly, especially in summer when food resources are scarcer. Interspecific competition was evident since the spatial variation in guanaco abundance was negatively related to sheep density and, at some sites, shifts to "sheep-empty" habitats by guanacos occurred after landowners moved the livestock to other paddocks (Baldi et al. 2001). Sheep densities were found to be positively related to the abundance of key forage plants for both herbivore species, while guanaco densities, presumably as a result of competition, were negatively related to the abundance of the same plants (Baldi et al. 2001). Subsequent studies found different lines of evidence on the negative effects of sheep ranching on guanacos at different spatial scale, both for Península Valdés and other sites across Patagonia. For example, a large-scale survey conducted across Santa Cruz province has shown that sheep densities affected negatively the probability of finding guanacos, which were restricted to the less productive areas (Pedrana et al. 2010). In Península Valdés, Nabte et al. (2013) found that the abundance of guanacos was inversely related to sheep-stocking rates and correlates of primary productivity in multiple paddocks across the whole area.

As expected from the evidence on *interespecific competition* (Baldi et al. 2001), the removal of all the sheep from a former ranch that was converted into a wildlife reserve (San Pablo de Valdés) led to the steady increase in guanaco abundance up to the highest value reported for a site in Península Valdés within 10 years (Marino et al. in press). Recent surveys of guanacos and sheep in Península Valdés following a model-based analysis are showing that the abundance of guanacos is negatively affected by sheep numbers, and to a lesser degree by the distance to windmills and to the nearest fences. Also, they are more abundant as paddock area increases, and where mean annual plant productivity is lower (Antún et al. 2015).

Although guanacos still face several threats in Península Valdés, their numbers have increased markedly over the last 20 years. Available estimates for the entire area gave account of an average density of 0.6 guanacos/km^2 (Baldi et al. 1997) after legal commercial hunting terminated in 1992 and before Península Valdés was declared World Heritage Site in 1999. During the period following the termination of legal hunting, the highest densities within the area were estimated to be around 3 guanacos/km^2 (Baldi et al. 2001). The abundance of guanacos continued to increase, as the average density for the whole area was 4 guanacos/km^2 10 years later (Baldi et al. 2009) and subsequently increased to 7 guanacos/km^2 recently (Antún et al. 2015).

The principal threat to carnivores is hunting, both retaliatory killing or to prevent predation or presumed predation on domestic animals, especially on lambs. However, there are no studies on the effects of sheep ranching and human activities on carnivores in Península Valdés. Predators like the culpeo fox are very rare in Península Valdés, and pumas are chased and killed when detected. Although the occasional presence of pumas does result in active persecution by ranchers allegedly to protect their sheep, the occurrence of attacks has not been documented in recent years. However, sheep were as important as guanacos in the diet of pumas at a coastal site in Golfo San Matías located at 60 km NW from Península Valdés, where *stocking rates* were similar to the density of guanacos. Within 3 years the sheep were removed from this ranch and pumas responded by increasing the number of smaller prey in their diet, while guanaco densities remained constant (Fernández and Baldi 2014). Some rural workers kill the smaller cats, foxes and mustelids, allegedly because they prey on lambs, although the impact has not been assessed in Península Valdés. Also, a common practice to kill predators is to spread poisoned meat and eggs which kills not only carnivores but also birds of prey, armadillos and scavengers.

For other sites across Patagonia, spatially heterogeneous hunting pressure affects the demography and dispersal patterns of Geoffroy's cats (Pereira and Novaro 2014) and culpeo foxes. Hunting pressure can affect recolonisation rates of the carnivores, and particularly for species with large home ranges and dispersal patterns like pumas which are under a regional system of bounty hunting in different Patagonian provinces (Novaro et al. 2005; Walker and Novaro 2010).

4 Perspectives and Future Work

4.1 Conservation and Management

Compared to other sites in the region, Península Valdés still harbours significant numbers of native herbivore species like the guanaco, the lesser rhea and the endemic mara. Also, two near-threatened felid species, the Geoffroy's and Pampas cats are present across the area, whereas the presence of pumas is occasional. Other small predators include the Grey fox, the lesser grison and, potentially, the very rare Patagonian weasel. Several arthropod species are endemic to the peninsula, as well as there are endemic birds including two endangered species. There are raptors such as eagles, harriers, kites and hawks, a seemingly diverse inland-wetland assemblage and several migrant species including passerines. Mammals have undergone substantial changes in species composition during the last 1000 years, including extirpations and colonisations which took place before the arrival of Europeans. The island-like geography of Península Valdés, connected to the mainland through the narrow isthmus, appears to contribute to shape its own dynamics in terms of biological diversity and also in the different ways humans interact with the environment.

The implementation of the management plan in Península Valdés must take into account how the different human activities interact with biodiversity and the environment. It needs to address the sustainable use and protection of wild species and habitats, and also to develop programmes for the coexistence of sustainable livestock production, responsible tourism and healthy wildlife populations. Land management based mostly on traditional sheep ranching is usually in conflict with wild, native species. Today, average-sized ranches are not economically viable to sustain a family as they did decades ago. Land degradation due to overstocking, coupled with the severe droughts and low prices of wool in the international market, has resulted in substantial sheep losses. In addition, the lack of alternative options led to increased conflicts with wildlife as many landowners aim to restock their ranches, even within this unfavorable context to compensate reduced income per sheep by increasing sheep numbers.

While a few ranches were able to develop ecotourism based on coastal wildlife—such as penguins and elephant seals—and keep their sheep stocks, others were closed as they cannot afford the costs of maintenance. However, over 90% of the land in Península Valdés is still devoted to sheep ranching, although the overall stock is probably the lowest in decades (Evolución Existencia de Ganado Ovino 2005–2014 2016). As a result, pressure by some landowners on wildlife authorities to reduce guanaco densities and kill predators in particular areas of the peninsula has increased during the last few years. At the same time, other landowners have recently certified the wool they produce given that they coexist with the wild herbivores and predators, in order to add value to their production and gain access to markets of responsible consumers (Wildlife Friendly Enterprise Network 2015). Therefore, opportunities for the development and implementation of economic activities compatible with the conservation of biodiversity are priorities which need to be undertaken to increase the benefits for wild species, habitats and the people.

4.2 Research Priorities

Well designed studies on diversity, distribution and abundance of different taxa throughout Península Valdés are a key to make decisions on conservation-oriented management. Also, the identification and assessment of the effects of the multiple variables—physical, climatic, biological and human-related—on the distribution and abundance of different taxa, communities and populations is required to understand the resulting biodiversity patterns.

It is necessary to carry out a systematic survey of reptile diversity and its association with different habitats, as the information available is outdated and results from occasional observations. Aspects of the ecology and estimates of the abundance of birds such as raptors and inland-wetlands birds, are all needed to understand the dynamics of these diverse wildlife assemblages in Península Valdés.

It is also a priority to describe the composition of the mammalian predator assemblage across different sites within the peninsula, as well as to gain knowledge on the processes driving colonisation and extirpation of mammal populations.

Given the particular morphology and location of Península Valdés, we have the opportunity to study its animal diversity, distribution and ecological interactions in order to understand the ecological processes operating at different scales. It is desirable to coordinate survey and monitoring efforts to generate and maintain a biodiversity data-base for Península Valdés, essential to support proper management actions and decision-making about the protected area.

It is necessary to prioritize applied research and monitoring on the effects of human activities on different taxa. For example, it is known that the arthropods play multiple roles as detritivorous, herbivores and predators, hence they are key in nutrient cycles and the energy flux at multiple levels of the trophic web (Polis 1991; Flores 1998; Ayal 2007). Many arthropod species show high habitat fidelity and respond rapidly to changes in the surrounding environment (Cheli 2009; Cheli et al. 2010; Carrara et al. 2011; Flores et al. 2011; Martínez 2013). Thus, they can be good indicators of ecosystem changes. Variation in abundance, species composition and richness due to disturbance can be expressed either at the taxonomic or functional group level.

Future work needs to assess the consequences of habitat loss on the distribution and abundance of reptiles, birds and small mammals throughout Península Valdés. The consequences of habitat loss on wild populations inhabiting the Monte ecoregion remain unknown, but it may lead to species range contractions as it was documented for birds (Pezzola et al. 2004) and probably to local and/or regional population extirpations (Llanos et al. 2011). For reptiles, impacts of human population settlements and recreational activities could be significant, especially along sandy, coastal habitats which people tend to select. Increased vehicle traffic on the main roads results in roadkills for middle-sized mammals such as foxes, cats, maras, grisons and armadillos, and for birds like tinamous. The opening of new internal, dirt roads results in habitat loss to smaller species, and access to poachers. Therefore, it is important to assess the different impacts of humans on the habitat and wildlife and to implement actions to mitigate the effects.

Accurate estimates of the impact of predators on sheep are necessary to quantify the magnitude of the conflict, hence to plan and implement appropriate mitigation actions. Alternative, non-lethal methods to control predators using guarding animals such as dogs and donkeys are worth to explore as they can be effective (Andelt 2004; Novaro et al. 2016). As it was already described (see Sect. 4.1), human activities compatible with the persistence of wildlife populations can result in added value to local production. Although Península Valdés is internationally renowned for its marine and coastal biodiversity, terrestrial wildlife is diverse and attractive hence there is a potential to develop responsible, low-impact tourism activities. Also, sustainable use of guanaco fibre after herding, shearing and subsequent release of the animals is possible as an opportunity to diversify production and increase the value of wild species. Shearing of wild populations of guanacos has been developed and implemented at other sites in Argentina following protocols of animal welfare and monitoring the effects on wild populations (Baldi et al. 2010; Rey et al. 2012). Therefore, work on specific actions to promote responsible, sustainable human activities compatible with biodiversity conservation needs to be continued and implemented on the basis of sound scientific knowledge.

4.3 Final Remarks

As it was suggested, the implementation of the management plan is a priority for the conservation of biodiversity and the ecosystems of Península Valdés. Among other steps to be taken, we find necessary to (a) assess the main, spatially explicit interactions between biodiversity and human activities; (b) assess specific threats to biodiversity and draft the priority actions to mitigate those threats; (d) identify a number of users willing to get involved in making their activities compatible with the conservation of biodiversity; and (e) develop standards for the different activities, promoting practices that benefit wildlife, while providing increased benefits to responsible users and promoting goodwill. The Provincial authorities, research institutions, nongovernmental organisations and the private sector are all part of the Península Valdés Administration. Therefore, there is a genuine opportunity to move forward towards a goal in which sustainable management and monitoring will be a key to achieve both the conservation of healthy wildlife and the improvement of human livelihoods.

Acknowledgements We are indebted to the Centro Nacional Patagónico—CONICET for financial and logistical support to the different projects referred to in this chapter, as well as to the Agencia de Promoción Científica y Tecnológica, Fundación Vida Silvestre Argentina, the Wildlife Conservation Society and Idea Wild. We thank all our colleagues and students who participated in data collection, processing, data analysis and species determination, through the numerous projects conducted in the Península Valdés area over the years. Thanks to Darío Podestá and Hernán Povedano for kindly allowing their photographs to illustrate the chapter. We appreciate the critical comments and suggestions made by the editors and the reviewers Andrés Novaro and Federico Kacoliris, as they contributed to improve the original manuscript. We are grateful to the Dirección de Fauna y Flora Silvestre and the Ministerio de Turismo y Áreas Protegidas de Chubut for the permits to work in Península Valdés, and to all the landowners and rural workers who allowed us to work at their sites.

Glossary

Amphisbaenian	A lizard clade with elongated, distinctly annulated body, limbs are not present as they evolved to reduced internal organs. Fossorial, secretive life style

Breed communally	When more than one pair raise their offspring in the same nest or den. In this case it is referred to the mara, the monogamous mammal endemic to Argentina. Several pairs can share the same den to raise their pups during the breeding season in Península Valdés
Co-dominance patterns	Refers to two or more species, or taxonomical entities, which are equally dominant—in terms of abundance—within a biotic community
Endemic	Inhabiting only a specific geographic area
Grasshopers	Insects of the order Orthoptera with powerful hind legs which enable them to escape from threats by leaping vigorously
Intermediate selectivity	It refers to the herbivore's ability to select particular parts of plants while foraging. For the case of guanacos and sheep mentioned in the text, they can select both grasses and parts of woody plants. Thus, they are *intermediate* rather than *non selective* (grazers) or *highly selective* (browsers) feeders
Interspecific competition	Is a form of competition in which individuals of different species compete for the same resource in an ecosystem (e.g. food or living space). If the resource cannot support populations of both species, then lowered fecundity, growth, or survival may result in at least one of the competing species
IUCN criteria	A globally used system for classifying species in terms of their risk of extinction. It provides the framework to elaborate the Red List, regularly updated by the International Union for the Conservation of Nature (IUCN) organized in specialists groups worldwide
Stocking rate	Relationship between the livestock (sheep for the case of Península Valdés) and the forage available. In other words, is the relationship between secondary production (animal biomass) and primary production (forage biomass) per unit of land
Springtail	Insects members of Collembola, usually found in leaf litter and other decaying material. They are primarily detritivores and one of the main biological agents responsible for the control and the dissemination of soil microorganisms
Solifuges	Although they belong to the class Arachnida and look like spiders, solifuges an order on their own as there are not true spiders or scorpions. Moderate to large-sized, solifuges or "sun spiders" are omnivorous, opportunistic feeders displaying an aggressive hunting behaviour

Appendix 1: List of Arthropods Recorded at Península Valdés

Subphylum	Class	Order	Family	Species
Chelicerata	Arachnida	Araneae	Amphinectidae	
			Gnaphosidae	
			Linyphiidae	
			Lycosidae	
			Philodromidae	
			Prodidomidae	
			Salticidae	
			Sicariidae	
			Sparassidae	
			Theriididae	
			Thomisidae	
			Trachelidae	
			Zodariidae	
		Pseudoscorpiones	Family 1	
		Scorpiones	Bothriuridae	*Bothriurus burmeisteri*
				Urophonius martinezi
				Brachistosternus alienus
				Brachistosternus angustimanus
		Solifuga	Ammotrechidae	
			Mummusidae	
Atelocerata	Insecta	Archaeognatha	Machilidae	
		Coleoptera (beetles)	Anobiidae	*Anobiidae* sp1
				Anobiidae sp1
			Anticidae	*Anthicidae* sp1
				Anthicidae sp2
				Anthicidae sp3
				Anthicidae sp4
			Apionidae	*Apion* sp1
			Carabidae	*Trirammatus (P.) vagans*
				Metius malachiticus
				Metius latemarginatus
				Metius caudatus
				Cnemalobus litoralis

(continued)

(continued)

Subphylum	Class	Order	Family	Species
				Metius harpaloides
				Metius sp1
				Notiobia sp1
				Pseudoanisotarsus nicki
		Coleoptera (beetles)	Carabidae	*Metius* sp2
				Trirammatus (F) striatula
			Cerambycidae	*Cerambycidae* sp1
			Chrysomelidae	*Cryptocephalus patagonicus*
			Cleridae	*Cleridae* sp1
			Coccinellidae	*Coccinellidae* sp2
			Curculionidae	*Entiminae* sp1
				Eurymetopus oblongus
				Pantomorus ruizi
				Listroderes costrirrostris
Atelocerata	Insecta			*Chryptorhynchinae* sp1
			Elateridae	*Conoderus* sp1
				Conoderinae sp3
				Conoderus sp2
			Heteroceridae	*Efflagitatus* sp1
			Histeridae	*Euspilotus lacordaiere*
				Euspilotus sp3
				Euspilotus sp4
			Meloidae	*Epicauta* sp1
			Nitidulidae	*Nitidulidae* sp1
			Pselaphidae	*Pselaphidae* sp1
				Pselaphidae sp2
			Scaphidiidae	*Scaphidiidae* sp1
			Scarabaeidae	*Alidiostoma* sp1
				Scarabeidae sp2
				Eucranium dentrifrons
				Scylophagus lacordaire

(continued)

(continued)

Subphylum	Class	Order	Family	Species
				Scylophagus patagonicus
			Staphylinidae	*Carpelimus* sp1
				Staphilinidae sp1
				Staphilinidae sp3
				Staphilinidae sp5
			Tenebrionidae	*Hylithus tentyroides*
				Epipedonota cristallisata
				Mitragenius araneiformis
				Nyctelia circumundata
				Nyctelia darwini
				Nyctelia dorsata
				Nyctelia nodosa
				Patagonogenius collaris
				Patagonogenius quadricollis
				Psectracelis sulcicollis
				Pimelosomus sphaericus
		Coleoptera	Tenebrionidae	*Calymmophorus patagonicus*
				Calymmophorus peninsularis
				Plathestes kuscheli
				Praocis argentina
				Praocis sellata granulipennis
				Praocis sellata peninsularis
				Praocis (*Hemipraocis*) sp.
				Ecnomoderes bruchi
				Salax lacordairei
Atelocerata	Insecta			*Blapstinus punctulatus*
				Emmallodera crenatocostata
				Emmallodera hirtipes

(continued)

(continued)

Subphylum	Class	Order	Family	Species
				Leptynoderes strangulata
				Rhypasma cuadricoldis
			Trogidae	*Polynoncus* sp1
		Collembola	Family 1	
			Sminthuridae	
		Dictyoptera	Blatellidae	
			Blattidae	
			Mantidae	
		Heteroptera–Hemiptera	Blissidae	*Blissus parasitaster*
				Blissus sp1
			Cydnidae	*Cydnidae* sp2
				Cydnidae sp1
			Lygaeidae	*Nysius simulans*
				Lygaeus alboornatus
			Miridae	*Miridae* sp1
				Miridae sp2
				Miridae sp7
				Miridae sp3
				Valdesiana curiosa
				Miridae sp5
				Miridae sp6
			Nabidae	*Pagasa* sp.
			Oxycarenidae	*Anomaloptera patagonica*
			Pentatomidae	*Pentatomidae* sp1
			Reduviidae	*Reduvidae* sp3
				Reduvidae sp4
			Rhopalidae	*Rhopalidae* sp2
				Rhopalidae sp1
			Rhyparochromidae	*Erlacda argentinensis*
				Rhyparochromidae sp1
		Heteroptera	Rhyparochromidae	*Lethaeini* sp1
			Scutelleridae	*Scutelleridae* sp1
		Hymenoptera	Formicidae (ants)	*Pheidole aberrans*
				Acromyrmex striatus
				Pheidole bergi

(continued)

(continued)

Subphylum	Class	Order	Family	Species
				Solenopsis patagonica
				Pheidole cf. *P. spininodis*
				Solenopsis sp1
				Acromyrmex sp4
				Acromyrmex lobicornis
				Pheidole cf. *P. spininodis*
				Acromyrmex cf. *A. ambigeis*
Atelocerata	Insecta			*Mycetophyllax* sp1
				Solenopsis sp4
				Solenopsis sp6
				Pogonomyrmex rastratus
				Solenopsis sp7
				Forelius chalybaeus
				Dorymyrmex breviscapis
				Dorymyrmex cf. *D. ensifer*
				Dorymyrmex hexanguis
				Forelius cf. *F. grandis*
				Dorymyrmex cf. *D. silvestris*
				Dorymyrmex wolffhügeli
				Forelius sp2
				Camponotus punctulatus
				Brachymyrmex sp2
				Brachymyrmex sp1
			Mutillidae	
			Pompilidae	
			Several families non-determined	
		Isoptera	Kalotermitidae	
			Termitidae	
		Neuroptera	Myrmeleontidae	

(continued)

(continued)

Subphylum	Class	Order	Family	Species
		Orthoptera	Acrididae	
			Grillidae	
			Ommexechidae	
			Proscopidae	
		Phasmatodea	Phasmidae	
		Psocoptera	Family 1	
		Siphonaptera	Family 1	
		Thysanoptera	Phlaoeothripidae	

Appendix 2: List of Reptiles Recorded at Península Valdés

Order	Family	Name	Species
Squamata	Phyllodactylidae	Darwin's gecko	*Homonota darwinii*
	Leiosauridae	Great lizard	*Leiosaurus bellii*
	Liolaemidae	Black Head Lizard	*Liolaemus melanops*
		Darwin's lizard	*Liolaemus darwinii*
		Slender lizard	*Liolaemus gracilis*
	Viperidae	Snub-nosed yarará viper	*Bothrops ammodytoides*
	Dipsadidae	Patagonian green racer	*Phylodrias patagonicus*
		Mousehole snake	*Phylodrias trilineata*
		False yarará viper	*Pseudotomodon trigonatus*
		Brown Musurana	*Clelia rustica*

Appendix 3: List of Terrestrial and Inland-Wetlands Birds Recorded at Península Valdés

Order	Family	Name	Species
Rheiformes	Rheidae	Lesser Rhea	*Rhea pennata*
Tinamiformes	Tinamidae	Darwin's Nothura	*Nothura darwinii*
		Elegant Crested-Tinamou	*Eudromia elegans*
Anseriformes	Anatidae	Coscoroba Swan	*Coscoroba coscoroba*
		Black-necked Swan	*Cygnus melancoryphus*[ND]

(continued)

(continued)

Order	Family	Name	Species
		Upland Goose	*Chloephaga picta*[AUS]
		Ashy-headed Goose	*Chloephaga poliocephala*[AUS]
		Ruddy-headed Goose	*Chloephaga rubidiceps*[AUS]
		Crested Duck	*Lophonetta specularioides*
		Chiloe Wigeon	*Anas sibilatrix*[ND]
		Yellow-billed Pintail	*Anas geórgica*
		Yellow-billed Teal	*Anas flavirostris*
		Red Shoveler	*Anas platalea*
		Cinnamon Teal	*Anas cyanoptera*
		White-cheeked Pintail	*Anas bahamensis*
		Rosy-billed Pochard	*Netta peposaca*
		Black-headed Duck	*Heteronetta atricapilla*
		Lake Duck	*Oxyura vittata*
Podicipediformes	Podicipedidae	White-tufted Grebe	*Rollandia rolland*
		Silvery Grebe	*Podiceps occipitalis*
		Great Grebe	*Podiceps major*
Phoenicopteriformes	Phoenicopteridae	Chilean Flamingo	*Phoenicopterus chilensis*
Ciconiiformes	Ciconiidae	Maguari Stork	*Ciconia maguari*[a]
Pelecaniformes	Ardeidae	Great Egret	*Ardea alba*
		Snowy Egret	*Egretta thula*
		Cattle Egret	*Bubulcus ibis*
		Cocoi Heron	*Ardea cocoi*
		Black-crowned Night-Heron	*Nycticorax nycticorax*
	Threskiornithidae	Black-faced Ibis	*Theristicus melanopis*[AUS]
		White-faced Ibis	*Plegadis chihi*
		Roseate Spoonbill	*Platalea ajaja*[a]
Cathartiformes	Cathartidae	Turkey Vulture	*Cathartes aura*[ND]
		Black Vulture	*Coragyps atratus*[a]
Accipitriformes	Accipitridae	White-tailed Kite	*Elanus leucurus*
		Long-winged Harrier	*Circus buffoni*
		Cinereous Harrier	*Circus cinereus*
		Black-chested Buzzard-Eagle	*Geranoaetus melanoleucus*
		White-tailed Hawk	*Geranoaetus albicaudatus*

(continued)

(continued)

Order	Family	Name	Species
Accipitriformes	Accipitridae	Variable Hawk	*Geranoaetus polyosoma*
Gruiformes	Rallidae	Plumbeous Rail	*Pardirallus sanguinolentus*
		Austral Rail	*Rallus antarcticus*
		Red-gartered Coot	*Fulica armillata*
		White-winged Coot	*Fulica leucoptera*
		Red-fronted Coot	*Fulica rufifrons*
Charadriiformes	Recurvirostridae	Black-necked Stilt	*Himantopus mexicanus*
	Charadriidae	Southern Lapwing	*Vanellus chilensis*
		American Golden-Plover	*Pluvialis dominica*[NEA]
		Black-bellied Plover	*Pluvialis squatarola*[NEA]
		Tawny-throated Dotterel	*Oreopholus ruficollis*[AUS]
		Two-banded Plover	*Charadrius falklandicus*[AUS]
		Rufous-chested Dotterel	*Charadrius modestus*[AUS]
	Scolopacidae	Greater Yellowlegs	*Tringa melanoleuca*[NEA]
		Lesser Yellowlegs	*Tringa flavipes*[NEA]
		Pectoral Sandpiper	*Calidris melanotos*[NEA]
		Baird's Sandpiper	*Calidris bairdii* [NEA]
		White-rumped Sandpiper	*Calidris fuscicollis*[NEA]
		Hudsonian Godwit	*Limosa haemastica*[NEA]
		Wilson's Phalarope	*Phalaropus tricolor*[NEA]
	Thinocoridae	Least Seedsnipe	*Thinocorus rumicivorus*[AUS]
		Grey-breasted Seedsnipe	*Thinocorus orbignyianus*
	Laridae	Kelp Gull	*Larus dominicanus*
		Brown-hooded Gull	*Chroicocephalus maculipennis*
Columbiformes	Columbidae	Rock Pigeon	*Columba livia*[1]
		Eared Dove	*Zenaida auriculata*
		Picui Ground Dove	*Columbina picui*
Cuculiformes	Cuculidae	Guira Cuckoo	*Guira guira*
		Dark-billed Cuckoo	*Coccyzus melacoryphus*[NEO,a]
Strigiformes	Tytonidae	Barn Owl	*Tyto alba*
	Strigidae	Burrowing Owl	*Athene cunicularia*
		Great Horned Owl	*Bubo virginianus*
		Short-eared Owl	*Asio flammeus*

(continued)

(continued)

Order	Family	Name	Species
Caprimulgiformes	Caprimulgidae	Band-winged Nightjar	*Systellura longirostris*
Falconiformes	Falconidae	Southern Caracara	*Caracara plancus*
		Chimango Caracara	*Milvago chimango*
		Peregrine Falcon	*Falco peregrinus*
		Aplomado Falcon	*Falco femoralis*
		American Kestrel	*Falco sparverius*
Psittaciformes	Psittacidae	Burrowing Parakeet	*Cyanoliseus patagonus*
		Monk Parakeet	*Myiopsitta monachus*
Passeriformes	Furnariidae	Common Miner	*Geositta cunicularia*
		Scale-throated Earthcreeper	*Upucerthia dumetaria*[AUS]
Passeriformes	Furnariidae	Band-tailed Earthcreeper	*Ochetorhynchus phoenicurus*
		Buff-winged Cinclodes	*Cinclodes fuscus*[AUS]
		Rufous Hornero	*Furnarius rufus*
		Wren-like Rushbird	*Phleocryptes melanops*
		Sharp-billed Canastero	*Asthenes pyrrholeuca*[AUS]
		Cordilleran Canastero	*Asthenes modesta*
		Patagonian Canastero	*Pseudasthenes patagónica*
		White-throated Cacholote	*Pseudoseisura gutturalis*
		Plain-mantled Tit-Spinetail	*Leptasthenura aegithaloides*
	Tyrannidae	Grey-bellied Shrike-Tyrant	*Agriornis micropterus*[AUS]
		Lesser Shrike-Tyrant	*Agriornis murinus*[AUS]
		Chocolate-vented Tyrant	*Neoxolmis rufiventris*[AUS]
		Rusty-backed Monjita	*Xolmis rubetra*[AUS]
		White Monjita	*Xolmis irupero*
		Black-crowned Monjita	*Xolmis coronatus*[AUS]
		Austral Negrito	*Lessonia rufa*[AUS]
		Dark-faced Ground-Tyrant	*Muscisaxicola maclovianus*[AUS]
		Spectacled Tyrant	*Hymenops perspicillatus*
		White-winged Black-Tyrant	*Knipolegus aterrimus*
		Hudson's Black-Tyrant	*Knipolegus hudsoni*[AUS,a]
		Many-coloured Rush Tyrant	*Tachuris rubrigastra*
		Great Kiskadee	*Pitangus sulphuratus*

(continued)

(continued)

Order	Family	Name	Species
		Fork-tailed Flycatcher	*Tyrannus savana*^{NEO}
		Vermilion Flycatcher	*Pyrocephalus rubinus*^{NEO}
		Yellow-billed Tit-Tyrant	*Anairetes flavirostris*
		Tufted Tit-Tyrant	*Anairetes parulus*
		Straneck's Tyrannulet	*Serpophaga griseicapilla*^{a}
		White-crested Elaenia	*Elaenia albiceps*
	Cotingidae	White-tipped Plantcutter	*Phytotoma rutila*
	Hirundinidae	Barn Swallow	*Hirundo rustica*^{NEA,a}
		Southern Martin	*Progne elegans*^{NEO}
		Chilean Swallow	*Tachycineta meyeni*^{AUS}
		Blue-and-white Swallow	*Pygochelidon cyanoleuca*
	Troglodytidae	House Wren	*Troglodytes aedon*
	Turdidae	Austral Thrush	*Turdus falcklandii*
		Chiguanco Thrush	*Turdus chiguanco*
	Motacillidae	Short-billed Pipit	*Anthus furcatus*
		Correndera Pipit	*Anthus correndera*
		Hellmayr's Pipit	*Anthus hellmayri*
	Mimidae	White-banded Mockingbird	*Mimus triurus*^{AUS}
		Chalk-browed Mockingbird	*Mimus saturninus*
		Patagonian Mockingbird	*Mimus patagonicus*
	Thraupidae	Common Diuca-Finch	*Diuca diuca*^{AUS}
		Grassland Yellow-Finch	*Sicalis luteola*
Passeriformes	Thraupidae	Patagonian Yellow-Finch	*Sicalis lebruni*
		Grey-hooded Sierra-Finch	*Phrygilus gayi*
		Mourning Sierra-Finch	*Phrygilus fruticeti*
		Carbonated Sierra-Finch	*Phrygilus carbonarius*
		Yellow Cardinal	*Gubernatrix cristata*^{a}
	Emberizidae	Rufous-collared Sparrow	*Zonotrichia capensis*
	Icteridae	Shiny Cowbird	*Molothrus bonariensis*

(continued)

(continued)

Order	Family	Name	Species
		Greyish Baywing	*Agelaioides badius*
		Yellow-winged Blackbird	*Agelasticus thilius*
		Long-tailed Meadowlark	*Sturnella loyca*
	Fringillidae	Black-chinned Siskin	*Sporagra barbata*
		Hooded Siskin	*Sporagra magellanica*[a]
	Passeridae	House Sparrow	*Passer domesticus*[I]

NEA Neartic Migrant, *NEO* Neotropical Migrant, *AUS* Austral Migrant, *ND* Northward Dispersion, *I* Introduced
[a]Rarely observed

Appendix 4: List of Native Terrestrial Mammals Recorded at Península Valdés

Order	Family	Name	Species
Didelphimorphia	Didelphidae	Pallid fat-tailed opossum	*Thylamys pallidior*
Cingulata	Dasypodidae	Big hairy armadillo	*Chaetophractus villosus*
		Pichi	*Zaedyus pichiy*
Chiroptera	Vespertilionidae	Small big-eared brown bat	*Histiotus montanus*
	Molossidae	Brazilian free-tailed bat	*Tadarida brasiliensis*
Carnivora	Canidae	Culpeo fox	*Lycalopex culpaeus*
		South american grey fox	*Lycalopex gymnocercus*
	Felidae	Geoffroy's cat	*Leopardus geoffroyi*
		Pampas cat	*Leopardus colocolo*
		Puma	*Puma concolor*
	Mephitidae	Molina's hog-nosed skunk	*Conepatus chinga*
	Mustelidae	Lesser grison	*Galictis cuja*
		Patagonian weasel	*Lyncodon patagonicus*[a]
Artiodactyla	Camelidae	Guanaco	*Lama guanicoe*
Rodentia	Cricetidae	Dolores grass mouse	*Akodon dolores*
		Intelligent grass mouse	*Akodon iniscatus*
		Drylands vesper mouse	*Calomys musculinus*
		Eastern patagonian laucha	*Eligmodontia typus*
		Grey leaf-eared mouse	*Graomys griseoflavus*

(continued)

(continued)

Order	Family	Name	Species
		Bunny rat	*Reithrodon auritus*
	Caviidae	Common yellow-toothed cavy	*Galea leucoblephara*
		Southern mountain cavy	*Microcavia australis*
		Mara	*Dolichotis patagonum*
	Ctenomyidae	Tuco-tuco	*Ctenomys* sp.

[a]Presence needs to be confirmed

References

Abba AM, Nabte MJ, Udrizar Sauthier DE (2010) New data on armadillos (Xenarthra: Dasypodidae) for central Patagonia, Argentina. Edentata 11:11–17

Abba AM, Poljak S, Gabrielli M, Teta P et al (2014) Armored invaders in Patagonia: recent southward dispersion of armadillos (Cingulata, Dasypodidae). Mastozool Neotrop 21:311–318

Alonso Roldán V (2012) Patrones de distribución espacial de la mara (*Dolichotis patagonum*) a distintas escalas. Dissertation, Universidad Nacional del Sur. Bahía Blanca, Argentina

Andelt WF (2004) Use of livestock guarding animals to reduce predation on livestock. Sheep Goat Res J. Paper 3. http://digitalcommons.unl.edu/icwdmsheepgoat/3

Andrew MH (1988) Grazing impact in relation to livestock watering points. Trends Ecol Evol 3:336–339

Antún M, Baldi R, Bandieri L (2015) Análisis de la variación espacial de guanacos (Lama guanicoe) y ovinos (Ovis aries) mediante el uso de modelos de superficie de densidad en Península Valdés. Paper presented at the III Jornadas Patagónicas de Biología, Universidad Nacional de la Patagonia San Juan Bosco, Trelew, 23–25 Sept 2015

Avila LJ (2009) Reptilia, Squamata, Colubridae, *Liophis saggitifer saggitifer*: Distribution extension. Check List 5(3):712–713

Avila LJ, Kozykariski M, Feltrin N et al (2007) Geographic distribution: *Amphisbaena plumbea*. Herpetol Review 38(2):217

Avila LJ, Morando M, Perez DR (2001) New records and natural history notes for lizards and snakes from Patagonia, Argentina. Herpetol Rev 32(1):64

Ayal Y (2007) Trophic structure and the role of predation in shaping hot desert communities. J Arid Environ 68:171–187

Baldi R (2007) Breeding success of the endemic mara *Dolichotis patagonum* in relation to habitat selection: conservation implications. J Arid Environ 68:9–19

Baldi R, Campagna C, Saba S (1997) Abundancia y distribución del guanaco (Lama guanicoe) en el NE del Chubut, Patagonia Argentina. Mastozool Neotrop 4(1):5–15

Baldi R, Albon SD, Elston D (2001) Guanacos and sheep: evidence for continuing competition in arid Patagonia. Oecologia 129:561–570

Baldi R, Pelliza-Sbriller A, Elston D et al (2004) High potential for competition between guanacos and sheep in Patagonia. J Wildlife Manag 68(4):924–938

Baldi R, Burgi MV, Marino A (2009) Factores que afectan la distribución y abundancia de guanacos, maras y choiques en la Península Valdés: hacia un modelo de manejo y conservación de herbívoros silvestres en áreas protegidas de Chubut. Report submitted to the Ministerio de Educación and the Dirección de Conservación y Áreas Protegidas de la Provincia de Chubut

Baldi R, Novaro AJ, Funes M et al (2010) Guanaco management in Patagonian rangelands: a conservation opportunity on the brink of collapse. In: du Toit J, Kock R, Deutsch J (eds) Wild rangelands. Conserving wildlife while maintaining livestock in semi-arid ecosystems. Blackwell Publishing, Oxford, pp 266–290

Baldi R, Pirronitto A, Burgi MV et al (2015) Abundance estimates of the lesser rhea *Rhea pennata pennata* in the Argentine Patagonia: conservation implications. Front Ecol Evol 3(135). Doi:10.3389/fevo.2015.00135

Balmford A, Gaston KJ, Blyth S et al (2003) Global variation in terrestrial conservation costs, conservation benefits, and unmet conservation needs. Proc Natl Acad Sci 100:1046–1050

BirdLife International (2012) The IUCN Red List of Threatened Species 2012. http://www.iucnredlist.org/

Blanco PD, Rostagno CM, Del Valle HF et al (2008) Grazing impacts in vegetated dune fields: predictions from spatial pattern analysis. Rangeland Ecology and Management 61:194–203

Bromham L, Cardillo M, Bennett AF et al (1999) Effects of stock grazing on the ground invertebrate fauna of woodland remnants. Aust J Ecol 24:199–207

Burgi MV, Marino A, Rodríguez MV et al (2012) Response of guanacos to changes in land management in Península Valdés, Argentine Patagonia. Conservation implications. Oryx 46:99–105

Burkart R, Bárbaro NO, Sánchez RO et al (1999) Eco-regiones de la Argentina. Administración de Parques Nacionales

Carpintero DL, Dellapé PM, Cheli GH (2008) *Valdesiana curiosa*: a remarkable new genus and species of Clivinematini (Heteroptera: Miridae: Deraeocorinae) from Argentina and a key to Argentinean genera and species. Zootaxa 1672:61–68

Carrara R, Cheli GH, Flores GE (2011) Patrones biogeográficos de los tenebriónidos epígeos (Coleoptera: Tenebrionidae) del Área Natural Protegida Península Valdés, Argentina: implicaciones para su conservación. Rev Mex Biodiver 82:1297–1310

Carrera HM, Avila LJ (2008a) Natural history notes: predation/scavenging: *Oxyrhopus rhombifer bachmanni*. Herpetol Review 39(3):356–357

Carrera HM, Avila LJ (2008b) Geographic distribution: *Oxyrhopus rhombifer bachmanni*. Herpetol Rev 39(2):208–209

Carrera M, Udrizar Sauthier DE (2011) Los cánidos (Mammalia, Carnivora) del Holoceno tardío del noreste de la provincia del Chubut, Argentina. Paper presented at the II Jornadas Patagónicas de Biología, Universidad Nacional de la Patagonia San Juan Bosco, Trelew 20–24 June 2011

Carrera M, Nabte MJ, Udrizar Sauthier DE (2012) Distribución geográfica, historia natural y conservación del hurón menor *Galictis cuja* (Molina, 1782) (Carnivora, Mustelidae) en la Patagonia central, Argentina. Rev Mex Biodiv 83:1252–1257

Cei JM (1986) Reptiles del centro, centro-oeste y sur de la Argentina. Herpetofauna de las zonas áridas y semiáridas. Monografie IV. Museo Regionale di Scienze Naturali. Torino

Cheli GH (2009) Efectos del disturbio por pastoreo ovino sobre la comunidad de artrópodos epígeos en Península Valdés (Chubut, Argentina). Dissertation, Universidad Nacional del Comahue, Argentina

Cheli GH, Corley J, Bruzzone O et al (2010) The ground-dwelling arthropod community of Península Valdés (Patagonia, Argentina). J Insect Sci 10:50. www.insectsicence.org/10.50

Cheli GH, Flores GE, Martínez Román N et al (2013) Tenebrionid beetle's dataset (Coleoptera, Tenebrionidae) from Península Valdés (Chubut, Argentina). ZooKeys 364:93–108. http://www.pensoft.net/journals/zookeys/article/4761/abstract/a-tenebrionid-beetle

Codesido M, Beeskow AM, Blanco P et al (2005) Relevamiento ambiental de la "Reserva de Vida Silvestre San Pablo de Valdés". Fundación Vida Silvestre Argentina

Coronato (2010) Le rôle de l'élevage ovin dans la construction du territoire de la Patagonie. Dissertation, Ecole doctorale ABIES, Paris Institute of Technology

Crawford CS (1988) Nutrition and habitat selection in desert detritivores. J Arid Environ 14:111–121

Crespo JA (1974) Comentarios sobre nuevas localidades para mamíferos de Argentina y de Bolivia. Rev Mus Arg Cs Nat 11:1–31

Crespo JA, de Carlo JL (1963) Estudio ecológico de una población de zorros colorados *Dusicyon culpaeus culpaeus* (Molina) en el oeste de la Provincia de Neuquén. Rev Mus Arg Cs Nat 1(1):56

Cruz FB, Schulte JA II, Bellagamba P (1999) New distributional records and natural history notes for reptiles from southern Argentina. Herpetol Rev 30(3):182

D'Agostino RL, Llanos R, Udrizar Sauthier D (2015) Nuevos registros y conservación de mamíferos carnívoros en el Área Natural Protegida Península Valdés y alrededores. Paper presented at the III Jornadas Patagónicas de Biología, Universidad Nacional de la Patagonia San Juan Bosco, Trelew, 23–25 Sept 2015

Daciuk J (1977) Notas faunísticas y bioecológicas de Península Valdés y Patagonia. VI. Observaciones sobre áreas de nidificación de la avifauna del litoral marítimo patagónico (Provincias de Chubut y Santa Cruz, Rep. Argentina). Hornero 11:361–376

Daciuk J (1978) Notas faunisticas y bioecológicas de la Península Valdés y Patagonia. XXIII. Estudio bioecológico y etológico preliminar del ñandu petiso patagonico y de los tinámidos de la Península Valdés, Chubut, Argentina (Aves, Rheidae y Tinamidae). Physis 38(95):69–85

Daciuk J (1979) Notas faunísticas y bioecológicas de Península Valdés y Patagonia. XXII. Elenco sistemático de las aves colectadas y observadas en la Península Valdés y litoral marítimo de Chubut (República Argentina). Acta Zool Lilloana 25:643–666

Daciuk J, Miranda ME (1980) Notas faunísticas y bioecológicas de Península Valdés y Patagonia. XXV. Batraco—Herpetofauna de la Península Valdés y costas patagónicas (Río Negro, Chubut, Santa Cruz y Tierra del Fuego, República Argentina). Neotropica 26(75):99–115

De la Peña MR (2013) Citas, observaciones y distribución de aves argentinas: Edición ampliada. Serie Naturaleza, Conservación y Sociedad N° 7, Ediciones Biológica. Museo Provincial de Ciencias Naturales Florentino Ameghino

De Tommaso D, Formoso AE, Teta P et al (2014) Distribución geográfica de *Calomys musculinus* (Rodentia, Sigmodontinae) en Patagonia. Mastozool Neotrop 21(1):121–127

Dellapé PM, Cheli G (2007) First record of the genus *Terenocoris* (Heteroptera: Rhyparochromidae: Antillocorini) from Argentina and Bolivia. Rev Soc Entomol Arg 65 (3–4):87–88

Dennis P, Young MR, Gordon IJ (1998) Distribution and abundance of small insects and arachnids in relation to structural heterogeneity of grazed, indigenous grasslands. Ecol Entomol 23:253–264

Evolución Existencia de Ganado Ovino 2005–2014 (2016). Dirección General de Estadística y Censos de la Provincia de Chubut. http://www.estadistica.chubut.gov.ar/home/index.php?option=com_content&view=article&id=331&Itemid=276. Accessed 6 Apr 2016

Elissalde NO, Miravalles HR (1983) Evaluación de los campos de pastoreo de Península Valdés. Informe 70, Centro Nacional Patagónico (CONICET), Puerto Madryn, Argentina

Farji-Brener A, Corley JC, Bettinelli J (2002) The effects of fire on ant communities in northwestern Patagonia: the importance of habitat structure and regional context. Divers Distrib 8:235–243

Fernández C, Baldi R (2014) Hábitos alimentarios del puma (*Puma concolor*) e incidencia de la depredación en la mortandad de guanacos (*Lama guanicoe*) en el noreste de la Patagonia. Mastozool Neotrop 21(2):331–338

Flores GE (1998) Tenebrionidae. In: Morrone JJ, Coscarón S (eds) Biodiversidad de Artrópodos Argentinos, Volumen 1, Ediciones Sur, La Plata, Argentina, pp 232–240

Flores GE, Carrara R, Cheli GH (2011) Three new Praociini (Coleoptera: Tenebrionidae) from Península Valdés (Argentina), with zoogeographical and ecological remarks. Zootaxa 2965:39–50

Folgarait PJ, Sala OE (2002) Granivory rates by rodents, insects, and birds at different microsites in the Patagonian steppe. Ecography 25:417–427

Formoso AE, Martin GM, Teta P et al (2015) Regional extinctions and Quaternary shifts in the geographic range of *Lestodelphys halli*, the southernmost living marsupial: clues for its conservation. PlosOne. Doi:10.1371/journal.pone.0132130

Formoso AE, Udrizar Sauthier DE, Teta P et al (2011) Dense-sampling reveals a complex distributional pattern between the southernmost marsupials *Lestodelphys* Tate, 1934 and *Thylamys* Gray, 1843 in Patagonia, Argentina. Mammalia 75:371–379

Frutos N, Camporro L, Avila LJ (2005) Geographic distribution: *Cnemidophorus longicauda*. Herpetol Rev 36(3):336

Gardner SM, Cabido MR, Valladares GR et al (1995) The influence of habitat structure on arthropod diversity in Argentine semi-arid Chaco forest. J Veg Sci 6:349–356

Haene E (2004) La avifauna de las eco-regiones de la Patagonia y Antártida. In: Narosky T, Yzurieta D (eds) Aves de la Patagonia y Antártida, Guía para su reconocimiento. Vazquez Mazzini Editores, Buenos Aires, pp 17–29

Harris G (2008) Guía de aves y mamíferos de la costa patagónica. El Ateneo y Ecocentro, Buenos Aires

Jarman PJ (1974) The social organisation of antelope in relation to their ecology. Behaviour 48:215–267

Kovacs C, Kovacs O, Kovacs Z et al (2005) Manual ilustrado de las aves de la Patagonia, Antártida Argentina e Islas del Atlántico Sur. Authors edition

Lagos SJ (2004) Diversidad biológica de las comunidades epígeas de artrópodos en áreas pastoreadas y no pastoreadas del Monte (Argentina). Dissertation, Universidad Nacional de Cuyo, Argentina

León RJC, Bran D, Collantes M et al (1998) Grandes unidades de vegetación de la Patagonia extra andina. In: Oesterheld M, Aguiar MR, Paruelo JM (eds) Ecosistemas patagónicos. Ecología Austral, pp 125–144

Llanos FA, Failla M, García GJ et al (2011) Birds from the endangered Monte, the Steppes and Coastal biomes of the province of Río Negro, northern Patagonia, Argentina. Check List 7:782–797

Lopez de Casenave J (2001) Estructura gremial y organización de un ensamble de aves del desierto del Monte. Dissertation, Universidad de Buenos Aires, Argentina

López-Lanús B, Grilli P, Coconier E et al (2008) Categorización de las aves de la Argentina según su estado de conservación. Informe de Aves Argentinas/AOP y Secretaría de Ambiente y Desarrollo Sustentable. Buenos Aires, Argentina

Magurran AE (2004) Measuring Biological Diversity. Blackwell Publishing, Oxford, UK

Marino A, Rodríguez MV, Pazos GE (2016) Self-limitation of population density in resource-defense ungulates. Behav Ecol (in press)

Martínez FJ (2013) Estructura de las comunidades de artrópodos epigeos en ambientes representativos de Península Valdés. Dissertation, Universidad Nacional de la Patagonia San Juan Bosco, Argentina

Martínez Román N (2014) Composición taxonómica y estructura de las comunidades de artrópodos epígeos en áreas quemadas del noreste de Chubut. Dissertation, Universidad Nacional de la Patagonia, Argentina

Massoia E, Vetrano AS, La Rossa FR (1988) Análisis de regurgitados de *Athene cunicularia* de Península Valdez, Departamento Biedma, provincia de Chubut. APRONA 4:4–13

Mazar Barnett J, della Seta M, Imberti S et al (1998) Notes on the rediscovery of the Austral Rail *Rallus antarcticus* in Santa Cruz, Argentina. Cotinga 10:96–101

Mazía NC, Chaneton E, Kitzberger T (2006) Small-scale habitat use and assemblage structure of ground-dwelling beetles in a Patagonian shrub steppe. J Arid Environ 67:177–194

Milchunas DG, Lauenroth WK (1993) Quantitative effects of grazing on vegetation and soils over a global range of environments. Ecol Monogr 63:327–366

Minoli I, Morando M, Avila LJ (2015) Reptiles of Chubut province, Argentina: richness, diversity, conservation status and geographic distribution maps. ZooKeys 498:103–126

Molina SI, Valladares GR, Gardner S et al (1999) The effects of logging and grazing on the insect community associated with a semi-arid Chaco forest in central Argentina. J Arid Environ 42:29–42

Nabte MJ (2010) Desarrollo de criterios ecológicos para la conservación de mamíferos terrestres en Península Valdés. Doctorate thesis, Universidad Nacional de Mar del Plata, Argentina

Nabte MJ, Pardiñas UFJ, Saba SL (2008) The diet of the Burrowing Owl, *Athene cunicularia*, in the arid lands of northeastern Patagonia, Argentina. J Arid Environ 72:1526–1530

Nabte MJ, Andrade A, Monjeau JA et al (2009) Mammalia, rodentia, Sigmodontinae, *Akodon molinae* (Contreras, 1968): new locality records. Check List 5(2):320–324

Nabte MJ, Marino AI, Rodríguez MV et al (2013) Range management affects native ungulate populations in Península Valdés, a World Natural Heritage. PLoS One 8(2):e55655

Narosky T, Izurieta D (2004) Aves de la Patagonia y Antártida. Guía para su reconocimiento. Vazquez Mazzini Editores

Novaro AJ (1997) Pseudalopex culpaeus. Mammalian Species 518:1–8

Novaro AJ, Funes MC, Walker RS (2005) An empirical test of source-sink dynamics induced by hunting. J Appl Ecol 56:709–718

Novaro AJ, González A, Pailacura O et al (2016) Manejo del conflicto entre carnívoros y ganadería en Patagonia utilizando perros mestizos protectores de ganado. Mastozool Neotrop (in press)

Ojanguren-Affilastro AA, Cheli G (2009) New data on the genus *Urophonius* in Patagonia with a description of a new species of the *exochus* group (Scorpiones, Bothriuridae). J Arachnol 37:346–356

Oliver I, Beattie AJ (1993) A possible method for the rapid assessment of biodiversity. Cons Biol 7(3):562–568

Oliver I, Beattie AJ (1996) Invertebrate morphospecies as surrogates for species: a case study. Cons Biol 10(1):99–109

Oliver I, Beattie AJ (1997) Future taxonomic partnerships: reply to Goldstein. Cons Biol 11(2):575–576

Pardiñas UFJ (2009) El género *Akodon* (Rodentia: Cricetidae) en Patagonia: estado actual de su conocimiento. Mastozool Neotrop 135–151

Pardiñas UFJ, Cirignoli S, Podestá DH (2001) Nuevos micromamíferos registrados en la Península de Valdés (provincia de Chubut, Argentina). Neotrópica 47:101–102

Pedrana J, Bustamante J, Travaini A et al (2010) Factors influencing guanaco distribution in southern Argentine Patagonia and implications for its sustainable use. Biodivers Conserv 19:3499–3512

Pereira JA, Novaro AJ (2014) Habitat-specific demography and conservation of Geoffroy's cats in a human dominated landscape. J Mammal 95:1–10

Pérez CHF, Delhey K, Petracci PF (2006) Aves nuevas o poco frecuentes del norte de la Patagonia Argentina. Nuestras Aves 52:25–29

Pezzola A, Winschel C, Sánchez R (2004) Estudio multitemporal de la degradación del monte nativo en el partido de Patagones-Buenos Aires. INTA, Boletín Técnico 12

Pik AJ, Oliver IAN, Beattie AJ (1999) Taxonomic sufficiency in ecological studies of terrestrial invertebrates. Aust J Ecol 24(5):555–562

Plan de Manejo del Área Protegida Sistema Península Valdés (1999) Organismo Provincial de Turismo. Gobierno de la Provincia del Chubut, Argentina

Pol RG, Camín SR, Astié AA (2006) Situación ambiental en la Ecorregión del Monte. In: Brown A et al (eds) Situación ambiental argentina 2005. Fundación Vida Silvestre Argentina, Buenos Aires, pp 227–233

Polis GA (1991) The ecology of desert communities. University of Arizona Press, Tucson

Prevosti FJ, Ramírez M, Schiaffini M et al (2015) Extinctions in near time: New radiocarbon dates indicate a very recent disappearance of the South American fox *Dusicyon avus* (Carnivora, Canidae). Zool J Linn Soc 116(3):704–720

Pruscini F, Morelli F, Sisti D et al (2014) Breeding passerines communities in the Valdes Peninsula (Patagonia, Argentina). Ornitol Neotrop 25:13–23

Pugnali G, Pearman M, Escudero G et al (2004) New localities for the Austral Rail *Rallus antarcticus* in Argentina, and first record from the Falkland Islands. Cotinga 22:35–37

Remsen Jr JV, Areta JI, Cadena CD et al (2015) A classification of the bird species of South America, American Ornithologists' Union. http://www.museum.lsu.edu/~Remsen/SACCBaseline.html

Rey A, Novaro AJ, Sahores M et al (2012) Demographic effects of live shearing on a guanaco population. Small Ruminant Res 107(2):92–100

Reynolds JF, Stafford Smith DF (2002) Do humans cause deserts? In: Reynolds JF, Stafford Smith DF (eds) Global desertification: do humans cause deserts? Dahlem University Press, Berlin, pp 1–21

Roig FA, Roig-Juñent S, Corbalán V (2009) Biogeography of the Monte Desert. J Arid Environ 73:164–172

Schiaffini MI, Martin GM, Giménez AL et al (2013) Distribution of *Lyncodon patagonicus* (Carnivora, Mustelidae): changes from the Last Glacial Maximum to the present. J Mamm 94:339–350

Scolaro JA (1976) Lista sistemática de reptiles de la Península de Valdés (Chubut). I. Sauria. Physis 35(91):267–271

Scolaro A (2006) Reptiles patagónicos norte: una guía de campo. Universidad Nacional de la Patagonia San Juan Bosco, Comodoro Rivadavia, Argentina

Scolaro JA, Cei JM (1979) The southernmost population of *Elapomorphus bilineatus* in Argentine Patagonia. Copeia 1979(4):745–747

Seymour CL, Dean WRJ (1999) Effects of heavy grazing on invertebrate assemblages in the Succulent Karoo, South Africa. J Arid Environ 43:267–286

Soriano A (1956) Los distritos florísticos de la provincia Patagónica. Rev Invest Agrop 10:323–347

Taber AB, MacDonald DW (1992a) Communal breeding in the mara, *Dolichotis patagonum* (Rodentia: Caviomorpha). J Zool Lon 227:439–452

Taber AB, MacDonald DW (1992b) Spatial organization and monogamy in the mara *Dolichotis patagonum*. J Zool Lon 227:417–438

Trejo A, Lambertucci S (2007) Feeding habits of Barn Owls along a vegetative gradient in northern Patagonia. J Raptor Res 41:277–287

Udrizar Sauthier DE (2009) Los micromamíferos y la evolución ambiental durante el Holoceno en el río Chubut (Chubut, Argentina). Dissertation, Universidad Nacional de La Plata, Argentina

Udrizar Sauthier DE, Nabte MJ (2012) Buscado en la Península Valdés: historia del huroncito patagónico. Biológica 15:129–135

Udrizar Sauthier DE, Pardiñas UFJ (2006) Micromamíferos de Puerto Lobos, Chubut. Argentina. Mastozool Neotrop 13(2):259–262

Udrizar Sauthier D, Frutos N, Avila LJ (2007) Natural History Notes: *Leiosaurus belli*: Predation. Herpetol Rev 38(1):78–79

Udrizar Sauthier DE, Teta P, Formoso AE et al (2013) Bats at the end of the world: new distributional data and fossil records from Patagonia, Argentina. Mammalia 77(3):307–315

Udrizar Sauthier DE, Pardiñas UFJ (2014) Estableciendo límites: distribución geográfica de los micromamíferos terrestres (Rodentia y Didelphimorphia) de Patagonia centro-oriental. Mastozool Neotrop 21(1):79–99

Udrizar Sauthier DE, Formoso AE, Teta P et al (2015) Dense sampling provides a reevaluation of the southern geographic distribution of the cavies Galea and Microcavia (Rodentia). Mammalia. Doi:10.1515/mammalia-2014-0156

Veiga JO, López-Lanús B, Earnshaw A (2010) Expansión del Zorzal Chiguanco (*Turdus chiguanco*) al norte de la Patagonia Argentina: una revisión y aporte de nuevos registros. Nuestras Aves 55:23–25

Walker S, Novaro A (2010) The world's southernmost pumas in Patagonia and the southern Andes. In: Hornocker M, Negri S (eds) Cougar: ecology and conservation. University of Chicago Press, Chicago, pp 91–99

Wildlife Friendly Enterprise Network (2015) http://wildlifefriendly.org/wfen-welcomes-wildlife-conservation-society-argentina-first-certified-wool-producers-from-iconic-peninsula-valdes/

Yokes M, Morando M, Avila LJ et al (2006) Phylogeography and genetic structure in the Cnemidophorus longicauda complex (Squamata, Teiidae). Herpetologica 62(4):424–438

Index

A

Accommodation space, 9, 10, 13
Active Sand Dunes, 194. *See also* dunefields
Aeolian Erosion, 113, 114, 194, 203, 204
Aeolic deposits, 201
Alluvium and colluvium deposits, 38, 218
Amphissbaenian, 268, 270
Andean orogeny, 12
Andes, 1, 2, 6–8, 10, 12, 14–16, 39, 74, 87–91, 234
Animal
　adaptive responses, 132
　distribution, 284
　diversity, 278, 284
　predators, 265, 278, 283, 284
　threats, 263, 275, 279, 285
Annual
　forbs, 137
　grasses, 134, 137, 141
　herbs, 137, 140, 142
Anticyclone
　atlantic, 98
　subtropical, 85, 87, 88
Aquicludes, 216, 219
Aquifers
　confined, 215, 220
　high permeability aquifers, 216–218
　low permeability, 218, 219, 227
　moderate, 216
　phreatic, 215, 220, 227
　semi-confined, 215, 220
Aquifuges, 216
Aquitard, 216, 219
Argentine continental shelf, 1, 2, 5
Argillic horizons, 112, 175–177, 199, 237, 268
Arid lands, 197, 265, 267
Aridisols
　Aquisalids, 161
　Calciargids, 161, 169
　Haplocalcids, 161, 169
　Natrargids, 161, 169
　Natrigypsids, 161
　Petrocalcids, 161
Aridity index, 99
Armadillos, 239, 246, 254, 275–277, 282, 284
Arthropods, 263–268, 279, 280, 284
Assemblages, 28, 32, 51, 57, 61–63, 65, 66, 69, 72–74, 76, 143, 185, 264, 268, 275, 277, 279, 282–284

B

Bajada, 111, 115–118, 122, 123, 134, 149, 171
Barelands, 194
Bare soil matrix, 150
Barrier islands, 121
Basin, 2, 5, 6, 9, 12–15, 17, 18, 23, 24, 27, 28, 30, 31, 41, 42, 49, 54, 72, 91, 105–107, 109, 111–119, 123, 132, 134, 139, 143, 148–150, 163, 164, 168–172, 218, 219, 222–228, 243, 270, 272
Beach ridges, 35–37, 39, 107, 109, 118–121, 123, 142, 163, 171, 172, 175, 181, 182, 224, 237, 238, 242, 247
Biodiversity, xi, 263–265, 283, 284, 285
Biogeographical regions, 264
Biostratigraphic, 48, 72
Bioturbation
　development, 30, 53, 54, 78
　index, 52, 55, 56, 78
Birds, 47, 48, 66–68, 70–72, 75, 145, 241, 245, 246, 254, 263, 264, 271–275, 280, 282–284
Blowouts, 113, 148, 152, 203–205, 279
Brackish, 62, 75, 218, 222

C

Calcareous crusts, 24, 26, 32, 51, 120, 251
Caleta Valdés, 26, 34, 36, 37, 39
Carnivores, 275, 278, 281, 282

Carrying capacity, 195
Chon Aike magmatic province, 5
Chronological record
  datations, 234
  Equestrian Period, 242
  middle Holocene, 241, 242
  late Holocene, 242
Chronosequence, 167, 170, 174, 177, 187
Chubut river, 60, 74, 234, 235
Chuquiraga species, 132
Circulation variability, 88
Climate
  change, 9–11, 14, 73, 89, 91, 100, 177, 195, 210
  classification, 99, 105
  mediterranean-like, 91
  model, 11, 91, 207, 208
  transitional, 85, 99, 100
Closed basin, 9, 105, 109, 112, 114, 115, 123, 126
Cloud coverage, 88
Coarse fragments, 197, 198
Coastal
  aeolian dunes, 111, 114, 122
  bajada, 111, 118, 119, 122, 123, 161, 171, 218, 233, 237, 242, 245, 254
  dunes, 5, 11, 111, 113, 114, 122, 125, 194, 218, 222, 224, 228, 242, 254, 270, 271
  piedmont pediment, 111, 118, 119, 140, 141, 151, 161, 170, 172, 174, 201
  zone, 10, 56, 75, 105, 107, 110, 111, 114, 117, 119, 121–123, 125, 126, 140, 143, 161, 171–173, 175, 178, 219, 222, 224–226, 228, 233, 242, 254, 268, 270
Co-dominance, 265, 267, 286
Compatible, 264, 283–285
Conflict (anthropic), 283, 284
Conservation, 11, 13, 14, 77, 118, 121, 122, 131, 132, 143, 145, 147, 148, 153, 168, 169, 172, 205, 239, 263, 264, 268, 271, 274, 275, 277–279, 282, 283, 285, 286
Continentality index, 92, 94
Convergent margins, 12
Creep, 194, 272, 273, 295
Cryodisturbances, 35

D

Deposits
  ash, 53, 57
  bioclastic, 55–57, 59, 61
  conglomerates, 34, 54
  heterolithic, 52, 55, 56, 59
  inclined heterolithic stratification (IHS), 52, 55
  mudstones, 23, 51, 52, 54–56, 57
  pyroclastic, 15, 23, 27, 51, 53, 55, 56
  sandstones, 23, 27, 28, 52, 54–56, 237
  shell beds, 53, 54, 55, 57, 60, 61
  volcanic, 10, 15, 27, 52, 53, 55, 56
Desertification, 74, 88, 192, 207
Desert pavement, 109, 126, 149, 150, 197, 198, 201
Diet
  actualistic studies, 246
  archaeofaunal studies, 233, 246, 254
    guanacos, 240, 241, 245, 246, 254, 282
    sea lions, 240, 241, 245
    mollusks, 233, 241
  stable isotope analyses, 233
    plants, 233, 254
    high trophic level, 245
Discharge
  local, 215, 222, 223, 227
  regional, 215, 222, 225
Domestic herbivores, 147
Drifting, 4–6, 16
Drylands, 192, 277, 278, 288
Dunefields, 109, 111, 113, 114, 143, 151, 152, 164, 194, 202, 203, 217, 218, 220, 242, 244
Dune Mobility, 194
Dunes
  barchans, 115, 125
  complex, 114, 117, 125
  compound, 114, 122
  dome-shaped, 114, 126
  linear, 113, 122, 126
  parabolic, 114, 126
  simple sand, 126
  star, 114, 126
  transverse, 114, 126
Dynamic topography
  Subsidence, 12, 13
  uplift, 12, 13

E

Eccentricity (orbital), 16
Ecological
  Interactions, 144, 267, 284
  processes, 77, 278, 279, 284
Ecosystem
  functioning, 264
  conservation, 131, 264, 285
Ecotone, 132, 143, 153, 264
Electric conductivity (EC), 222, 223
Endangered, 274, 282
Endemic, 66, 71, 145, 265, 270, 272, 282, 286
Entisols
  Aquents, 162, 175

Index

Endoaquents, 161, 162
Fluvaquents, 161, 175
Hydraquents, 161
Psammaquents, 161
Sulfaquents, 161, 175
Torriorthents, 161
Torripsamments, 161
Entrerriense Transgression, 48, 49, 51
Equilibrium profile, 9, 10, 16
Equipotential map, 222
Erodibility, 193, 194, 210
Erosive processes, 55, 152
Erosivity, 193, 210
Eruption style, 10
Evaporation processes, 225
Evapotranspiration, 96, 98, 99, 101, 177, 226
Expert Judgment, 191, 196
Extinction, 66, 69, 286

F
Facies, 41, 52–57, 59, 60, 74, 78, 223
Faults, 2, 6, 12, 13, 17, 33, 38, 57, 123
Fauna, 11, 31, 61, 63, 65–67, 69, 70, 73–78, 197, 233, 239
Field Monitoring, 192, 196
Flooding, 13, 48, 49, 51, 52, 59, 60–62, 72, 74, 76, 78, 79, 105, 125, 143, 175
Floristic composition
  grassy vegetation, 131
  shrubby vegetation, 131
Floristic element, 132, 153
Flow velocity, 224
Fluvial terraces, 7, 109, 164
Fossil
  assemblage, 61
  group, 48, 51, 77, 78
  records, 68, 77, 91
  trace, 30, 53–57, 61, 75, 76
Framboidal pyrite, 181, 187
Freshwater, 48, 56, 70, 75, 76, 222, 227
Formation
  Caleta Valdés, 26, 34–37, 118, 119, 120, 121, 171, 172, 175, 182, 186, 247, 248
  Gaiman, 14–16, 26, 28, 30, 31, 34, 39, 47, 49, 51–53, 57, 59, 60, 62, 63, 66–69, 72–74, 215, 219, 220, 237
  Patagonian Shingle, 25, 109
  Puerto Madryn, 11–13, 15, 28, 31–36, 38, 39, 47, 48, 51, 53–55, 57, 59–75, 77, 110, 112, 114, 150, 169, 171, 215, 218–220, 224, 237, 238
  San Miguel, 36, 37, 39, 118, 119, 121, 171, 218, 247

G
Geohydrological regions, 225, 226
Geomorphic surface, 107, 112, 161, 162, 164, 167–170, 172, 174, 177, 182, 187, 197
Glaciation, 14, 35, 109, 112, 172, 186, 187
Glauberite, 26, 40, 42
Global meteoric water line, 225
Golfo
  Nuevo, 25, 30, 31, 33, 37, 39, 57, 66, 72, 106, 113, 114, 117, 118, 121–123, 125, 141, 151, 172, 222, 223, 235, 237, 240, 241, 244–246, 271
  San José, 24, 30–32, 36, 37, 39, 106, 115, 117, 118, 120–123, 125, 141, 172, 222, 235, 239, 241, 242, 244, 246
  San Matías, 30, 36, 38, 125, 239, 241, 245, 246, 282
Gondwana, 1, 3–5
Gran Salitral, 111, 115, 117, 149, 171, 221, 222, 227
Grass matrix, 148
Grasshoppers, 265
Grasslands, 150, 197, 264, 272, 274, 277, 281
Gravels, 11, 25, 26, 34–37, 39–41, 126, 165–167, 170, 173, 174, 178, 185, 201, 237, 238, 240, 247, 248
Great endorheic basins, 105, 107, 112, 114, 115, 118, 123, 134, 139, 168, 169, 171, 228
Greatest Patagonian Glaciation, 112
Greenhouse-world, 10, 17
Groundwater
  balance, 220
  flow, 221, 222
  quality, 227, 229
  resources, 98, 215, 216, 228, 237, 244
  salinity, 222
Guanaco, 148, 233, 239–241, 244–254, 276, 278, 281–284, 286, 297
Gullies, 106, 109, 111, 117, 149, 150, 193, 202, 203, 237
Gypsum, 26, 32, 38, 40, 53, 164, 177, 179, 180, 184, 187

H
Habitat
  modification, 270, 274, 278
  degradation, 263
  loss, 279, 284
Halite, 23, 26, 38, 40
Halophytic, 121, 143, 153, 175, 278
Haplocalcid soils, 150
Height, 18, 87, 111, 113, 137–142, 172, 242, 257

Herbaceous
 layer, 134, 137, 139, 141, 142, 147, 148
 perennial forbs, 134
 perennial species, 134
Herbivore preferred species, 147
Herpetofauna, 264, 268
Hiatus, 32, 60
Holocene Climatic Optimum, 121
Human
 activities, 153, 194, 263, 264, 277–280, 282–285
 infrastructure, 277
 related factors, 280
Hunting, 245, 246, 249, 257, 280– 282, 286
Hyalix argentea, 113, 134, 137, 139, 146–148, 151, 152, 156, 168
Hydraulic conductivity, 216, 229
Hydrochemistry, 216, 222, 227, 229
Hydrogeological
 system, 226
 units, 216, 218
Hydrogeology, 215, 216, 229
Hypersaline water, 228

I

Ice-sheets, 10, 13, 17
Ice-wedge casts, 11, 110, 112, 126
Ice-world, 11, 17
Ichnofacies
 Cruziana, 53, 54
 Glossifungites, 57
 Skolithos, 54
Illite, 182, 184, 251
Infiltration, 199, 201, 205, 210, 217, 220, 225–227, 228
Inland-wetlands, 264, 272, 274, 283, 292
Insects, 263–265, 286
Insolation, 11
Interdune wetlands, 217
Intermediate selectivity, 281
Interspecific competition, 281
Intertidal environments, 143
Invertebrates
 Bivalves, 53, 63, 65
 Brachipodos, 63
 Bryozoans, 63
 Corals, 63
 Decapods, 63
 Echinoids, 53, 63
 Gastropods, 53, 63, 65
 Molluscs, 57, 63, 237
 Oysters, 53, 56, 57
Islands of Fertility, 193
Isotopic composition, 162, 182, 224, 225, 228

Istmo Carlos Ameghino, 51, 52, 57, 59, 60, 67, 72, 121, 122, 134, 175, 222, 270
IUCN, 264, 274, 286

K

Köppen-Geiger, 99

L

La Pastosa-Punta Delgada, 52, 57, 59, 60, 61, 69, 71, 75
Land degradation, 149, 193, 195–197, 199, 203, 210, 283
Landforms hierarchically classification, 106
Landscape
 evolution, 35, 105, 106, 122–125
  bajadas, 115, 117, 118, 134, 139, 149, 150, 169, 171, 201, 218, 237, 242, 254
  clay minerals, 234
  Gullies, 106, 109, 111, 117, 149, 150, 202, 203, 237
  lithic raw material, 237, 240, 245, 247, 249, 251, 255
  restingas, 237
  sandy environments, 237, 242
  water supply, 237, 240, 244
Larrea species, 132
Lava flows, 15
Life form
 deciduous shrubs, 133
 dwarf shrubs, 134
 evergreen shrubs, 133
 forbs, 133
 grasses, 133, 134
 shrubs, 133, 134
Lithic Technology
 bipolar flaking, 247, 248, 255
 anvils, 248, 249, 255
  cores, 248
  flakes, 247, 248
  hammer stones, 249, 255
 direct percussion, 247, 248, 255
  blades, 248
  cores, 248
  flakes, 247, 248
  hammer stones, 248, 255
 dispersive X-Ray analyses, 248
 expedient artefacts, 248
 geochemical analyses, 248
 instruments
  bolas, 246, 249
  burins, 249, 255
  drills, 249, 255
  fishing weights, 246, 248, 249, 255

hand stones, 248, 249
knives, 246, 249, 255
manos, 248, 255
metates, 248
mortars, 249
proyectil points, 246, 249, 255
scrapers, 246, 248, 249, 255
lithic raw material
basalt, 237, 247–249, 255
chalcedony, 249, 255
obsidian, 248, 249, 255
porphyritic rocks, 248, 249
riolite, 247
silica, 237, 255
sandstone, 237, 248, 255
xylopal, 249, 255
macroscopic analyses, 251
nodules, 247, 248, 255
pebbles, 248, 249, 254, 255
sampling, 247
Scanning Electron Microscope, 248
Lithologic discontinuities, 164, 180
Lithostratigraphic unit, 49, 51, 164
Livestock, 148, 195, 203–205, 279, 281, 283
Lizards, 268, 270, 271
Llanquihue glaciation, 172, 186
Logging, 279, 280
Low marsh, 143, 145, 176

M

Major ionic composition, 222, 224
Mammals, 48, 66, 67, 70, 71, 73, 76, 145, 239, 241, 245, 246, 254, 263, 264, 271, 275–277, 280, 282, 284, 297
Management plan, 148, 264, 283, 285
Marine
Coast, 143, 234
terraces, 13, 14, 233, 254
Marine Isotope Stages
MIS 1, 121
MIS 5, 120
MIS 7, 120, 123
Maritime influence, 85, 92, 93, 94
Marshes, 75, 118, 119, 121, 122, 126, 132, 143–145, 161, 162, 170, 174–176, 178, 180, 181, 186, 222, 272
Marsupials, 272
Mass wasting processes, 112, 126
Maximum Flooding, 51, 52, 59, 60, 61, 62, 72, 74
Megapatches, 194, 202
Meteorological records, 100
Mid-Holocene Thermal Maximum, 121, 125
Milankovitch cycles, 11, 14

Miocene
early, 7, 14, 25, 30, 31, 39, 49, 51, 60, 68, 69, 71–74
late, 12, 23, 28, 31, 34, 36, 39, 47–49, 63–66, 68–76
middle, 7, 11, 31, 39, 48, 66, 69, 73, 74
Modelling, 229, 267, 280
Monitoring, 100, 196, 202, 209, 284, 285
Monte, 100, 132, 263, 264, 268, 270, 274, 277, 279, 284
Monteeco-region, 264, 274, 277, 278
Morphospecies, 265
Mosaic, 134, 137–139, 146–148, 193, 264
Mudstones, 23, 26–28, 30–33, 47, 51, 52, 54–57, 76, 219
Mulinum spinosum, 140, 146, 151, 155, 205

N

Nassella tenuis, 134, 137, 138, 147, 150, 151, 154, 164
Near threatened, 282
Nebkas, 109, 114, 126
Neogene, 12, 23, 25–28, 30, 38, 42, 57, 59, 61, 63, 65, 68, 73, 75, 77, 106, 109, 237
Non-saturated zone (NSZ), 215, 220, 230
North Patagonian Massif, 2–4

O

Obliquity, 11, 17
Oil exploration drill, 27
Orbital changes, 10, 11
Orogenic wedge, 12, 17
Overgrazing, 191, 194, 195, 199, 202, 203, 206, 210, 270, 279
Overland Flow, 193, 199, 200, 210

P

Paddocks, 195, 203, 204, 278, 281
Paleoenvironments
inner estuary, 55
nearshore, 47, 52, 54, 57, 59, 60, 61, 76
neritic, 62
offshore, 2, 27
outer estuary, 59
rivers, 15, 17, 35, 56, 60, 74
saltmarsh, 75
shallow marine, 9, 10, 27, 49, 51, 54, 55, 61, 76
shelf, 28, 39, 47, 52–55, 57, 59, 60, 61, 74
shoreface, 52, 54, 75
subtidal, 55, 56
swamp, 62, 75
tidal channel, 47, 52, 55, 59, 61, 75, 176
tidal flat, 47, 56, 59, 60

tidal-fluvial, 55, 61, 75
Paleosols, 172, 175, 177, 180, 182, 186
Palygorskite, 162, 182, 184, 186
Palynomorphs
  dinoflagellates/dinocysts, 61, 62
  forest, 62, 76
  herbs, 62
  neotropical, 62, 76
  palms, 62
  pollen, 61, 62, 76
  shrubs, 62, 76
  spore, 61, 62, 76
  trees, 62
Passerines, 274, 275, 280, 282
Patagonia
  eco-region, 263, 264, 271, 274, 277, 278
  terrane, 2, 4
Patagonian
  foreland, 1–3, 6, 7, 12, 15, 39, 99, 109, 234, 235
  steppe, 99, 239, 264, 268, 270, 271, 275
Patagoniense Transgression, 14, 48, 49, 72
Patchy, 77, 132, 134, 138, 193
Patterns, 11, 53, 73, 87, 88, 92, 96, 106, 118, 120, 182, 186, 193, 201, 202, 207, 254, 265, 267, 272, 277–279, 282, 283
Pediments, 115, 117, 118, 122, 123, 134, 139, 141, 149–151, 161, 168, 169, 171, 172, 201, 242
Pedogenic carbonates
  alpha-type micro-fabric, 177
    circum-granular cracks, 177, 179
      floating coarse mineral components, 177, 179
        lenticular gypsum aggregates, 177, 179
        matrix nodules, 177, 179
        micritic calcite, 177
        ooids, 177, 179
        pellets, 177
        pendant, 177, 179
        rhombohedral micrite, 177, 179
  beta-type micro-fabric, 177
    calcified filaments, 179
    calcispheres, 177
    needle-fiber calcite, 177, 179
  calcification, 161, 177, 179
  gypsification, 161, 179
  illuviation, 161, 176
  stable isotopic composition, 182
    d13C, 185, 186
    d18O, 162, 182, 185, 186, 225, 228
  sulfidization, 161
Perennial grass steppes, 133, 134
Permeability, 216–219, 226, 227

Phreatic aquifer, 215, 220, 230
Physiognomic, 131, 133, 153
Phytogeographical
  districts, 100, 132
    central, 132
  provinces, 132, 153, 277
    Monte, 132, 264
    Patagonian, 132
Piedmont Pediment, 111, 115, 116, 118, 119, 139–141, 149–151, 161, 168–170, 201
Pilot Studies, 192, 196
Plant canopy, 139, 142, 143
Plant community zonation, 143
Plant cover, 134, 137–141, 146, 148, 153, 193, 197, 210, 267
Playa lake, 26, 38, 111, 115–117, 134, 149, 169–172, 222, 223, 225, 272
Pleistocene glaciations, 109
Poaching, 263, 279
Polygenetic soils, 188
Populations, 158, 228, 233, 254–256, 264, 270, 271, 278, 280, 283
Porosity, 218, 219, 229
Pottery Technology
  actualistic studies, 246
  building, 251, 280
    pinching, 251
    ring building, 251
  ceramic matrix, 251
    aggregates, 251, 254
    clay, 233, 234, 237, 251, 254, 255
    soils, 168, 251, 269, 270, 278
  decoration, 253, 257
    engobe, 253, 257
    painting, 253
    polishing, 253
  3-D digital reconstructions, 251, 253
  exchange, 249, 253, 254, 255
    alloctonous pieces, 254
  functionality, 242, 248
    domestic activities, 233, 255
    dynamic pieces, 253
  gas chromatography, 246, 251
  isotope analyses, 233, 243, 244, 246, 251
  local development, 251
  morphology, 251, 254, 268, 277, 284
    bases, 233, 251
    handle, 251, 253
    mouths, 251
    necks, 251
    shapes, 251
  macroscopic analyses, 251
  petrography, 248, 251
  volume, 253

Index 311

weight, 246, 248–250, 253, 255, 257
  x-ray, 248, 251
    diffraction, 251
    images, 251
Precession (axial), 17
Precipitation
  gradient, 76, 87, 88, 94, 193, 203, 204, 228
  interannual variability, 89, 90, 94
  seasonality, 133, 186
Predictive models
  optimal foraging theory, 240
    central place model, 240
    diet breath model, 240
    patch choice model, 240
  regional preliminary model, 240
    average diet, 241
    coastal productivity, 237, 240, 241
    hierarchy, 241
    water abailability, 241
Present beach, 36, 111, 119, 121, 125
Pressure
  cells, 101
  gradient, 87
  systems, 88, 98
Priority sites, 118, 121, 132, 143, 153, 168, 169, 172, 239, 268, 271, 277, 279
Protected area, 263, 264, 265, 284
Proximal areas, 9, 10, 15, 17
Proxy, 8, 17, 162
Psammophile plant species, 168
Puerto Pirámide, 41, 52, 57, 59–61, 67, 72, 75, 90, 94, 216, 222, 223, 278
Punta
  Buenos Aires, 32, 51, 70, 71
  Cero, 32
  Cono, 32, 37
  Delgada, 32, 52, 57, 59–61, 69, 70, 75, 92, 93, 97–100, 236, 240, 241, 244, 245
  Hércules, 33
  Ninfas, 37, 198
  Norte, 32, 240
  Pardelas, 37, 241
Pyroclastic content, 53

R
Rainfall, 11, 85, 87–95, 97, 99–101, 106, 162, 177, 193–195, 199–201, 208, 268, 272
Rain shadow, 11, 74, 85, 99
Rainsplash erosion, 197
Rangeland degradation, 143, 191–199, 201, 279
Raptors, 272, 274, 282, 283
Ratites, 272
Rawson Basin, 2, 5, 6

Recharge
  mechanism, 226, 228
  zone, 215, 217, 224
Recreational activities, 279, 280, 284
Redoximorphic features
  nodules, 171, 176, 177, 180
  rhizo-concretions, 180, 181
  stratified sediments, 179, 180
Regressive secuence, 32, 48, 51, 59–61, 75, 79
Remote Sensing, 191, 196, 197, 202, 209, 210
Reptiles, 263, 264, 268–270, 280, 283, 292
Retroarc foreland basins, 12, 15, 17
Riacho San José, 37, 121, 122, 144, 145, 175, 176, 181, 222
Rifting, 3–6, 17
Rills, 106, 149, 193, 202, 203
Río de la Plata craton, 2–4
Rodados Patagónicos, 11, 15, 16, 25, 34, 109, 110, 112, 123, 161, 164, 168, 182, 186, 218, 222, 226, 227, 237, 247
Rodents, 70–72, 197, 246, 275, 277, 278
Rooting depth, 133
Runoff, 35, 106, 112, 169, 175, 193, 197, 199, 201

S
Salina
  Chica, 23, 38, 40, 41, 111, 115–117, 123, 221, 222, 224, 225, 227, 228, 237, 240–242, 254
  Grande, 23, 40, 41, 111, 115–117, 123, 221–223, 225, 227, 228, 240, 242, 243, 254, 256
Salt
  flats, 40
  marshes, 118, 119, 121, 126, 132, 143–145, 161, 162, 170, 174–176, 178, 181, 272
Saltation, 194
Sandstones, 4, 23, 26–28, 31–33, 47, 51, 52, 54–56, 215, 219, 220, 233, 237, 248, 255
Sandy
  beaches, 114, 151
  soils, 133, 188, 270, 278
Sarcocornia marshes, 145
Sea-level
  eustacy, 13, 16
    glacioeustasy, 13
    tectonoeustasy, 13
  relative, 10, 13, 14
Sedimentary
  basin, 9, 14, 15, 17, 41
  environment, 10, 15, 16, 35
  flux, 18
  routing system, 1, 9, 16, 17

structures, 51, 53, 54, 75, 78, 79
  cross stratification/bedding, 54
  current ripples, 56
  hummocky cross-stratification, 54
  massive, 54, 75
  mud drapes, 55
  parallel stratification, 54
  planar lamination, 55
  ripple lamination, 55
  sand-mud couplets, 55
  through cross stratification, 54
  wave ripples, 55
Sediments, 1, 9, 10, 13, 15–17, 23–28, 30, 36, 38, 39, 41, 47, 54, 56, 60, 61, 66, 68, 69, 72, 77, 109, 112, 114, 115, 117, 118, 143, 169, 172, 176, 180, 181, 192, 194, 218, 219, 220, 227, 228
Sepiolite, 162, 182, 184, 186
Sheep
  ranching, 274, 278–283
  grazing, 146, 148, 149, 195, 198, 205, 206, 279
Sheet erosion, 117, 193, 199
Shoreline, 7, 13, 16, 36, 54, 114, 118, 120, 121, 194
Shrubby
  layer, 137–139, 141, 142, 172
  matrix, 133
  steppes, 133, 134, 138
Shrublands, 150, 264, 268, 272, 274, 278, 281
Site diversity
  base camps, 242, 244, 251
  human burials, 237, 242, 243
  locations, 149, 151, 242, 243
  palimpsests, 242
  post-depositional, 242
  shellmiddens, 242, 243
  transient camps, 242, 243
Slab
  pull, 12
  window, 7, 12, 18
Slope
  gradient, 193
  length, 193
Smectite, 162, 182, 184, 186, 251
Snakes, 268, 271
Sodium
  bicarbonate, 223, 224, 227, 228
  chloride, 223, 224, 228
  chloride-sulfate, 181

Soil
  compaction, 88, 193, 279
  complex, 161, 169, 172, 188
  degradation, 113, 143, 147, 186, 191–197, 201, 209
    factors, 196, 197
    processes, 191, 192, 197, 209, 210
  detachment, 193, 199
  fertility, 192
  moisture, 133, 161, 186, 202, 222
  organic matter, 198
  properties, 165, 168, 170, 173, 178, 192
  roughness, 194
  structure, 192
  texture, 194, 210
Solifuges, 265, 286
Southern Andes, 1, 2, 8, 14, 89
Southern Volcanic Zone (SVZ), 15
Spartina marshes, 145, 155, 176
Species
  arrangement, 131, 133, 138
  composition, 69, 140, 143, 206, 275, 277, 280, 282, 284
  richness, 66, 265, 267, 270, 274, 279, 284
Spiders, 263–265
Sporobolus rigens, 113, 137, 146, 151, 168
Spring, 89, 94, 116, 133, 194, 195, 208, 223, 237, 239, 245, 275
Springtail, 265
Squamata, 270, 292
Stabilized Aeolian Field, 109, 111, 113, 134, 137, 139, 143, 147, 148, 151, 152, 161, 164, 168, 217, 218, 220, 226
Stocking rate, 195, 206, 279, 281, 282
Stratum, 133, 134, 139–142, 150–152, 205, 206
Streams, 89, 118, 151, 172, 220
Strict reserves, 147
Structural high, 39
Subsidence, 9, 10, 12–14, 16, 87
Sulfidic materials, 122, 176, 180, 181
Surface
  hydrological processes, 193, 197, 227
  runoff, 35, 106, 112, 169, 193, 197, 199
Survey, 105, 109, 148, 162, 175, 181, 196, 197, 209, 220, 256, 268, 275, 279, 281, 283, 284
Suspension, 56, 74, 194
Sustainable, 203, 205, 210, 264, 283–285

T

Taxa
  abundance, 62, 263, 265, 267, 279, 283–284
  definition, 264, 279
Temperature
  gradient, 76, 87–88, 92, 94
  interannual variability, 89, 94
  paleo, 8
  range, 75, 99
Terrace levels, 109, 123, 147, 149, 150, 164, 182, 201, 224, 226, 227
Terrain Tertiarie Patagonien, 24
Terrestrial
  arthropods, 263
  birds, 263
  mammals, 263
Tertiary beds, 24, 25, 28
Tidal
  amplitude, 143
  plain, 121
Tourism, 216, 283, 284
Trace fossil
  asterosoma, 53
  chondrites, 53
  helicodromites, 53
  planolites, 53
  rosselia, 53
  teichichnus, 53
  thalassinoides, 53
Transgressive sequence, 32, 78
Trelew, 90, 94, 100
Two phase mosaic, 137, 138, 141

U

UNESCO, 77, 147, 264, 278
Ungulate, 71, 278, 281
Unsaturated
  definition, 223, 225
  zone, 225, 227
Uplands and Plains, 105, 107, 109, 111, 132–134, 143, 146, 147, 149, 164, 165, 168, 225–227
Uplift
  foreland, 7
  fault-block, 12

V

Valdés Basin, 5, 6, 27, 28
Vegetation
  community, 118, 145, 150–153, 279
  condition, 106, 117, 131, 133
  ground Cover, 193, 279
  physiognomy, 264
  resources, 191, 192
    edible, 239
    medicinal, 239
    woody species, 239
  spatial Pattern, 193
  structure, 267, 277, 280
  units, 161, 168
Vertebrates
  birds, 48, 66, 67, 241, 245
  cetacean, 48, 61, 66, 244
  elasmobranchs, 66, 67
  fish, 48, 61, 66, 67, 74, 76, 241, 244, 245
  mammals, 48, 67, 241, 245
  osteichthyans, 66, 67
  penguin, 47, 61
  pinnipeds, 47, 66, 244
Vesicular horizon, 164
volcanic edifices
  composite volcano, 14
  calderas, 14
Volcanic explosivity index (VEI), 8, 15
Volcanism, 6, 7, 9, 10, 14–16

W

Water
  bearing, 219, 229, 230
  erosion, 88, 106, 192, 193, 197, 199, 202, 242, 280
  table, 117, 171, 176, 218, 223, 228
Waterbirds, 272
Watersheds, 220, 222
Wave-cut platforms, 118, 119, 122, 237
Westerly winds, 74, 98, 114, 152, 194
Wildfires, 194, 195, 207, 279
Wildlife, 147, 263, 264, 278, 279, 281, 283–285
Wind
  chill, 88
  erosion, 123, 143, 191–194, 197, 201–203, 206, 209
  severity, 193
  speed, 88, 194, 205
  storms, 194
  trend, 98, 114, 203
World Natural Heritage Site, 264

X

Xeric Calciargid, 166, 168, 170, 173, 176, 198

Z
Zoological resources
   terrestrial vertebrates, 239
      armadillos, 239
      choiques, 239
      guanacos, 239
   marine fauna, 239
      birds, 239
      fish, 239
      mollusks, 239
      sea lions, 239